MOLECULAR MECHANISMS IN CELLULAR GROWTH AND DIFFERENTIATION

MOLECULAR MECHANISMS IN CELLULAR GROWTH AND DIFFERENTIATION

Edited by
Anthony R. Bellvé
Henry J. Vogel

College of Physicians and Surgeons
Columbia University
New York, New York

ACADEMIC PRESS, INC.
Harcourt Brace Jovanovich, Publishers
San Diego   New York   Boston
London   Sydney   Tokyo   Toronto

This book is printed on acid-free paper. ∞

Copyright © 1991 by Academic Press, Inc.
All Rights Reserved.
No part of this publication may be reproduced or transmitted in any form or by any means, electronic or mechanical, including photocopy, recording, or any information storage and retrieval system, without permission in writing from the publisher.

Academic Press, Inc.
San Diego, California 92101

United Kingdom Edition published by
Academic Press Limited
24–28 Oval Road, London NW1 7DX

Library of Congress Cataloging-in-Publication Data

Molecular mechanisms in cellular growth and differentiation / [edited
  by] Anthony R. Bellvé, Henry J. Vogel.
    p. cm.
  Includes index.
  ISBN 0-12-085360-4 (alk. paper)
  1. Cells–Growth–Regulation. 2. Growth factors. 3. Paracrine mechanisms. 4. Autocrine mechanisms.   I. Bellvé, Anthony R.
II. Vogel, Henry J. (Henry James), Date.
QH604.M65 1991
574.87'61–dc20   90-1034
                                                                        CIP

Printed in the United States of America
91 92 93 94   9 8 7 6 5 4 3 2 1

Contents

Preface   xv

PART I   POLYPEPTIDE GROWTH FACTORS AND OTHER SIGNALS: CELL MEMBRANE RECEPTORS, TRANSDUCTION MECHANISMS, AND PROLIFERATIVE RESPONSE

1   A Dual Binding Site Hypothesis of the Type I Insulin-Like Growth Factor Receptor

JUDSON J. VAN WYK AND SAMUEL J. CASELLA

Introduction	3
Which Receptor Mediates the Actions of IGF-II?	4
References	8

2   Insulin Action on Membrane Protein Recycling

MICHAEL P. CZECH, SILVIA CORVERA, MARK A. TEPPER, CAROL M. BROCKLEBANK, AND RICHARD G. MACDONALD

Introduction	9
Membrane Protein Recycling	9
Conclusions	17
References	18

3 Growth-Factor and Oncogene-Mediated S6 Phosphorylation during the Mitogenic Response: A Novel S6 Kinase

PAUL JENÖ, L. M. BALLOU, AND GEORGE THOMAS

Stimulation of S6 Phosphorylation	19
Activation of S6 Kinase	20
Phosphatases	21
Characteristics of the S6 Kinase	21
Pathways Involved in S6 Phosphorylation	22
References	23

4 Phosphorylation of pp60$^{c\text{-}src}$

TONY HUNTER, CHRISTINE A. CARTWRIGHT, JAMES R. WOODGETT, AND KATHLEEN L. GOULD

Introduction	25
Serine 12	27
Serine 17	29
Other Amino-Terminal Phosphorylations	30
Tyrosine 416	31
Tyrosine 527	32
Platelet-Derived Growth Factor Treatment	34
Neuronal pp60$^{c\text{-}src}$	35
pp60$^{c\text{-}src}$ and Polyomavirus Middle T Antigen	37
Vanadate Treatment	39
Elevation of pp60$^{c\text{-}src}$ Protein Kinase Activity in Transformed Cells	39
Summary	40
References	41

5 G Proteins May Exist as Polymeric Proteins: A Basis for the Disaggregation Theory of Hormone Action

MARTIN RODBELL

Introduction	45
Depolymerization Theory	47
Current Thoughts on G Proteins	47
Some Structural Aspects of G Proteins	49
Pertussis Toxin Actions	51
Effects of Detergents on G Proteins	52

Signal Transduction and Cytoskeletal Elements	55
Summary and Speculations	55
References	57

6 Pleiotropic Functions of the Insulin Receptor

ALAN R. SALTIEL

Introduction	59
Purification and Biological Characterization of the Insulin-Sensitive Enzyme Modulators	61
The Enzyme Modulators Are Inositol Phosphate Glycans	62
The Inositol Phosphate Glycan Enzyme Modulators Are Produced by an Insulin-Sensitive Phosphatidylinositol Glycan-Specific Phospholipase C Activity	65
Turnover of Glycosylphosphatidylinositol in Insulin Action	66
References	67

7 Cooperative Regulation of Gene Expression in Liver Cells by Hormones and Extracellular Matrix

LOLA M. REID, ISABEL ZVIBVEL,
TOHRU WATANABE, YASU FUJITA, MARIA
AGELLI, KATE MONTGOMERY, ANDREA KRAFT,
AND ANDREAS OCHS

Introduction	69
Cell Culture	71
Gene Expression in Normal Liver	80
Gene Expression Studies in Hepatoma Cells	90
Hypotheses on the Mechanisms of Matrix and Hormonal Regulation of Gene Expression: Delivery of Protein Hormones to the Nucleus by Proteoglycan/Glycosaminoglycan Carriers	94
Summary	95
References	98

8 Receptor–Ligand Interactions: Role in Cancer Invasion and Metastasis

MARK E. SOBEL AND LANCE A. LIOTTA

Introduction	107
Basement Membrane and the Three-Step Hypothesis of Invasion	108
Components of Basement Membrane	108
Laminin Receptor	109

Role of Laminin and Laminin Receptor in Tumor Cell Metastasis	109
Monoclonal Antibodies to Human Laminin Receptor	110
Isolation of Human Laminin Receptor cDNA	110
Overlapping Domain of Laminin Receptor cDNA Inserts	111
Nucleotide and Deduced Amino Acid Sequence of the 2H5 Epitope of Human Laminin Receptor	113
Synthetic Peptides and the Putative Ligand Binding Domain of Laminin Receptor	116
Expression of Laminin Receptor mRNA	116
Conclusions	118
References	119

9 Control of Cell Proliferation by Transforming Growth Factors

H. L. MOSES, J. BARNARD, C. C. BASCOM,
R. D. BEAUCHAMP, R. J. COFFEY,
R. M. LYONS, AND N. J. SIPES

Introduction	121
Transforming Growth Factor-α	121
Transforming Growth Factor-β	123
Autocrine Regulation of Epithelial Cells by Transforming Growth Factors-α and -β	125
References	126

PART II POLYPEPTIDE GROWTH FACTORS AND THEIR CELL MEMBRANE RECEPTORS: RELATION TO ONCOGENES

10 The EGF Receptor Protooncogene: Structure, Evolution of Properties of Receptor Mutants

JOSEPH SCHLESSINGER

Introduction	131
Structure of the EGF Receptor	132
Properties of EGF Receptor Mutants	133
References	134

11 Biological Effects of the v-*erbA* Oncogene in Transformation of Avian Erythroid Cells

BJÖRN VENNSTRÖM, HARTMUT BEUG, KLAUS DAMM, DOUGLAS ENGEL, ULLRICH GEHRING, THOMAS GRAF, ALBERTO MUÑOZ, JAN SAP, AND MARTIN ZENKE

Biological Effects of the v-*erbA* and v-*erbB* Oncogenes	138
Comparison of the v-*erbA* and c-*erbA*	138
Effects of Mutant v-*erbA* Genes on Erythroblasts	140
How Does v-*erbA* Contribute to Transformation of Erythroblasts?	141
Conclusions	145
References	147

12 *ras* Transformation of 3T3 Cells Induces Basic Fibroblast Growth Factor Synthesis *in Vitro* and *in Vivo*

MICHAEL KLAGSBRUN, PAUL FANNING, AND SNEZNA ROGELJ

Introduction	149
Results	150
Discussion	151
References	154

13 Cellular Oncogenes Conferring Growth Factor Independence on NIH 3T3 Cells

MITCHELL GOLDFARB

Introduction	157
Results	158
Discussion	171
References	172

14 Potential Function of the *mos* Protooncogene in Germ Cell Differentiation and Early Development

GEOFFREY M. COOPER

Introduction	175
Expression of *mos* in Male Germ Cells	177

Expression of *mos* in Oocytes	178
Posttranscriptional Processing of *mos* in Eggs	179
References	181

PART III POLYPEPTIDE GROWTH FACTORS AND THEIR CELL MEMBRANE RECEPTORS: ROLE IN DIFFERENTIATION AND DEVELOPMENT

15 Pleiotypic Actions of the Seminiferous Growth Factor

ANTHONY R. BELLVÉ AND WENXIN ZHENG

Introduction	187
Expression of Mitogenic Activity	188
Biochemical Characterization	189
Two Testicular Cell Lines: TM_3 Leydig and TM_4 Sertoli Cells	191
Proliferation of TM_3 Leydig and TM_4 Sertoli Cells	192
Extracellular Accumulation of Sulfated Glycoprotein	194
Synthesis of Specific Peptides	196
Immunological Properties of Heparin-Binding Growth Factors	200
Discussion	201
References	203

16 Regulation of Blood Vessel Growth and Differentiation

WERNER RISAU, HANNES DREXLER, HANS-GÜNTER ZERWES, HARALD SCHNÜRCH, URSULA ALBRECHT, AND RUPERT HALLMANN

Introduction	207
Growth and Development	207
Embryonic Stem Cells as an Experimental Model System for Early Vascular Development	211
Endothelial Cell Differentiation	217
Concluding Remarks	221
References	221

17 Vascular Endothelium: Role in Albumin Transport

G. E. PALADE AND A. J. MILICI

Multiplicity of Endothelial Functions	223
General Organization	224
Continuous Microvascular Endothelium	225
Physiological Data	226
Attempts at Structural–Functional Integration	227
Perspectives	245
References	246

18 A Growth Factor Homologous Gene Controlling Pattern Formation in *Drosophila*

WILLIAM M. GELBART

Introduction	249
Overview of the Organization of the *decapentaplegic* Gene	249
Contribution of the *Hin* Region to *dpp* Function	250
Nature of the *dpp* Polypeptide	254
References	257

PART IV SEQUENCE-SPECIFIC DNA-BINDING PROTEINS: ROLE IN DIFFERENTIATION AND DEVELOPMENT

19 The Role of Master Regulatory Genes in Controlling Development

E. B. LEWIS

Text	261
References	265

20 *engrailed*, A Gene for All Segments

THOMAS KORNBERG, CHIHIRO HAMA, NICHOLAS J. GAY, AND STEPHEN J. POOLE

Introduction	267
The *engrailed* Gene Promoter	268
The *engrailed* Protein Is a Phosphoprotein	269
References	278

21 Sequence-Specific DNA-Binding Activities of *Even-Skipped* and Other Homeo Box Proteins in *Drosophila*

TIMOTHY HOEY, MANFRED FRASCH, AND MICHAEL LEVINE

Introduction	279
Results	281
Discussion	296
Materials and Methods	301
References	303

22 Identification and Characterization of Two Mouse Genes, *EN-1* and *EN-2*, with Sequence Homology to the *engrailed* and *invected* Genes of *Drosophila*

GAIL R. MARTIN AND ALEXANDRA L. JOYNER

Introduction	305
Results	308
Discussion	315
References	317

23 Structure and Developmental Expression of Murine Homeo Box Genes

PETER GRUSS, CAROLA DONY, BERND FÖHRING, AND MICHAEL KESSEL

Introduction	319
Results and Discussion	320
References	325

PART V   SMALL MOLECULES IN DIFFERENTIATION AND DEVELOPMENT

24 Molecular Dissection of Pattern Formation in Vertebrate Limbs: Concepts and Experimental Approaches

GREGOR EICHELE AND CHRISTINA THALLER

Background	329
Retinoic Acid Mimics ZPA Tissue and Induces Pattern Duplications in a Highly Specific Way	331

Retinoic Acid Can Induce Digit Formation in the Absence of ZPA Tissue	334
Limb Buds Contain Endogenous Retinoic Acid	336
Future Problems	339
References	340

25 Differentiation of a Multipotent Human Intestinal Cell Line: Expression of Villin, a Structural Component of Brush Borders

D. LOUVARD, M. ARPIN, E. COUDRIER, B. DUDOUET, J. FINIDORI, A. GARCIA, O. GODERFROY, C. HUET, E. PRINGAULT, S. ROBINE, AND C. SAHUQUILLO MERINO

The Intestinal Epithelium: A Model System for Study of Cell Diffferentiation and the Development of a Polarized Cytoskeleton	341
A Pluripotent Human Intestinal Cell Line That Can Be Induced to Differentiate in Culture	343
Organization of Intestinal Brush Border Cytoskeleton	344
Villin, a Calcium-Regulated Actin-Binding Protein	347
Villin Is an Organ-Specific Marker in Adults and Embryos	348
Villin Expression during Intestinal Differentiation	350
Summary and Conclusions	351
References	351

Index   **355**

Preface

If to genetics, embryology, and molecular biology, we add biochemistry, physiology, and pathology, and if we focus on the region in which all six of these fields overlap, we are looking at the general area underlying this volume. Within this interdisciplinary area, we find three processes of broad import for all the life sciences: mitogenesis, oncogenesis, and differentiation. These processes are interrelated, partly because all three characteristically can involve a sequence of three major types of mechanisms, namely, cell–cell communication, signal transduction, and regulation of gene expression. Overall, this sequence provides a means by which information from a cell's chemical environment can modify the expression of the cell's genetic endowment. Mitogenesis, oncogenesis, and differentiation are also interrelated because all three can additionally involve another mode of regulation of gene expression, which depends on intracellular factors, such as sequence-specific DNA-binding proteins, rather than on cell–cell communication.

A significant type of cell–cell communication involves the interaction of a polypeptide growth factor, secreted by one cell, with a corresponding specific membrane receptor on the surface of a neighboring cell. Polypeptide growth factors are involved not only in cell growth, as their name suggests, but also in the control of multiple cell functions during embryogenesis, development, and adulthood. Such factors thus can cause discrete populations of cells to proliferate or to differentiate, or both. Proliferation and differentiation can occur in spatially and temporally regulated patterns to bring about coordinated interactions and development of the cell populations forming a tissue. In cell proliferation as well as in differentiation, growth factors can act singly or in concert or in opposition to coordinate the expression of multiple functions.

Molecules such as polypeptide growth factors can be regarded as signals that, through interaction with membrane receptors, are capable of triggering a series of signal transduction events culminating in the regulation of gene expression. Signal transduction tends to be specific, complex, and versatile, and may be pleiotypic, i.e., productive of more than one specific effect. Characteristically, signal transduction involves phosphorylation mechanisms in which membrane-associated kinases and guanine nucleotide-binding proteins may participate. Between the triggering of transduction and gene expression, various translocation-type processes must occur through which the information supplied by the signal is delivered to appropriate cytoplasmic or nuclear sites.

Signals may be (a) autocrine, i.e., produced by a cell and active on the same cell, or on a neighboring sister cell; (b) paracrine, i.e., secreted by a cell and active on a neighboring cell of different lineage (as in some types of cell–cell communication); or (c) endocrine, i.e., produced by a source cell and active on a target cell, at some distance. In this volume, autocrine and paracrine signals are emphasized. Because of their role in cell–cell communication, paracrine signals are discussed first.

An important alternative to the role of paracrine signals, such as growth factors, in the regulation of gene expression is the functioning of autocrine regulatory elements, e.g., sequence-specific DNA-binding proteins. Such regulatory elements include proteins that are capable of interacting, in receptor–ligand fashion, with small effector molecules, e.g., steroid hormones.

Gene expression may be regulated (a) directly, by the positive or negative control of transcription, through mechanisms including induction, repression, and derepression, or (b) indirectly, at any of several posttranscriptional levels up to and including translation. Induction (positive control) and repression (negative control) represent an increase or decrease in the rate of transcription of a structural gene and hence in the rate of synthesis of the protein encoded by that gene, respectively. Both of these mechanisms typically involve interactions between sequence-specific DNA-binding proteins and part of the promoter region for the structural gene. In induction, the transcription-accelerating effect is brought about, in the promoter region, by the DNA-binding protein in conjunction with a small-molecule inducer, such as an estrogen. The DNA-binding protein thus can function as an intracellular receptor for the small-molecule inducer. In repression, the transcription-decelerating effect is produced, in the promoter region, by the DNA-binding repressor protein, in conjunction with a small repressive molecule, e.g., in some cases, a glucocorticoid. In the absence of the small repressive molecule, derepression occurs. However, the regulation of gene expression is not unrestricted. For instance, in cells that have undergone commitment during embryogenesis, regulation is limited to the expression of those genes that are within the regulable repertoire of the committed cells. Thus, a particular signal may elicit different responses from different cell types.

Among the polypeptide growth factors are the transforming growth factors, which are so designated because, for appropriate cell types, they can bring about not only proliferation, but also morphological transformation, in monolayer culture, and anchorage-independence. Their special characteristics relate the transforming growth factors to neoplastic transformation, oncogenesis, and oncogenes, although such growth factors can occur in normal cells.

Oncogenes, which are capable of inducing neoplastic transformation, bear a remarkable relationship to the growth factor–signal transduction–gene expression sequence method above. Thus, oncogene-encoded proteins include polypeptide growth factors, growth factor receptors, membrane-associated kinases, and guanine nucleotide-binding proteins, and other proteins that can function in various steps of signal transduction pathways, with consequent autocrine production of cellular abnormalities, including stimulation of the growth of cultured cells and of solid tumors. Oncogenes are activated derivatives of protooncogenes, which are oncogene homologs present in normal cells. Mechanisms of activation include (a) changes in regulation of gene expression; (b) point mutations; and (c) formation of recombinant fusion proteins. There are indications that some protooncogenes encode growth factors and receptors, which can function in the control of normal cell proliferation.

Protooncogenes, with or without involvement of proliferative functions, can play key roles in normal processes of differentiation. A number of such roles have been identified for specific protooncogenes. Thus, the *src* protooncogene, which is the cellular progenitor of the tyrosine kinase oncogene of Rous sarcoma virus, can be involved in neuronal differentiation in the absence of cell proliferation.

In line with the various connections between protooncogenes and oncogenes, on the one hand, and peptide growth factors, on the other (Parts I and II of this volume), these growth factors importantly participate in differentiation phenomena. In the present treatment of differentiation and development, emphasis is placed on pattern formation, i.e., the genesis of spatial relationships among the parts of an organism, embryonic or adult. Vertebrate (mammalian and avian) and invertebrate (dipteran) organisms are discussed. Parts III to V of this volume are arranged according to molecular mechanisms that are exhibited in illustrative organ systems: paracrine mechanisms in gonadal and vascular systems and one case of a segmentation system (Part III); autocrine mechanisms in other cases of segmentation and segmentation-like systems (Part IV); and autocrine-type mechanisms with responsiveness to an exogenous small-molecule morphogen or effector in limb and intestinal cell systems (Part V).

Part I deals with major steps leading from growth factor–receptor interactions, via transduction and modulation mechanisms, to the proliferative response. Insulin and insulin-like growth factors and their receptors, which are of broad relevance to differentiation, are considered from the points of view of ligand–receptor binding (Van Wyk and Casella) and membrane protein recycling (Czech *et al.*). Insulin and other growth factors, including epidermal growth factor, can trigger, among others, a transduction mechanism with widespread consequences, namely, the multiple phosphorylation of the 40 S ribosomal protein S6, which can lead to a considerable increase in the rate of general protein synthesis, as a significant step in the transition of animal cells from the quiescent to the proliferative state of growth (Jenö, Ballou, and Thomas). The complexity of phosphorylation events at the cytosolic face of cell membranes is exemplified by the functioning of pp60$^{c\text{-}src}$, a protein product of the c-*src* gene, which is thought capable of participating in the phosphorylation of other proteins. In pp60$^{c\text{-}src}$, there is a multiplicity of phosphorylation sites, and there is a multiplicity of protein kinases that can phosphorylate these sites, with the possibility of interactive effects, either synergistic or counteractive, allowing pp60$^{c\text{-}src}$ to act as a node in a regulatory circuit, which receives inputs through the different protein kinases able to phosphorylate it, and which provides an integrated output by phosphorylating its substrates (Hunter *et al.*). In signal transduction, a large number of receptors can interact with GTP-binding proteins whose properties resemble those of certain cytoskeletal proteins. GTP-binding proteins appear to have higher order polymeric structures which are coiled and springlike, and which are adapted to rapid transmission of information provided by subtle changes in receptor structure (Rodbell). Transduction can be accompanied by the appearance of modulators capable of regulating the activity of enzymes relevant to the transduction signal. Thus, in response to insulin, plasma membranes can release modulators identified as inositol phosphate glycans (Saltiel).

In addition to soluble signals, such as polypeptide growth factors, certain insoluble signals, such as extracellular matrix components, can function as regulators of gene expression. One such component, a heparan sulfate proteoglycan, bound to the surface of hepatocytes by inositol phosphate, can be internalized, and the heparan sulfate chains can be translocated to the cell nucleus. Since heparins are capable of binding growth factors, the possibility is considered that heparins or related molecules can serve as vehicles that would permit the growth factors to reach the nucleus (Reid). One of the components of the extracellular matrix is the basement membrane, and a component of this membrane is laminin, which is a large, complex, and versatile glycoprotein. Laminin can interact with tumor cells through a cell membrane receptor. This interaction plays a significant role in the attachment of tumor cells to the basement membrane. Once attached, the tumor cell can secrete, or induce host cells to secrete, enzymes that can locally degrade the basement membrane. Such degradation is an essential step in tumor cell invasiveness and in metastasis (Sobel and Liotta).

A polypeptide growth factor, transforming growth factor-β, can counteract the degradation of the basement membrane not only by decreasing protease production and stimulating protease inhibitor, but also by increasing the synthesis of extracellular matrix components. This growth factor is mitogenic for fibroblastic and certain other mesenchymal cells, but is growth-inhibitory for many cell types. Transforming growth factor-α, however, is a potent mitogen for most cell types. Cell proliferation and its positive or negative control, under the influence of these two growth factors, are considered (Moses *et al.*).

Part II is concerned with the relation of growth factors and their receptors to oncogenes and to protooncogenes. Transforming growth factor-α shows partial sequence and structural homology to epidermal growth factor, and appears to function through interaction with epidermal growth factor receptor. The major structural elements of the mature receptor are an extracellular epidermal growth factor-binding domain, a transmembrane region of hydrophobic amino acids, and a cytoplasmic domain. The binding of the growth factor to its receptor activates the receptor tyrosine kinase. A truncated form of the epidermal growth factor receptor is encoded by the v-*erbB* oncogene of the avian erythroblastosis virus, and the v-*erbB* protein appears to transform by functioning as an activated growth factor receptor (Schlessinger). The v-*erbB* oncogene can transform erythroblasts and fibroblasts *in vitro*. A second oncogene, v-*erbA*, having an enhancing function, is carried by the same virus, which can induce erythroblastosis and sarcomas *in vivo*. The cellular protooncogene c-*erbA*, corresponding to this viral oncogene, encodes a high-affinity nuclear receptor for thyroid hormones, presumably acting by direct regulation of transcription. The v-*erbA* protein, however, is ligand-independent. Nucleotide sequences, effects of mutations, and the mechanism of transformation are discussed (Vennström *et al.*).

Transformation of NIH 3T3 cells with the *ras* oncogene induces the synthesis of basic fibroblast growth factor. Tumors produced by *ras*-transformed cells express this growth factor. *In vitro*, the expression of the growth factor in *ras*-transformed cells is increased by a factor of 25, in comparison with that in the parental cells (Klagsbrun *et al.*). The finding that NIH 3T3 cells, cultured in a defined medium, die in the absence of fibroblast growth factor or platelet-derived growth factor forms the basis of a new assay for the detection of oncogenes, which has led to the identification of an oncogene encoding a novel fibroblast growth factor (Goldfarb). In connection with the normal cellular functioning of several protooncogenes and with their relationships to peptide growth factors or growth factor receptors, the expression of the *mos* protooncogene in murine male germ cells and oocytes is described (Cooper).

Part III pertains to roles of growth factors and receptors in differentiation and development, particularly, in pattern formation. In male gonadal development and functions, a number of polypeptide growth factors and their receptors are involved. The growth factors include seminiferous growth factor, which appears to be related to acidic fibroblast growth factor; basic fibroblast growth factor; insulin-like growth factor-I; transforming growth factors α and β; β-nerve growth factor; and, probably originating from extragonadal sites, epidermal growth factor and insulin-like growth factor-II. Growth factors of local origin may act primarily in an avascular environment, whereas those present in serum may act within the interstitial or peritubular regions where appropriate cell membrane receptors are accessible. In the developing murine seminiferous epithelium, the expression of seminiferous growth factor is consistent with levels of mRNA encoded by the c-*myc* protooncogene. Seminiferous growth factor and the acidic and basic fibroblast growth factors are related, and have significant effects, which may or may not be concerted, on the regulation of the growth and function of Leydig and Sertoli cells. All three of these growth factors can bind heparin, can induce the proliferation of endothelial cells (among others), and are thought to be involved in neovascularization of the developing gonads (Bellvé and Zheng). Endothelial cells, which form the inner surface of all functional blood vessels, are polarized: the luminal surface has nonthrombogenic properties, whereas abluminally the cells, early in development, produce an extracellular matrix, which is subsequently remodeled into a basement membrane. Laminin expression appears to be an early marker for vascular maturation. Two heparin-binding endothelial growth factors, which are also angiogenic *in vivo*, are structurally similar to the fibroblast growth factors. Endothelial cells cultured on nitrocellulose membranes secrete a platelet-derived growth factor-like chemotactic factor into the abluminal compartment. Such polarized secretion would be expected for a factor involved in the development of the vascular wall. These growth factor-dependent mechanisms are discussed within a broader view of blood vessel growth and differentiation (Risau *et al.*). The vascular endothelium, especially, the endothelium of the capillaries, is involved in the transport not only of small molecules but also of macromolecules such as albumin. The transport of macromolecules, from the lumen of the blood capillaries to the interstitial fluid, is effected by plasmalemma vesicles. In such transcytosis, albumin can serve as a vehicle for fatty acids and other passenger molecules as well as for metal ions. The vesicles functioning in transcytosis are differentiated microdomains, which can be maintained only in the polarized endothelium, presumably with the aid of cell–cell interactions and of serum factors (Palade and Milici).

Unlike the preceding topics, which deal with vertebrate differentiation, the next topic regards a *Drosophila* gene, *decapentaplegic* (*dpp*) which is involved in pattern formation. Genes like *dpp* sometimes are called homeotic since, when mutated, they can cause one part of an organism to transform toward a corresponding homologous part. The protein specified by *dpp*, to date, is the only member of the transforming growth factor-β family of proteins discovered in invertebrates. The nearest vertebrate relative of the *dpp* product is the bone morphogenesis protein BMP-2A. The *dpp* product appears to function, at the cell–cell level, through interaction with a transforming growth factor-type receptor (Gelbart). Frequently, the protein products of homeotic genes function through quite a different mechanism at the level of a single cell.

Part IV discusses homeotic systems regulated intracellularly, with the characteristic involvement of sequence-specific DNA-binding proteins. Homeotic genes in *Drosophila* have a long history beginning with the discovery of *bithorax*, but a detailed understanding of their expression had to await the availability of suitable molecular biological methods. Since homeotic genes typically act early in the programming of the embryonic development of an organism, they can be regarded as master regulatory genes. Three such genes, *bithorax*, *antennapedia*, and *engrailed*, and their expression are discussed. Certain regions of these and related genes share a remarkable degree of homology in the form of the homeo box. Similar sequences are found in other invertebrate genomes as well as in avian and mammalian DNAs. The original mutant of *engrailed* has a striking homeotic effect in that the posterior wing margin transforms toward a mirror image of the anterior wing margin. The *engrailed* gene contributes to the control of segmentation, a graphic instance of pattern formation in embryogenesis (Lewis). The unusual structures of the *engrailed* gene promoter and of the phosphoprotein gene product have been determined. The gene product of *engrailed*, like those of similar genes, is thought to act intracellularly as a sequence-specific DNA-binding protein that regulates transcription. It is suggested that the *engrailed* gene directly orchestrates a subset of other genes whose role is to elaborate the developmental program of the cell (Kornberg *et al.*). Another gene, *even-skipped*, which affects segmentation in *Drosophila*, controls morphogenesis by regulating the expression of *engrailed* and by also regulating its own expression (Hoey, Frasch, and Levine).

In the mouse, there are two genes with sequence homology to the *engrailed* gene and the related *invected* gene of *Drosophila*. The region of homology, containing a centrally located homeo box, can code for 107 amino acids, of which 78 are identical in all four genes. There is evidence that the two mouse genes are expressed early in embryogenesis during periods when fundamental cell lineage decisions are made and when the basic embryo body plan is established (Martin and Joyner). Structural and functional analyses of murine homeo box genes have revealed highly distinct temporal and spatial expression profiles. The intriguing possibility exists that similar molecular mechanisms underlie the generation of cellular diversity and pattern formation in many different organisms (Gruss *et al.*).

Part V describes two differentiation systems thought to involve sequence-specific DNA-binding proteins in conjunction with small molecules. In the developing chick limb, pattern formation depends on a gradient of a diffusible morphogen released from the zone of polarizing activity. This morphogen is concluded to be all-*trans*-retinoic acid. Implantation, at the appropriate site *in ovo*, of all-*trans*-retinoic acid-containing Dowex beads can mimic the effects of the zone of polarizing activity. The morphogen is indicated to act, in a transcriptional induction process, together with a sequence-specific DNA-binding protein belonging to the superfamily of steroid receptors (Eichele and Thaller).

An *in vitro* differentiation system that seems to depend on effects of small molecules is represented by a pluripotent human intestinal epithelial cell line that can be caused to differentiate in culture. Phenotypes resembling terminally differentiated absorptive cells appear in highly polarized confluent cell monolayers when galactose is substituted for glucose as the main carbon source in the culture medium. Apparently, a readily utilizable carbon source (glucose) does not permit such differentiation, whereas a restrictive carbon source (galactose) does. During differentiation of cells grown in the presence of galactose, there is a considerable increase in the rate of synthesis of villin, a calcium-regulated actin-binding cytoskeletal protein present in the brush border of the absorptive cells. The increased rate of villin synthesis is indicated to result from an increased rate of transcription, suggestive of derepression, although a possible stabilization of villin mRNA has not been ruled out (Louvard *et al.*).

It is hoped that the paracrine and autocrine molecular mechanisms considered in this volume contribute to an appreciation of the warp and woof of differentiation as well as of normal and abnormal cellular growth and to an integration of the findings flowing from the various fields of this interdisciplinary area. As documented here, the research is in a highly active state, and accelerating advances are foreshadowed.

It is a pleasure to acknowledge the advice and help of Dr. G. Blobel, Dr. H. Green, Dr. H. Hanafusa, Dr. E. B. Lewis, Dr. D. Louvard, Dr. H. L. Moses, Dr. G. E. Palade, Dr. M. Rodbell, and Dr. J. J. Van Wyk.

We are grateful for the fine support of the College of Physicians and Surgeons (P & S) of Columbia University, without which this volume would not have reached fruition, and for the continued interest of Dr. D. F. Tapley. This volume was developed from a P & S Biomedical Sciences Symposium held at Arden House on the Harriman Campus of Columbia University.

Anthony R. Bellvé
Henry J. Vogel

PART I

POLYPEPTIDE GROWTH FACTORS AND OTHER SIGNALS: CELL MEMBRANE RECEPTORS, TRANSDUCTION MECHANISMS, AND PROLIFERATIVE RESPONSE

1
A Dual Binding Site Hypothesis of the Type I Insulin-Like Growth Factor Receptor

JUDSON J. VAN WYK AND SAMUEL J. CASELLA[1]
Department of Pediatrics
The University of North Carolina at Chapel Hill School of Medicine
Chapel Hill, North Carolina 27599

INTRODUCTION

A continuing interest of our laboratory is to determine which receptors mediate the growth promoting effects of insulin and the insulin-like growth factors IGF-I and IGF-II. For this purpose we and others have utilized human placental membranes, which are rich in type I and type II somatomedin receptors as well as in insulin receptors (1). The type I receptor is a well-characterized disulfide linked tetramer consisting of two 90-kDa β subunits and two 130-kDa α subunits (2–4). This receptor has higher affinity for IGF-I than for IGF-II and binds insulin weakly. The insulin receptor has similar structural characteristics but has highest affinity for insulin and weakly binds IGF-II and IGF-I (4). The type II receptor migrates as a 220-kDa band under nonreducing conditions and as a 260-kDa band after reduction. The type II receptor has higher affinity for IGF-II than IGF-I and does not bind insulin (2,3) (Table I). The type II receptor has no protein kinase activity, and it remains to be demonstrated whether any of the actions of the somatomedins are mediated through this receptor. The type II receptor has recently been

---

1. Present affiliation: Department of Pediatrics, The Johns Hopkins University School of Medicine, Baltimore, Maryland 21205.

TABLE I
Physical Characteristics and Hormonal Specificities of the Receptors for Insulin and the Somatomedin/Insulin-Like Growth Factor

Receptor	Chemical structure	Specificity for		
		Insulin	SM-C/IGF-I	IGF-II/MSA
Insulin	Unreduced: Heterotetramer 300K Reduced subunits: α (binding): 130K β (tyrosine kinase): 90K	++++	+	++
Type I (Sm-C/IGF-I)	Unreduced Heterotetramer 300K Reduced subunits: α (binding): 130K β (tyrosine kinase): 90K	+	++++	++
Type II (IGF-II/MSA)	Unreduced: 220K Reduced: 260K No known subunits	[a]	++	++++

[a] Insulin stimulates increase in type II receptors on plasma membrane.

cloned and found to have near identity with the mannose 6-phosphate receptor (5). Thus, it remains to be clarified through which receptor(s) IGF-II produces its biological effects.

WHICH RECEPTOR MEDIATES THE ACTIONS OF IGF-II?

Hypotheses Based on Competitive Binding Studies

Radioreceptor assays for the somatomedins employing human placental membranes have poor specificity because of the structural homologies that exist between insulin and the somatomedins and their complex pattern of cross-reactivities. Based on competitive binding studies, Nissley and Rechler (6) proposed that such assays primarily measure interactions with the type I receptor, even when iodinated IGF-II is used as the tracer. Evidence to the contrary has been reported by Hintz *et al.* (7), who concluded that IGF-II binds not only to the type I and II receptors, but also to a distinct receptor which they termed the type III receptor. These workers pointed out that since a considerable portion of the binding of ^{125}I-IGF-II to IM-9 lymphoblasts was inhibited by insulin, this binding could not be to the classical type II receptor since insulin does not cross-react with the type II receptor. Because the binding was more sensitive to IGF-II than IGF-I, however, they argued that there might be a third class or "type III" receptor that had high affinity for IGF-II and that also recognized insulin (8). This postulate was supported by the work of Jonas et al., who reported that an atypical insulin receptor with high affinity for IGF-I and IGF-II copurified with the insulin receptor on insulin agarose affinity columns (9).

Subtypes of the Type I IGF Receptor

Jonas and Harrison described two subgroups of IGF type I receptors in human placentas based on whether they cross-reacted with an autoantibody to the insulin receptor (B-2) (10). Those receptors that were recognized by the antibody B-2 had lower affinity for IGF-I, whereas receptors with high affinity for IGF-I were not recognized by the antibody. They subsequently reported that the low affinity immunoreactive species could be converted by disulfide reduction to a high affinity nonimmunoreactive species (11). It is not clear, however, whether these subclasses represent different receptor subtypes *in vivo* or simply partially degraded receptors.

Perhaps the most convincing evidence for the existence of subtypes of the type I receptor was provided by Morgan and Roth (12), who developed a monoclonal antibody to the insulin receptor (5D9) that also blocks IGF-I binding. Although the antibody blocked the binding of ^{125}I-IGF-I to human placental membranes and IM-9 receptors, it did not inhibit binding to HEP G-2 cells. Thus, there are clearly immunological differences between type I receptors in different tissues. The authors also contend that there are subclasses of both types in placenta because they could not completely inhibit IGF-I binding to placental membranes with their monoclonal antibody. The primary difficulty with the above studies is that they are based primarily on binding studies and immunoreactivities. No one has yet succeeded in isolating the putative subtypes, nor have any unique biological functions been attributed to any of them.

Studies with α-IR3, a Monoclonal Antibody to the Type I Receptor

Kull et al. (13) developed a battery of monoclonal antibodies to a preparation of insulin receptor that had been purified from human placental membranes. Recognizing that type I IGF receptors might have been copurified, they screened their clones for an antibody that might preferentially recognize the IGF-I receptor. This effort led to the isolation of α-IR3, a monoclonal antibody that binds IGF-I but not insulin.

We have used α-IR3 in an attempt to further define the nature of the receptors that bind the respective peptides in human placental membranes. We found that the antibody inhibited ^{125}I-IGF-I binding to human placental membranes by 60–70% but inhibited the binding of ^{125}I-IGF-II by less than

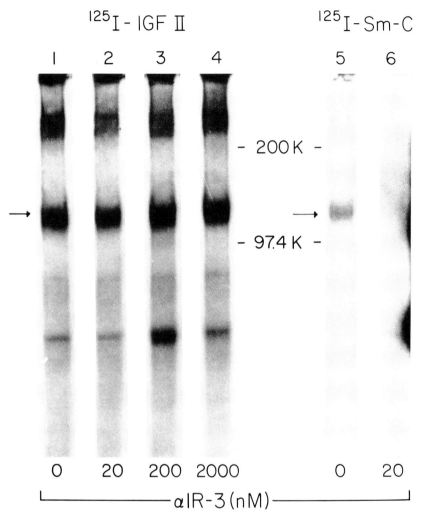

Fig. 1. Effect of the anti-type I receptor antibody α-IR3 on binding of ^{125}I-IGF-II and ^{125}I-Sm-C/IGF-I to the 130K binding site in human placental membranes. Particulate human placental membranes were incubated overnight at 4°C with ^{125}I-IGF-II (lanes 1–4) or ^{125}I-Sm-C (lanes 5 and 6) with graded concentrations of α-IR3, as shown below the lanes, then affinity cross-linked with 0.3 mM disuccinimidyl suberate. Samples were solubilized, reduced, and analyzed by SDS–PAGE (3–14% gradient gel), then autoradiographed. The antibody blocked most of ^{125}I-Sm-C/IGF-I binding at a concentration of 20 nM, whereas concentrations up to 2000 nM failed to block binding of ^{125}I-IGF-II.

15% (14). Affinity cross-linking studies demonstrated that, in this tissue, ^{125}I-IGF-II binds both to the type II receptor and to a 130K moiety that has an electrophoretic pattern that is identical to that of the type I receptor. Binding to neither band, however, was inhibited by α-IR3 (Fig. 1). We postulated either that IGF-II was binding to a very similar receptor (type III receptor as suggested by Hintz *et al.*) or, alternatively, that IGF-II binds to an immunologically distinct site on the type I receptor.

To address these two possibilities, type I receptors were purified from human placental membranes using an α-IR3 immunoaffinity column. We predicted that any and all binding of ^{125}I-IGF-I and ^{125}I-IGF-II to this highly purified population of receptors, all of which shared an epitope to α-IR3, would be completely inhibited by the antibody. We found that both ^{125}I-IGF-I and ^{125}I-IGF-II bound to the purified receptor preparation with high affinity. Surprisingly, however, the antibody inhibited ^{125}I-IGF-II binding by less than 15% and the binding of ^{125}I-IGF-I by only 60%.

The Two Site Hypothesis for the Type I Receptor

These studies at the very least revealed that IGF-I and IGF-II bind at immunologically distinct sites. Because the two ligands both bind with high affinity and each can effectively compete for binding sites with the other, we have postulated that there are two binding sites on the same receptor molecule. The 1A site has highest affinity for IGF-I and is blocked by the antibody α-IR3. The 1B site has higher affinity for IGF-II and is not blocked by the antibody. This model would also explain the inability of α-IR3 to block all of the IGF-I binding because IGF-I could presumably bind to the 1B site, but with slightly lower affinity.

Although our purified receptors appeared homogeneous, and Scatchard plots were linear, we cannot exclude the possibility that our preparation contained two very similar forms of receptors. Affinity cross-linking studies showed that the 130-kDa binding subunit of the purified receptor is slightly smaller than that of the insulin receptor; furthermore, competitive binding studies with radiolabeled insulin revealed no hint that our receptor preparation might contain insulin receptors or the "atypical" insulin receptors described by Hintz and Jonas *et al*. It is conceivable, however, that the type 1A and type 1B receptors described by Morgan *et al.* share an epitope recognized by α-IR3 and copurified on our affinity column. The resolution of these various possibilities must await the isolation of the receptor subtypes. The cloning of the type I receptor and the derivation of its primary sequence may make it possible to specifically characterize the binding sites on the type I receptor and to disprove or support the theory that two binding sites exist on a single receptor.

ACKNOWLEDGMENTS

The authors wish to acknowledge the assistance of Burton J. Balfour and Lori Ann Bono in the preparation of the manuscript. This work was supported by U.S. Public Health Service Research Grants 2 R01 DK01022 (J.J.V.W.) and 1R29 DK38542 (S.J.C.).

REFERENCES

1. Marshall, R. N., Underwood, L. E., Viona, S. G., Foushee, D. B., and Van Wyk, J. J. (1974). *J. Clin. Endocrinol. Metab.* **39**, 283–292.
2. Kasuga, M., Van Obberghen, E., Nissley, S. P., and Rechler, M. M. (1981). *J. Biol. Chem.* **256**, 5305–5308.
3. Massague, J., and Czech, M. P. (1981). *J. Biol. Chem.* **257**, 5038–5045.
4. Chernausek, S. D., Jacobs, S., and Van Wyk, J. J. (1981). *Biochemistry* **20**, 7345–7350.
5. Morgan, D. O., Edman, J. C., Standring, D. N., Fried, V. A., Smith, M. C., Roth, R. A., and Rutter, W. J. (1987). *Nature (London)* **329**, 301–307.
6. Nissley, S. P., and Rechler, M. M. (1984). *Clin. Endocrinol. Metab.* **3**, 43–67.
7. Hintz, R. L., Thorsson, A. V., Enberg, G., and Hall, K. (1984). *Biochem. Biophys. Res. Commun.* **118**, 774–782.
8. Misra, P., Hintz, R. L., and Rosenfeld, R. G. (1986). *J. Clin. Endocrinol. Metab.* **63**, 1400–1405.
9. Jonas, H. A., Newman, J. D., and Harrison, L. C. (1986). *Proc. Natl. Acad. Sci. U.S.A.* **83**, 4124–4128.
10. Jonas, H. A., and Harrison, L. C. (1985). *J. Biol. Chem.* **260**, 2288–2294.
11. Jonas, H. A., and Harrison, L. C. (1986). *Biochem. J.* **236**, 417–423.
12. Morgan, D. O., and Roth, R. A. (1986). *Biochem. Biophys. Res. Commun.* **138**, 1341–1347.
13. Kull, F. C., Jacobs, S., Su, Y.-F., Svoboda, M. E., Van Wyk, J. J., and Cuatrecasas, P. (1983). *J. Biol. Chem.* **258**, 6561–6566.
14. Casella, S. J., Han, V. K., D'Ercole, A. J., Svoboda, M. E., and Van Wyk, J. J. (1986). *J. Biol. Chem.* **261**, 9268–9273.

2
Insulin Action on Membrane Protein Recycling

MICHAEL P. CZECH, SILVIA CORVERA, MARK A. TEPPER,
CAROL M. BROCKLEBANK, AND RICHARD G. MACDONALD

Department of Biochemistry
University of Massachusetts Medical School
Worcester, Massachusetts 01655

INTRODUCTION

The activation of hexose transport in muscle and adipose cells in animals and man represents one of the major actions of insulin on cell metabolism. Experiments *in vivo* and *in vitro* have clearly demonstrated that this effect of insulin is rapid in onset and completed in a few minutes after exposure of tissue to the hormone. Insulin-sensitive glucose transporters in these tissues operate by a facilitated diffusion mechanism (1). Activation of hexose transport leads to enhanced glycolytic rates and glycogen deposition. Thus, this effect of insulin contributes in large part to its hypoglycemic action *in vivo*. Furthermore, impairment of insulin action on glucose transport appears to be associated with some forms of type II diabetes mellitus (2). Thus, the detailed molecular mechanism by which insulin modulates hexose transport in target cells has generated much scientific and clinical interest.

MEMBRANE PROTEIN RECYCLING

The facilitated diffusion of glucose across cell membranes could be hormonally regulated either by an increase in the number of glucose transport systems present in the plasma membrane or by an increase in the intrinsic activity of a similar number of transporters, or both. Convincing evidence was reported several years ago that the former possibility occurs. Cushman and Wardzala (3) as well as Suzuki and Kono (4) simultaneously reported that in isolated fat cells insulin caused an increased number of glucose transporters in adipocyte plasma membrane preparations at the expense of a pool of low density microsome transporter proteins. This conclusion was made by measuring specific cytochalasin B binding to adipocyte membrane fractions or by reconstituting transport activity from such fractions. Subsequently, using affinity labeling techniques, we demonstrated that, indeed, increased numbers of glucose transporters appear on the cell surface of intact adipocytes in response to insulin (5). Although these experiments established this phenomenon convincingly, they do not eliminate the possibility that enhanced intrinsic activity of glucose transporters also occurs upon addition of insulin. Indeed, recent published work has supported this concept. Baly and Horuk (6) have found that increased hexose transport activity in response to insulin can be dissociated from increased numbers of glucose transporters in the cell surface membrane using the protein synthesis inhibitor cycloheximide. These data suggest the possibility that modulation of the intrinsic activity of glucose transporters may contribute in a major way to the action of insulin on hexose transport. Other experiments will be needed to clarify this issue.

The action of insulin to cause redistribution of glucose transporter proteins from intracellular membrane pools to the cell surface membrane now appears to reflect a more general paradigm in cell biology. Experiments in our laboratory have shown that at least four other membrane proteins undergo a similar membrane redistribution reaction in response to insulin. These include the type II insulin-like growth factor (IGF) receptor (7), the transferrin receptor (8), the low density lipoprotein (LDL) receptor (9), and the α_2-macroglobulin receptor (10). In each case, an increased number of these receptor proteins can be directly monitored on the surface of intact target cells upon addition of the hormone. Furthermore, other hormones that activate specific receptor kinases also appear to have somewhat similar effects on some of these proteins, as depicted schematically in Fig. 1. We found that addition of insulin-like growth factor I or insulin to A431 cells in culture caused a rapid expression of transferrin receptors on the cell surface as monitored by specific antireceptor antibody (11). The effects of these growth factors on the number of transferrin receptors expressed on the cell surface exhibit a similar time course to that of insulin action on hexose transport. The membrane receptors modulated by insulin are known to recycle constitutively between the plasma membrane and intracellular endosomal membranes. Thus, it would appear that insulin and the other growth factors modulate one or more steps in the recycling process. Whether glucose transporters recycle in a similar manner is unknown at present. The

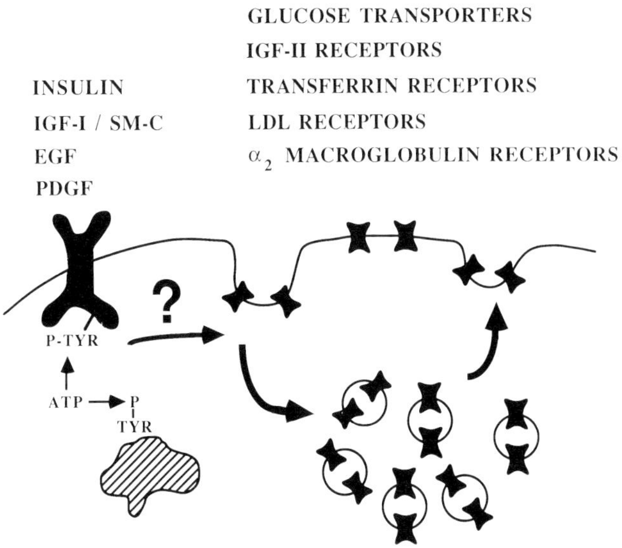

GLUCOSE TRANSPORTERS
IGF-II RECEPTORS
INSULIN   TRANSFERRIN RECEPTORS
IGF-I / SM-C   LDL RECEPTORS
EGF α_2 MACROGLOBULIN RECEPTORS
PDGF

Fig. 1. Activation of receptor tyrosine kinases causes the increased cell surface expression of several membrane proteins. Addition of epidermal growth factor (EGF), platelet-derived growth factor (PDGF), insulin-like growth factor I (IGF-I), or insulin to a number of rat, mouse, and human cell lines (A431, WI-38, H-35, 3T3-L1, rat adipocytes) elicits transient (EGF, PDGF) or sustained (IGF-I, insulin) increases in the cell surface expression of at least five distinct proteins.

detailed molecular events which occur subsequently to activation of the growth factor receptor tyrosine kinases and the modulation of the nutrient receptor recycling pathway are unknown.

We have also been able to demonstrate that membrane redistribution of IGF-II and transferrin receptors is associated with increased uptake of the respective ligands (7,8,11). Thus, similarly to the glucose transport system, these nutrient receptors appear to be regulated by insulin and the other growth factors for the purpose of enhancing cellular uptake of a variety of nutrients and other molecules.

We have studied the regulation of the transferrin and the IGF-II receptor systems in order to gain more insight into the steps of the recycling pathway affected by insulin. The transferrin receptor system offers an advantage in evaluating the recycling process in that transferrin remains bound to its receptor throughout the entire process of internalization and recycling back to the plasma membrane (12,13). Thus, reliable estimates of receptor internalization (endocytosis) and exocytosis rates can be obtained by measuring labeled transferrin uptake and release under certain carefully controlled experimental conditions. Using such methods, we were able to show that membrane redistribution of transferrin receptors due to IGF-I action results from both an increased receptor exocytosis rate and decreased receptor endocytosis when normalized per cell surface number of receptors (14). Thus, while the absolute number of receptors internalized per unit time is increased somewhat in response to IGF-I, the rate of receptor internalization normalized to the increased number of available cell surface receptors is lower after hormone treatment.

An interesting finding in our studies was that IGF-I also caused an increased percent occupancy of cell surface receptors by ligand at steady state, presumably due to the increased time each receptor spends on the cell surface in IGF-I-treated cells. This increased percent ligand occupancy of transferrin receptors appears to contribute to the increased ligand uptake by insulin- or IGF-I-treated cells at low ligand concentrations. Taken together, these experiments indicate that reciprocal control by insulin on transferrin receptor recycling occurs such that both the endocytotic and exocytotic pathways are affected in opposite directions.

In attempting to understand the molecular mechanisms whereby receptor recycling steps are modulated by insulin and other growth factors, two extreme models or hypotheses were considered. First, it is possible that insulin acts directly on the cellular machinery that operates to internalize and externalize membrane receptors during the recycling process. While some cellular elements involved in this process (e.g., clathrin-coated pits) are known, these processes are poorly understood at the molecular level at present. An alternative hypothesis is that the receptor proteins are themselves directly changed in structure in response to insulin, thereby modulating the rate at which they pass through one or more steps of the recycling process. In our studies, we have focused on the latter hypothesis, concentrating on changes in the phosphorylation state of IGF-II receptors. Phosphorylation/dephosphorylation reactions are important regulatory mechanisms, especially related to hormone action. We chose the IGF-II receptor system for these studies because these receptors show a particularly large response to insulin in isolated fat cells and because particularly potent polyclonal antibodies against purified receptor from rat placenta are available in our laboratory.

The type II IGF receptor is also an excellent model for studying possible phosphorylation changes because its primary amino acid sequence is now known. Amino acid sequences deduced from cDNAs isolated from human (15) and rat (16) libraries surprisingly indicated a high degree of identity with the primary structure of the bovine cation-independent mannose 6-phosphate (Man-6-P) receptor. Definitive biochemical experiments performed in our laboratory using affinity chromatography techniques showed that the same receptor protein indeed binds both IGF-II- and Man-6-P-linked lysosomal enzymes (16). These results indicate that the type II IGF receptor is a multifunctional protein that binds at least one other physiologically important ligand. In addition, it is known from previous studies using Man-6-P-derivatized ligands that this receptor is involved in the transport of newly synthesized Man-6-P-linked lysosomal enzymes from the Golgi to the lysosomes (17,18). We were also able to demonstrate that Man-6-P binding to this receptor increases its affinity for ^{125}I-IGF-II (16). The physiological consequence of this effect is not understood at present.

The deduced primary structure of the IGF-II/Man-6-P receptor provides the basis for the general receptor structure schematically depicted in Fig. 2. The receptor consists of a large extracellular domain containing over 2000 amino acid residues. This extracellular domain is composed of 15 segments containing multiple cysteines that can be aligned with each other. The spe-

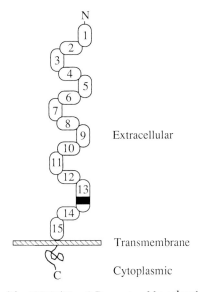

Fig. 2. Structure of the IGF-II/Man-6-P receptor. More than 90% of the receptor is located in the extracellular space. The extracellular structure is constituted by 15 domains, each of which contains multiple cysteine residues at similar positions within each segment. Domain 13 contains a sequence which exhibits homology to the type II repeat in the collagen binding domain of fibronectin. A 22 amino acid membrane-spanning sequence is followed by a small cytoplasmic portion, analyzed in detail in Fig. 4.

cific amino acid sequences within each of the 15 segments do not in themselves show a high degree of identity. Each of the 15 segments is approximately 150 amino acids long. One segment contains an amino acid sequence which appears similar to that found in a number of proteins including fibronectin. The receptor contains multiple N-linked glycosylation consensus sequences located predominantly in the amino-terminal half of the extracellular domain. The receptor appears to contain one putative transmembrane domain of about 22 amino acids followed by a relatively small cytoplasmic carboxy-terminal tail consisting of 167 (rat), 164 (human), or 163 (bovine) amino acids. The cytoplasmic domain is devoid of sequences that would suggest a kinase activity.

Figure 3 depicts results in experiments designed to assess the phosphorylation state of IGF-II/Man-6-P receptors in ^{32}P-labeled control and insulin-treated adipocytes (19). In these experiments, intact cells were incubated for 2 hr with ^{32}P prior to addition of insulin and further incubation for the indicated times. The cells were homogenized at various time points and fractionated into plasma membranes (Fig. 3A) and low density intracellular microsomes (Fig. 3B). IGF-II/Man-6-P receptors were immunoprecipitated from the membrane fractions and quantified by Western blotting. The ^{32}P

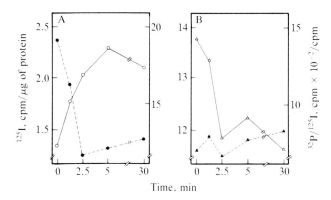

Fig. 3. Time course of the effect of insulin to produce redistribution of IGF-II receptors and decrease the phosphorylation state of the receptor in the plasma membrane. Adipocytes were incubated for 2 hr with ^{32}P at 1 mCi/ml. Insulin (10 nM) was added to each sample, and incubations were continued for the times indicated. The incubation medium was removed as quickly as possible, and the cells were homogenized. The earliest time point that could be recorded, approximately 1.2 min, was the shortest period of time between the addition of insulin and homogenization of the cells. Plasma membranes (A) and low density microsomes (B) were solubilized, and the IGF-II receptor was immunoprecipitated, electrophoresed, and immunoblotted. ^{125}I radioactivity is expressed as a function of the amount of protein initially used for immunoprecipitation. ^{125}I radioactivity per microgram of protein (○ and △) and ^{32}P/^{125}I ratios (● and ▲) for the IGF-II receptor are plotted. (Reprinted from Corvera and Czech, 19.)

content of the receptors was also determined, and an estimate of the specific activity (^{32}P/^{125}I) was calculated.

Three important findings were made in these experiments (19). First, IGF-II/Man-6-P receptors present in adipocyte plasma membranes have a much higher ^{32}P content than receptors in intracellular membranes. Second, addition of insulin to adipocytes dramatically decreases the estimated specific activity of [^{32}P]phosphate in IGF-II/Man-6-P receptors present in the plasma membrane concomitant with increasing numbers of receptors in this membrane fraction. Third, insulin action is not accompanied by a change in the specific activity of [^{32}P]phosphate incorporation in receptors in the low density microsome fraction. Thus, the rapid action of insulin to inhibit phosphorylation of IGF-II receptors in these cells was localized to the plasma membrane domain. More recently, similar experiments have been performed in H-35 hepatoma cells, showing that insulin decreases the phosphorylation of IGF-II/Man-6-P receptors concomitant with the membrane redistribution effect. The effect of insulin on [^{32}P]phosphate content of type II IGF receptors does not appear to be secondary to the increased numbers of receptors which appear in the plasma membrane. This conclusion is based on an experiment showing that the effect of insulin to decrease phosphorylation of type II IGF receptors occurs even when the membrane redistribution effect is inhibited at low temperature (19).

Phosphoamino acid analysis of IGF-II/Man-6-P receptors isolated from ^{32}P-labeled adipocytes or H-35 hepatoma cells shows that the phosphorylated residues are serine and threonine (20). Insulin decreases the phosphorylation of both residues in these experiments. Tryptic peptide mapping of ^{32}P-labeled IGF-II receptors derived from H-35 hepatoma cells revealed three major phosphorylated peptides, each of which appeared to be decreased in ^{32}P content in insulin-treated cells. These data are consistent with the interpretation that multiple phosphorylation sites on the cytoplasmic domain of the IGF-II/Man-6-P receptor are phosphorylated in intact insulin-sensitive cells and that insulin action decreases the phosphorylation or increases the dephosphorylation of these multiple sites.

In order to investigate which kinase or kinases might be responsible for the phosphorylation of IGF-II/Man-6-9 receptors in intact cells, experiments were performed with purified receptors and preparations of various purified kinases (cyclic AMP-dependent protein kinase, casein kinase II, protein kinase C, and phosphorylase kinase). Significant phosphorylation of purified receptors was catalyzed by casein kinase II but not the other kinases tested (20). Both threonine and serine residues on the receptor were phosphorylated *in vitro* by purified casein kinase II. Furthermore, bidimensional peptide mapping revealed that the casein kinase II-catalyzed phosphorylation of the IGF-II/Man-6-9 receptor involved a tryptic phosphopeptide which comigrated with the main tryptic phosphopeptide found in receptors obtained from cells labeled *in vivo* with [^{32}P]phosphate. Importantly, IGF-II/Man-6-P receptors isolated from insulin-treated H-35 cells or adipocytes were phosphorylated *in vitro* by casein kinase II to a greater extent than receptors isolated from control cells. Thus, the insulin-regulated phosphorylation sites on the IGF-II receptor appear to serve as substrates *in vivo* for casein kinase II or an enzyme with similar substrate specificity (20).

Careful analysis of the deduced amino acid sequence of the cytoplasmic portion of the rat IGF-II/Man-6-P receptor reveals the presence of two striking consensus sequences for casein kinase II substrates (Fig. 4, denoted by plus symbols). These two serine residues are flanked on both the amino-terminal and carboxy-terminal sides with repetitive aspartate and glutamate residues. In addition, an acidic stretch of amino acids between these serines and the transmembrane portion of the receptor contains two threonines which might serve as substrates for casein kinase II (Fig. 4, denoted by asterisks). These latter threonine residues seem less specific as casein kinase II substrates, judging from data obtained with synthetic peptides as substrates, but could be substrates in the native protein. In any case, the structural data obtained from cDNA sequences of the receptor reinforce the biochemical data obtained that indicate receptor phosphorylation occurs by casein kinase II or a similar enzyme in intact cells. Experiments in progress are devoted to determining definitively which specific receptor residues are phosphorylated and regulated by insulin in intact cell plasma membranes.

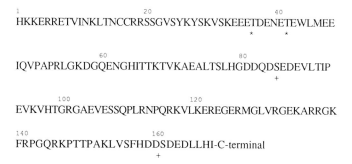

```
         1                    20                          40
         HKKERRETVINKLTNCCRRSSGVSYKYSKVSKEEETDENETEWLMEE
                                            *         *

                    60                         80
         IQVPAPRLGKDGQENGHITTKTVKAEALTSLHGDDQDSEDEVLTIP
                                                     +

                  100                     120
         EVKVHTGRGAEVESSQPLRNPQRKVLKEREGERMGLVRGEKARRGK

         140                160
         FRPGQRKPTTPAKLVSFHDDSDEDLLHI-C-terminal
                               +
```

Fig. 4. Primary amino acid structure of the cytoplasmic portion of the rat IGF-II/Man-6-P receptor. Residues are numbered from the first amino acid (H) following the putative membrane-spanning region in the receptor sequence. Salient features of the cytoplasmic portion of the receptor are (a) its acidic nature, (b) the presence of serine (+) and threonine (*) residues within acidic amino acid stretches (which comprise potential phosphorylation sites for casein kinase II), and (c) the absence of consensus sequences for phosphotransferase activities. (Data are from MacDonald *et al.*, 16.)

CONCLUSIONS

The information reviewed briefly here suggests that a major physiological action of insulin and other growth factors is to stimulate nutrient uptake in concert with membrane redistribution of transporters and receptors involved in nutrient transport. Although the molecular basis for these actions of insulin on receptor redistribution and ligand uptake is not understood, recent data obtained in our laboratory indicate three novel conclusions.

1. Studies on the transferrin receptor indicate that insulin action on this system involves reciprocal control of both exocytosis (stimulation) and relative endocytosis rates (inhibition). These combined effects account for steady-state receptor redistribution and increased ligand occupancy and uptake (14).
2. IGF-II/Man-6-P receptors are more highly phosphorylated in cell surface membranes than intracellular membranes in adipocytes and are rapidly dephosphorylated in response to insulin. Inhibition of receptor phosphorylation by insulin appears to occur on sites that are specific for casein kinase II or similar enzyme activity and correlates with membrane redistribution of this receptor.
3. The primary structure of the type II IGF receptor deduced from cDNA sequences indicates that it is a multifunctional receptor which also binds Man-6-P. The deduced amino acid sequences of the cytoplasmic domain of this insulin-regulated receptor contain consensus sequences for highly specific casein kinase II phosphorylation sites. These appear to be excellent candidates for the insulin-regulated phosphorylation sites observed in intact cells.

Is there a direct relationship between regulation of IGF-II/Man-6-P receptor phosphorylation and the membrane redistribution of this receptor in response to insulin? Data available to date do not allow a definitive answer to this question. Our results indicate only a high degree of correlation between insulin-inhibited IGF-II/Man-6-P receptor phosphorylation and increased receptor number in the plasma membrane. A rigorous approach toward determining whether changes in phosphorylation of the receptor in response to insulin actually participate in driving the membrane redistribution effect would be to express mutant forms of the receptor devoid of the regulated phosphorylation sites. The availability of the rat IGF-II/mannose 6-phosphate receptor cDNA in our laboratory should allow such experiments to answer this important question. It is also possible that actions of insulin on the cellular machinery involved in the endocytosis and exocytosis might occur. Clearly, this hypothesis also deserves careful attention in the future.

REFERENCES

1. Czech, M. P. (1980). *Diabetes* **29**, 399–409.
2. Reaven, G. M., Bernstein, R., Davis, B., and Olefsky, J. (1976). *Am. J. Med.* **60**, 80.
3. Cushman, S. W., and Wardzala, L. S. (1980). *J. Biol. Chem.* **225**, 4758–4762.
4. Suzuki, Y., and Kono, T. (1980). *Proc. Natl. Acad. Sci. U.S.A.* **77**, 2542–2545.
5. Oka, Y., and Czech, M. P. (1984). *J. Biol. Chem.* **255**, 10382–10386.
6. Baly, D. L., and Horuk, R. (1987). *J. Biol. Chem.* **262**, 21–24.
7. Oka, Y., Rozek, L. M., and Czech, M. P. (1985). *J. Biol. Chem.* **260**, 9435–9442.
8. Davis, R. J., Corvera, S., and Czech, M. P. (1986). *J. Biol. Chem.* **261**, 8708–8711.
9. Czech, M. P., Corvera, S., Tepper, M. A., Brocklebank, C. M., and MacDonald, R. G. Unpublished observations.
10. Czech, M. P., Corvera, S., Tepper, M. A., Brockelbank, C. M., and MacDonald, R. G. Unpublished observations.
11. Davis, R. J., and Czech, M. P. (1986). *EMBO J.* **5**, 653–658.
12. Ciahanover, A., Schwartz, A. L., Deurtry-Varsat, A., and Lodish, H. F. (1983). *J. Biol. Chem.* **258**, 9681–9689.
13. Klausner, R. D., Van Renswoude, J., Ashwell, G., Kempf, C., Schechter, A. N., Dean, A., and Bridges, K. R. (1983). *J. Biol. Chem.* **258**, 9681–9689.
14. Davis, R. J., Faucher, M., Kuck-Racaniello, L., Carruthers, A., and Czech, M. P. (1987). *J. Biol. Chem.* **262**, 13126–13134.
15. Morgan, D. O., Edman, J. C., Standring, D. N., Fried, V. A., Smith, M. C., Roth, R. A., and Rutter, W. J. (1987). *Nature (London)* **329**, 301–307.
16. MacDonald, R. G., Pfeffer, S. R., Coussens, L., Tepper, M. A., Brocklebank, C. M., Mole, J. E., Anderson, J. K., Chen, E., Czech, M. P., and Ullrich, A. (1988). *Science* **239**, 1134–1137.
17. von Figura, K., and Haslik, A. A. (1986). *Annu. Rev. Biochem.* **55**, 167–193.
18. Kornfeld, S. (1987). *FASEB J.* **1**, 462–468.
19. Corvera, S., and Czech, M. P. (1985). *Proc. Natl. Acad. Sci. U.S.A.* **82**, 7314–7318.
20. Corvera, S., Roach, P. J., DePaoli-Roach, A. A., and Czech, M. P. (1988). *J. Biol. Chem.* **263**, in press.

3
Growth-Factor and Oncogene-Mediated S6 Phosphorylation during the Mitogenic Response: A Novel S6 Kinase

PAUL JENÖ, L. M. BALLOU, AND GEORGE THOMAS
Friedrich Meischer Institut
Basel, Switzerland

STIMULATION OF S6 PHOSPHORYLATION

Within minutes of addition of a number of mitogens to quiescent Swiss mouse 3T3 cells, there is a rapid stimulation of 40 S ribosomal protein S6 phosphorylation and, depending on the mitogen employed, as much as a two- to threefold activation of protein synthesis (1,2). The increased rate of protein synthesis is due to an increase in the rate of initiation of protein synthesis (3), which is usually measured as a shift of inactive 80 S ribosomes and a large pool of stored mRNA into actively translating polysomes (4). In the case of epidermal growth factor (EGF) treatment of quiescent Swiss mouse 3T3 cells, 10^{-12} to 10^{-10} M EGF has little effect on S6 phosphorylation or on the rate of initiation of protein synthesis. However, as the concentration of EGF is raised to 10^{-9} M, there is a sharp burst in S6 phosphorylation and the rate of protein synthesis. The increase in protein synthesis is about 1.5-fold, as measured by the incorporation of [^{35}S]methionine into trichloroacetic acid (TCA)-insoluble material or as the percentage of ribosomes present in polysomes (2). Furthermore, these changes are accompanied by a number of marked alterations in the pattern of translation (5). These studies, supported by a number of other experiments *in vivo* and *in vitro*, have lead to the hypothesis that the multiple phosphorylation of S6 is either a prerequisite for or facilitates the activation of protein synthesis during the transition of animal cells from the quiescent to the proliferative state of growth.

The extent of S6 phosphorylation can be measured on two-dimensional polyacrylamide gels by the change in electrophoretic mobility of the protein. The more phosphorylated the protein becomes, the slower its mobility in both dimensions of electrophoresis. At 10^{-8} M EGF, S6 phosphorylation reaches its maximum after 1 hr, with most of S6 shifting to derivatives S6d and S6e, which contain 4 and 5 mol of phosphate per mole of S6, respectively. Between 1 and 2 hr postinduction, the extent of S6 phosphorylation begins to decrease, though a portion of the protein population remains highly phosphorylated up to 6 hr following the addition of EGF (A. Olivier and G. Thomas, unpublished). The activation of protein synthesis is slightly delayed, reaching maximum after 4 hr but remaining persistently elevated for at least 6 hr (6). Thus, S6 phosphorylation seems to be important for triggering some early response in the activation of protein synthesis but does not appear to be required for its long-term maintenance. The maximally phosphorylated form of S6 contains 10 or 11 tryptic phosphopeptides. All the phosphopeptides contain phosphoserine, and 2 contain traces of phosphothreonine. When the 5 increasingly phosphorylated derivatives of S6 were separated by two-dimensional polyacrylamide gel electrophoresis and digested with trypsin, it was shown that the phosphopeptides appeared in a specific order (7). The results above would suggest that the phosphates are added to S6 in a specific order, either by one or a number of kinases. If sequence studies substantiate these findings, it will be of interest to determine whether ordered phosphorylation of S6 is dictated by site phosphorylation or the sequential activation of a number of kinases (8).

ACTIVATION OF S6 KINASE

Having the phosphopeptide maps above and knowing the EGF dose–response curve for S6 phosphorylation in the intact cells made it possible to begin the search for the S6 kinase. High speed supernatants from EGF-stimulated 3T3 cells were found to contain an S6 kinase, which is 25-fold more active in phosphorylating S6 than extracts from quiescent cells (9). The activation of the kinase(s) is transient, reaching maximum within 15 min and then slowly declining to basal levels within 2 hr (6). Phosphopeptide analysis of *in vitro* phosphorylated S6 revealed an identical pattern to that observed *in vivo* (9). The EGF dose–response curve for S6 kinase activation closely follows the extent of S6 phosphorylation in cells: no activation up to 10^{-10} M EGF, full activation between 5×10^{-10} and 10^{-8} M EGF. Comparison of EGF binding and down regulation of EGF receptors to the time course of kinase activation shows that these two events roughly parallel one another. Thus, the level to which the S6 kinase and S6 phosphorylation are activated appears to be closely linked to the number of ligand-occupied EGF receptors on the cell surface. To test this possibility we began to look for receptor-independent routes of kinase activation. It was found that sodium orthovanadate (vanadate), an agent which has been reported to mimic the mitogenic effects of EGF (10,11), elevates the level of S6 kinase activity at concentrations as low as 10^{-6} M, with maximal activation at 4 mM. Vanadate activation, however, is not so rapid as with EGF (maximal activation at 60 min), and activation of the kinase is persistent, rather than transient. In cells in which down regulation of the EGF receptor has been induced by a 2-hr pretreatment with EGF, the S6 kinase can be fully reactivated by vanadate but not by EGF. This result argues that vanadate can bypass the down regulation of the EGF receptor.

PHOSPHATASES

In searching for an activated S6 kinase in 3T3 cells, it was noted that phosphatase inhibitors are necessary to preserve enzymatic activity during the extraction procedure. Phosphotyrosine, p-nitrophenyl phosphate, β-glycerophosphate, and phosphoserine are all effective in protecting the activity, yielding extracts which have 5- to 20-fold higher S6 kinase activity than quiescent cells. Furthermore, extracts from activated cells show a time-dependent loss of activity when incubated at 30°C, which can be prevented by the presence of p-nitrophenyl phosphate (6). Similar results have been obtained by others who employed β-glycerophosphate in the extraction buffer (12,13). These findings suggested that the activity of S6 kinase is regulated by one or more phosphatases. Indeed, recent results from our laboratory demonstrate that a type 2A phosphatase is the agent responsible in whole cell extracts for inactivating the S6 kinase. Present studies are aimed at determining whether the activity of this enzyme is also modulated during the mitogenic response. In the future the extreme sensitivity of this kinase to phosphatase 2A can be used as a diagnostic tool to discriminate the growth factor- and oncogene-stimulated S6 kinase from other kinases, which also phosphorylate ribosomal protein S6 (14).

CHARACTERISTICS OF THE S6 KINASE

Along with EGF, vanadate, and serum, a number of other mitogens [platelet-derived growth factor (PDGF), insulin, phorbol myristate acetate (PMA), and prostaglandin $F_{2\alpha}$ ($PGF_{2\alpha}$)] and oncogenes (*src* and *ras*) have been shown to induce S6 phosphorylation (15). Extracts prepared from cells, under conditions where the agents above lead to S6 phosphorylation, all contain activated S6 kinase. When the extracts were analyzed by anion-exchange chromatography, all contained a single kinase species which eluted at 0.34 M NaCl (15). As pointed out above, a number of kinases have been reported to phosphorylate S6 *in vitro*, including cAMP-dependent protein kinase (16), type II casein kinase (17), a trypsin-activated protein kinase from reticulocytes (18,19). Ca^{2+}- and phospholipid-activated protein kinase C (20,21), and a 92,000 molecular weight enzyme from unfertilized *Xenopus* eggs (13). The activity of the mitogen/oncogene-activated S6 kinase was not affected by the heat-stable inhibitor of cAMP-dependent protein kinase or by the addition of Ca^{2+} and phospholipids. In addition, low concentrations of trypsin lead to immediate inactivation of the enzyme (15). These results argue that the mitogen/oncogene-activated kinase is not equivalent to the first four enzymes listed above (16–21). Furthermore, recent purification studies from our laboratory, which are summarized below, argue that it is distinct from the *Xenopus* enzyme. Indeed, results to date suggest that it is a novel kinase.

Recently, we succeeded in purifying the kinase from vanadate-stimulated 3T3 cells. The enzyme shows an apparent molecular weight of 66,000–70,000 on polyacrylamide gel electrophoresis in the presence or absence of reducing agents [including sodium dodecyl sulfate (SDS)], suggesting a single polypeptide chain. It undergoes autophosphorylation when incubated in the absence of 40 S subunits, and it incorporates up to 5 mol of phosphate into S6. Moreover, the phosphopeptide maps obtained from *in vitro* phosphorylation of S6 match those from *in vivo* labeled S6 (22). Finally, the purified enzyme is inactivated by phosphatase 2A, arguing that activity of the enzyme itself and not a regulatory subunit is controlled by phosphorylation.

PATHWAYS INVOLVED IN S6 PHOSPHORYLATION

Previously it was demonstrated that EGF, insulin, $PGF_{2\alpha}$, and serum induce the phosphorylation of the same set of S6 peptides (7,23), leading to the conclusion that all four agents may be converging on a common regulatory mechanism. The results above show that all mitogens and oncogenes tested to date converge on the same S6 kinase. However, since these agents are thought to act through distinct regulatory pathways, a number of signaling systems have been implicated in the activation of the S6 kinase (25–30). Activation of an enzyme by such diverse pathways could be explained by multiple phosphorylation sites on the kinase, all of which differently regulate its activity. Such a hypothesis is supported by the fact that S6 kinase activity in crude extracts and in its purified form is labile in the presence of phosphatases, pointing to a regulation mechanism in which phosphorylation renders the molecule active. On the other hand, regulation via dephosphorylation can occur at the level of S6, and results from our laboratory suggest that the phosphatases involved in regulating the kinase and S6 are distinct enzymes (31). Knowledge of the molecular characteristics of the S6 kinase will greatly facilitate the understanding of mechanisms involved in S6 phosphorylation, protein synthesis, and cell division.

REFERENCES

1. Thomas, G., Siegmann, M., and Gordon, J. (1979). *Proc. Natl. Acad. Sci. U.S.A.* **76**, 3952.
2. Thomas, G., Martin-Perez, J., Siegmann, M., and Otto, A. M. (1982). *Cell (Cambridge, Mass.)* **30**, 235–242.
3. Stanners, C. P., and Becker, H. (1971). *J. Cell. Physiol.* **77**, 31–42.
4. Bandman, E., and Gurney, T. (1975). *Exp. Cell Res.* **90**, 159–168.
5. Thomas, G., Thomas, G., and Luther, H. (1981). *Proc. Natl. Acad. Sci. U.S.A.* **78**, 5712–5716.
6. Novak-Hofer, I., and Thomas, G. (1985). *J. Biol. Chem.* **260**, 10314–10319.
7. Martin-Perez, J., and Thomas, G. (1983). *Proc. Natl. Acad. Sci. U.S.A.* **80**, 926–930.
8. Martin-Perez, J., Rudkin, B. B., Siegmann, M., and Thomas, G. (1986). *EMBO J.* **5**, 725–731.
9. Novak-Hofer, I., and Thomas, G. (1984). *J. Biol. Chem.* **259**, 5995–6000.
10. Carpenter, G. (1981). *Biochem. Biophys. Res. Commun.* **102**, 1115–1121.
11. Smith, J. B. (1983). *Proc. Natl. Acad. Sci. U.S.A.* **80** 6162–6166.
12. Tabarini, D., Heinrich, J., and Rosen, O. (1985). *Proc. Natl. Acad. Sci. U.S.A.* **82**, 4369–4373.
13. Erikson, E., and Maller, J. (1985). *Proc. Natl. Acad. Sci. U.S.A.* **82**, 742–746.
14. Ballou, L. M., Jenö, P., and Thomas, G., submitted for publication.
15. Novak-Hofer, I., Luther, H., Siegmann, M., Friis, R. R. and Thomas, G., submitted for publication.
16. DelGrande, R. W., and Traugh, J. A. (1982). *Eur. J. Biochem.* **123**, 421–428.
17. Cobb, M. H., and Rosen, O. (1983). *J. Biol. Chem.* **258**, 12472–12481.
18. Perisic, O., and Traugh, J. A. (1983). *J. Biol. Chem.* **258**, 9589–9592.
19. Perisic, O., and Traugh, J. A. (1983). *J. Biol. Chem.* **258**, 13998–14002.
20. LePeuch, C. J., Ballester, R., and Rosen, O. (1983). *Proc. Natl. Acad. Sci. U.S.A.* **80**, 6585–6862.
21. Parker, P. J., Katan, M., Waterfield, M., and Leader, D. P. (1985). *Eur. J. Biochem* **148**, 579–586.
22. Jenö, P., Jaeggi, N., Ballou, L. M., Luther, H., Siegmann, M., and Thomas, G., submitted for publication.
23. Martin-Perez, J., Siegmann, M., and Thomas, G. (1984). *Cell (Cambridge, Mass.)* **36**, 287–294.
24. Habenicht, A. J. R., Glomset, J. A., King, W. C., Nist, C., Mitchell, C. D., and Ross, R. (1981). *J. Biol. Chem.* **256**, 12329–12335.

25. Blackshear, P. J., Witters, L. A., Girard, P. R., Kuo, J. F., and Quamo, S. N. (1985). *J. Biol. Chem.* **260**, 13304–13315.
26. Blenis, J., Spivack, J. G., and Erikson, R. L. (1984). *Proc. Natl. Acad. Sci. U.S.A.* **81**, 6408–6412.
27. Trevillyan, J. M., Kulkarni, R. K., and Byus, C. V. (1984). *J. Biol. Chem.* **259**, 897–902.
28. Moolenaar, W. H. (1986). *Annu. Rev. Physiol.* **48**, 363–376.
29. Decker, S. (1981). *Proc. Natl. Acad. Sci. U.S.A.* **78**, 4112–4115.
30. Maller, J. L., Foulkes, J. G., Erikson, E., and Baltimore, D. (1985). *Proc. Natl. Acad. Sci. U.S.A.* **82**, 272–276.
31. Olivier, A., Ballou, L., and Thomas G. unpublished.

4
Phosphorylation of pp60[c-*src*]

TONY HUNTER[1], CHRISTINE A. CARTWRIGHT[2],
JAMES R. WOODGETT[3], AND KATHLEEN L. GOULD[4]
1. *Molecular Biology and Virology Laboratory*
The Salk Institute
San Diego, California 92138
and
2. *Division of Gastroenterology S 069*
Department of Medicine
Stanford University
Stanford, California 94305
and
3. *Ludwig Institute for Cancer Research*
Middlesex Hospital
University College Branch
London, W1P 8BT, England
and
4. *Microbiology Unit*
Department of Biochemistry
University of Oxford
Oxford, OX1 3QU, England

INTRODUCTION

pp60[c-*src*], the 60-kDa product of the c-*src* gene, is a protein-tyrosine kinase that is not only highly conserved throughout the vertebrate kingdom but is also found in organisms as simple as *Drosophila* and the sea urchin (for reviews see 1–5). In vertebrates pp60[c-*src*], which is expressed in most cell types, is localized to the cytosolic face of cytoplasmic membranes. pp60[c-*src*] is present at the highest level in neurons and platelets, and it has recently been shown that there is a neuronal specific form of pp60[c-*src*], which differs from the normal pp60[c-*src*] by the presence of an additional 6 amino acid exon in its N-terminal half (see Fig. 1). In simpler organisms pp60[c-*src*] is also enriched in the nervous system.

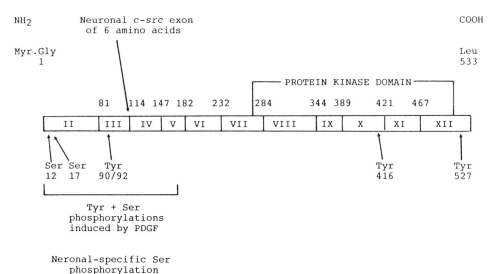

Fig. 1. Phosphorylation sites in pp60[c-src].

The c-src gene is one member of a closely related protein-tyrosine kinase gene family, which currently has seven members (the src, yes, fgr, lck, hck, fyn, and lyn genes) (4). All the src family genes encode protein-tyrosine kinases, of about 525 amino acids in length, with very similar structural organizations. These proteins are about 80% homologous over their C-terminal 450 amino acids, which includes the catalytic domain, but diverge almost completely in their N-terminal 80 amino acids. However, all these enzymes are predicted to have an N-terminal Gly, which in the case of pp60[c-src] and p56[lck] is known to be modified by the addition of a myristyl group that is essential for membrane association.

The functions of pp60[c-src] and the other src family members remain largely unknown, although it is presumed that they act to phosphorylate other proteins on tyrosine. It also is becoming apparent that these enzymes are normally in an inactive state in the cell as a consequence of a negative regulatory phosphorylation. They must be activated transiently by unknown stimuli. Individual members of the src gene family are present at highest levels in specific differentiated cell types, and some are expressed exclusively in particular hematopoietic lineages. The unique N-terminal sequences could target the individual enzymes to different substrates or subcellular locations. The expression of pp60[c-src] in neurons and platelets coupled with its membrane association has prompted the speculation that pp60[c-src] could play a role in exocytosis and secretion. An alternative hypothesis is that pp60[c-src] regulates the opening of an ion channel.

The genomes of three acutely transforming retroviruses, Rous sarcoma virus (RSV) and the S1 and S2 avian sarcoma viruses, contain sequences recognizably derived from the chicken c-*src* gene (3–5). These co-opted c-*src* sequences encode the viral transforming protein, pp60$^{v\text{-}src}$. Since high level expression of pp60$^{c\text{-}src}$ fails to transform cells (6–8), there must be critical differences between pp60$^{c\text{-}src}$ and pp60$^{v\text{-}src}$. All three pp60$^{v\text{-}src}$ species contain nearly the intact coding sequence of the c-*src* gene, but in each case the normal pp60$^{c\text{-}src}$ C terminus is missing, having been replaced by unique C-terminal sequences (5). In addition to the C-terminal alteration, there are scattered single amino acid mutations in pp60$^{v\text{-}src}$. By making chimeric pp60$^{v\text{-}src}$/pp60$^{c\text{-}src}$ molecules it has been shown that the C-terminal alteration by itself is sufficient to confer transforming activity on pp60$^{c\text{-}src}$. However, certain of the single amino acid substitutions are also capable of activating pp60$^{c\text{-}src}$ to transform cells (5). Thus, pp60$^{v\text{-}src}$ has been converted to a transforming protein by a series of independent mutations. It seems likely that the C-terminal truncation was the initial change and that subsequent mutations were selected as a consequence of their leading to a more tumorigenic virus.

Given the difference in biological activity between pp60$^{v\text{-}src}$ and pp60$^{c\text{-}src}$, one would predict that there must be discernible biochemical differences between these proteins. pp60$^{v\text{-}src}$ does, in fact, prove to have considerably greater protein kinase activity than pp60$^{c\text{-}src}$ both *in vitro* and *in vivo* (9,10), and there is every reason to believe that the increased protein kinase activity is responsible for its transforming activity. This difference in activity is now realized to be due in part to the C-terminal truncation of pp60$^{c\text{-}src}$ and the resultant loss of a negative regulatory tyrosine phosphorylation site present in pp60$^{c\text{-}src}$ (5). This observation has engendered new interest in the regulation of pp60$^{c\text{-}src}$ by phosphorylation.

pp60$^{c\text{-}src}$ is phosphorylated at multiple sites (see Fig. 1). It is the purpose of this chapter to review what is known about the location of these sites in pp60$^{c\text{-}src}$ and the effects of their phosphorylation on the function of pp60$^{c\text{-}src}$. Since there is no obvious logical order in which to describe the characterized phosphorylation sites, we will start from the N terminus.

SERINE 12

In untreated quiescent cells there is barely detectable phosphorylation of Ser-12 in pp60$^{c\text{-}src}$. However, in response to treatment of intact chicken and mammalian cells with tetradecanoylphorbol acetate (TPA) or teleocidin as well as synthetic diacylglycerols, all of which are activators of protein kinase C, there is rapid and nearly stoichiometric phosphorylation of Ser-12, an effect that is maximal within 2–5 min (11,12). We obtained proof that this phosphorylation is mediated by protein kinase C by showing that purified protein kinase C phosphorylates pp60$^{c\text{-}src}$ at Ser-12 *in vitro* (11). None of five other purified protein-serine kinases, including the cAMP-dependent protein kinase, was able to phosphorylate Ser-12 *in vitro*. Serine 12 is embedded in a highly basic region, Lys·Pro·Lys·Asp·Pro·**Ser**·Gln·Arg·Arg·Arg, which is characteristic of protein kinase C phosphorylation sites in general. For example, the sequence surrounding Ser-12 is homologous to that around Thr-654, the protein kinase C phosphorylation site in the epidermal growth factor (EGF) receptor (13). In chicken pp60$^{c\text{-}src}$ there is a second minor site of protein kinase C phosphorylation, Ser-48, in the sequence His·Arg·Thr·Pro·**Ser**·Arg·Ser. While there is an equivalent Ser in mammalian pp60$^{c\text{-}src}$, this is not phosphorylated, possibly because it lacks the Arg on the C-terminal side present in chicken pp60$^{c\text{-}src}$, which we have proposed is an essential element of *in vivo* protein kinase C phosphorylation sites (14). Serines 12 and 48 in pp60$^{v\text{-}src}$ are also phosphorylated when RSV-transformed cells are treated with TPA. However, phosphorylation of Ser-12 is not observed in cells infected by mutant pp60$^{v\text{-}src}$ that lack the myristylation signal and fail to associate with membranes (15). This indicates that membrane affiliation is essential for phosphorylation by protein kinase C, in keeping with the membrane localization of the activated enzyme.

Despite the rapid and extensive phosphorylation of Ser-12, there are no measurable effects on the ability of pp60$^{c\text{-}src}$ or pp60$^{v\text{-}src}$ to phosphorylate added substrates, at least when immunoprecipitates from untreated and TPA-treated cells are compared (11,12). Ideally this question should be reexamined with purified protein, but it may not be easy to obtain the high stoichiometry of pp60$^{c\text{-}src}$ phosphorylation by protein kinase C required for this. Another imponderable is the prior state of phosphorylation of pp60$^{c\text{-}src}$, which may well influence the ability to detect changes in activity. For example, if the C-terminal regulatory tyrosine phosphorylation site is fully phosphorylated, then pp60$^{c\text{-}src}$ may be maximally repressed, making it impossible to detect the effects of other phosphorylations which lead to decreased activity.

There are certainly good reasons for believing that phosphorylation of Ser-12 can regulate pp60$^{c\text{-}src}$ function. First, Ser-12 is conserved in all known vertebrate pp60$^{c\text{-}src}$ proteins, although it should be noted that there is no equivalent serine in *Drosophila* pp60$^{c\text{-}src}$ encoded by the locus at 64C. Second, there is striking homology between the Ser-12 site and other functional protein kinase C sites, both with regard to sequence and apposition to the membrane (14). Third, phosphorylation of Ser-12 is induced in quiescent cells by physiological agents which activate the turnover of phosphatidylinositol (59). This suggests that pp60$^{c\text{-}src}$ might be an important target for the protein kinase C limb of this signal pathway.

If phosphorylation of Ser-12 does not alter protein kinase activity, it could instead affect some other property of pp60$^{c\text{-}src}$ such as subcellular localization. We have been unable to obtain evidence that TPA treatment causes release of pp60$^{c\text{-}src}$ from membranes. However, phosphorylation of Thr-654 in the EGF receptor leads to internalization of the receptor in endocytotic vesicles, so phosphorylation of Ser-12 might likewise alter the distribution of pp60$^{c\text{-}src}$ between the plasma membrane and internal membranes. We are currently testing this possibility. In addition, the consequences of mutating Ser-12 to Ala, a nonphosphorylatable residue, and to Glu, which being negatively charged might mimic phosphoserine, are being investigated.

SERINE 17

In vivo, Ser-17 is constitutively phosphorylated in both pp60$^{c\text{-}src}$ and pp60$^{v\text{-}src}$ to a variable stoichiometry, with estimates ranging from 10 to 60% occupancy (16–18). The cAMP-dependent protein kinase phosphorylates Ser-17 in purified pp60$^{v\text{-}src}$ *in vitro* (19), and the sequence upstream of Ser-17 is Arg·Arg·Arg·**Ser,** which conforms with the deduced consensus sequence for phosphorylation by cAMP-dependent protein kinase, Arg·Arg·X·Ser. Moreover, in some RSV-transformed cells the phosphorylation of Ser-17 is increased significantly by elevating intracellular cAMP levels (20), which implies that the cAMP-dependent protein kinase can phosphorylate pp60$^{v\text{-}src}$ *in vivo*. To date the same experiment has not been done with pp60$^{c\text{-}src}$. Thus, it seems likely that the major Ser-17 protein kinase in the cell is cAMP-dependent protein kinase, but we cannot rule out that other protein kinases can phosphorylate Ser-17. For instance, the cGMP-dependent protein kinase has rather similar specificity to cAMP-dependent protein kinase and might be able to phosphorylate Ser-17. It would clearly be interesting to determine whether Ser-17 is phosphorylated in mutant cells which lack cAMP-dependent protein kinase activity.

The effect of Ser-17 phosphorylation on pp60^{src} is not clear-cut. When Ser-17 in purified pp60$^{v\text{-}src}$ is phosphorylated by cAMP-dependent protein kinase, there is a small increase in its ability to phosphorylate casein or the *in vivo* substrate p36 (A. F. Purchio and R. L. Erikson, personal communication). Similar experiments have not yet been performed with pp60$^{c\text{-}src}$. Mutants of pp60$^{v\text{-}src}$ in which Ser-17 or Ser-12 and Ser-17 have been deleted have essentially unaltered transforming ability (17), but it is possible that any deleterious effects of these mutations are overridden by the multiple activating mutations in pp60$^{v\text{-}src}$. Clearly, similar mutations need to be made in pp60$^{c\text{-}src}$.

OTHER AMINO-TERMINAL PHOSPHORYLATIONS

There are a number of other phosphorylation sites in the N-terminal 34 kDa of pp60^{c-src}, which can be readily demonstrated by tryptic peptide mapping of an N-terminal 34-kDa fragment of pp60^{c-src} generated by partial cleavage with *Staphylococcus aureus* protease (see Fig. 2). Two of these sites are minor phosphoserine sites which are present constitutively in pp60^{c-src} in all cell types. Other serine and tyrosine phosphorylation sites are inducible or present only in pp60^{c-src} isolated from specific cell types. With one exception these sites have not been precisely mapped. These N-terminal phosphorylations will be discussed under specific examples where the phosphorylation pattern of pp60^{c-src} is known to differ from the normal.

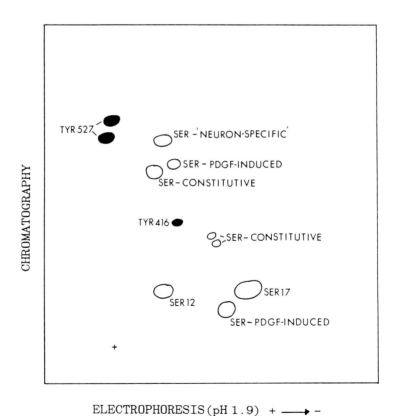

Fig. 2. Schematic phosphopeptide map of pp60^{c-src}.

TYROSINE 416

Tyrosine 416 is the major site of autophosphorylation *in vitro* for both pp60$^{c\text{-}src}$ and pp60$^{v\text{-}src}$ (17,21,22). However, when pp60$^{c\text{-}src}$ is isolated from normal cells, only a small fraction (1–5%) of molecules are phosphorylated at Tyr-416. In contrast 40–60% of pp60$^{v\text{-}src}$ molecules are found to be phosphorylated at Tyr-416. This difference correlates well with the enzymatic activities of the two proteins when isolated from the cell, in that pp60$^{v\text{-}src}$ has a specific protein kinase activity about 10 times that of pp60$^{c\text{-}src}$ (9,10). Given that the phosphate on Tyr-416 is known to turn over rapidly in the cell, one would expect the occupancy of Tyr-416 to be roughly proportional to the activity of pp60^{src}. As discussed below, the difference in activity can be accounted for by the fact that the majority of pp60$^{c\text{-}src}$ molecules are phosphorylated at Tyr-527, a negative regulatory phosphorylation site that has been deleted in pp60$^{v\text{-}src}$.

The exact function of Tyr-416 phosphorylation is not clear. Autophosphorylation appears to be an intramolecular reaction (23). There is an equivalent tyrosine in the catalytic domain of every bona fide protein-tyrosine kinase (1), and in almost every case this tyrosine is autophosphorylated *in vitro*. However, initial experiments in which Tyr-416 was changed to Phe in pp60$^{v\text{-}src}$ showed little difference in protein kinase activity or transforming potency between the wild-type and mutant proteins (24), although the mutant-transformed cells showed reduced tumorigenicity (25). In contrast, mutation of the equivalent tyrosine, Tyr-1073, to Phe in the P140$^{gag\text{-}fps}$ transforming protein-tyrosine kinase showed a significant diminution of both protein kinase activity and transforming potential (26). More recent experiments with the Phe-527 transforming mutant of pp60$^{c\text{-}src}$ prove that the presence of a Tyr at 416 is essential for both maximal protein kinase and transforming activities (27,28). The Phe-416/Phe-527 mutant pp60$^{c\text{-}src}$ has about 2-fold lower protein kinase activity and about 10-fold lower transformation efficiency. This implies that autophosphorylation at Tyr-416 results in enhanced enzymatic activity, as has been shown to be the case for the autophosphorylated insulin receptor protein-tyrosine kinase. With hindsight, the nearly wild-type properties of the Phe-416 mutant of pp60$^{v\text{-}src}$ can be explained by the presence of multiple independent activating mutations, which largely counteract the relatively small negative consequence of Phe-416.

Although one cannot exclude that Tyr-416 has been conserved for structural purposes, which cannot be served by phenylalanine, Tyr-1073 in P140$^{gag\text{-}fps}$ has been mutated to Phe, Ser, Thr, Gly, or Glu, and none of these amino acids restores full wild-type activity (29). This suggests that the ability to phosphorylate this residue is important for maximal enzymatic activity. This region of the protein-tyrosine kinases is believed to lie near the substrate binding site, based on work with an affinity peptide substrate for the cAMP-dependent protein kinase which becomes cross-linked to a Cys in the equivalent regions of the catalytic domain near the threonine autophosphorylation site (30). This suggests that this region of the catalytic domain could form a flexible loop that can fold into the active site and thus be autophosphorylated. If this were true, then Tyr-416 would act as a pseudosubstrate and compete for incoming exogenous substrates. Autophosphorylation of Tyr-416 would convert it to a pseudoproduct, which would have a lower affinity for the active site, and thus be a mechanism for relieving the competitive inhibition imposed by Tyr-416. The true solution to this problem may not be learned until three-dimensional structures of the phosphorylated and unphosphorylated forms of pp60^{c-src} are available.

TYROSINE 527

A difference between the major tyrosine phosphorylation sites of pp60^{c-src} and pp60^{v-src} was noted as soon as the two proteins were characterized. pp60^{c-src} and the form of pp60^{v-src} encoded by Rous sarcoma virus are very similar in structure between residues 1 and 514, differing only by a few scattered single amino acid substitutions (3,5). However, the C termini of the two proteins diverge completely, owing to the deletion of the normal C terminus of pp60^{c-src} from pp60^{v-src}. The analysis of chimeric pp60^{c-src}/pp60^{c-src} molecules created to test for the changes that are required for the transforming activity of pp60^{v-src} suggested that the C-terminal alteration might be critical (5). Subsequently, pp60^{v-src} was found to have a 10-fold higher specific protein kinase activity than pp60^{c-src}, which seemed likely to be important for its transforming activity. These two facts were neatly dovetailed by a series of observations which led to the understanding that the protein kinase activity of pp60^{c-src} is negatively regulated by phosphorylation of Tyr-527 (31–34). As this model would predict, enzymatic dephosphorylation of pp60^{c-src} *in vitro* leads to a 10-fold increase in protein kinase activity (34). Moreover an antiserum directed against the C-terminal 19 amino acids of pp60^{c-src} is able to stimulate enzymatic activity (34). Formal proof for the function of Tyr-527 comes from mutants of pp60^{c-src} in which Tyr-527 has been replaced by Phe. The Phe-527 mutant of pp60^{c-src} has about 10-fold higher protein kinase activity than the wild-type pp60^{c-src} (27,28,35). All three different pp60^{v-src} species lack the region containing Tyr-527 and are therefore not subject to negative regulation (5). This would explain their increased protein kinase activity, which in turn is essential for the transforming activity of pp60$^{v\text{-}src}$. Likewise, the increased protein kinase activity of the Phe-527 mutant of pp60$^{c\text{-}src}$ is presumably responsible for its ability to transform NIH 3T3 cells.

This negative regulation model for pp60$^{c\text{-}src}$ raises the critical question of how Tyr-527 is phosphorylated and how it inhibits protein kinase activity. Tyr-527 does not act as a good autophosphorylation site, since when pp60$^{c\text{-}src}$ is fully dephosphorylated with phosphatase the majority of phosphate incorporated into the protein is still at Tyr-416 (34). When pp60$^{c\text{-}src}$ is expressed in yeast, where the level of endogenous protein-tyrosine kinase activity is vanishingly low, again one finds the preponderance of phosphate at Tyr-416, although there is a trace of phosphate at Tyr-527 (64,65). In contrast, a kinase-negative mutant of pp60$^{c\text{-}src}$ shows no phosphorylation at Tyr-527 in yeast (66,67). This implies that yeast lacks a protein-tyrosine kinase that can phosphorylate Tyr-527. Thus, the phosphate detected at Tyr-527 when wild-type pp60$^{c\text{-}src}$ is expressed in yeast is due either to a low level of autophosphorylation or to transphosphorylation between pp60$^{c\text{-}src}$ molecules. When the kinase-negative mutant is overexpressed in chick cells, it is fully phosphorylated at Tyr-527 (66). At the very least this indicates that Tyr-527 phosphorylation cannot be an obligate autophosphorylation event. Moreover, given the low endogenous protein kinase activity of pp60$^{c\text{-}src}$ in chick cells, it seems likely that another protein-tyrosine kinase is normally responsible for phosphorylating Tyr-527. The identification of this protein-tyrosine kinase will obviously be critical for an understanding of the regulation of pp60$^{c\text{-}src}$ in the cell. Of equal importance will be the characterization of the phosphatase which dephosphorylates Tyr-527.

How phosphorylated Tyr-527 inhibits pp60$^{c\text{-}src}$ phosphotransferase activity is not yet settled. One proposal based on a comparative kinetic analysis of the phosphorylated and unphosphorylated forms of pp60$^{c\text{-}src}$ is that Tyr-527 in the phosphorylated state acts as a product analog (34). Another possibility is that phosphorylation of this residue induces a conformational change in the catalytic domain, resulting in increased activity. In principle this not unreasonable, since there are point mutations (e.g., Thr-338 to Ile) within the catalytic domain, which presumably alter conformation, that activate pp60$^{c\text{-}src}$ both as an enzyme and a transforming protein (5,36,37). Once again, we may not learn the answer to this question until a three-dimensional structure of pp60$^{c\text{-}src}$ is available.

Tyrosine 527 brings us to the end of the list of identified phosphorylation sites in pp60$^{c\text{-}src}$. We now turn to systems in which the phosphorylation of pp60$^{c\text{-}src}$ differs from that detected in untreated fibroblasts either constitutively or else in an experimentally manipulatable fashion.

PLATELET-DERIVED GROWTH FACTOR TREATMENT

In response to PDGF treatment of quiescent mammalian fibroblasts, a small fraction (about 5%) of the pp60^{c-src} population is found to migrate more slowly on sodium dodecyl sulfate (SDS) gel electrophoresis (38,59). We find that this new species (p60$^+$) is detectable within 2 min, is maximal by 5 min, and decays over the next 2 hr. A 2- to 3-fold increase in the protein kinase activity of pp60^{c-src} both for autophosphorylation and enolase phosphorylation is observed in immunoprecipitates isolated from PDGF-treated cells. p60$^+$ appears to be a more highly phosphorylated derivative of the form of pp60^{c-src} present in untreated cells. Two-dimensional gel analysis of [^{35}S]-methionine-labeled immunoprecipitates indicates that pp60^{c-src} normally occurs in the di- or triphosphorylated state (i.e., 2 or 3 mol of phosphate per mole of protein). In contrast p60$^+$ has 4 or 5 mol of phosphate per molecule. Treatment of p60$^+$ with phosphatase converts it to a species identical to that seen when the fibroblast form is dephosphorylated.

Mapping with *S. aureus* V8 protease shows that these additional phosphates are located in the N-terminal half of pp60^{c-src}, and more specifically in the N-terminal 18 kDa (38, 59). Phosphoamino acid analysis of the N-terminal fragments of the modified form reveals phosphoserine as well as phosphotyrosine in a ratio of about 4:1. In our hands, tryptic digestion of the N-terminal 18 kDa of pp60^{c-src} generates four phosphoserine-containing peptides. These comprise Ser-17, Ser-12, and two unidentified serine residues. The presence of the Ser-12 peptide is expected because PDGF induces the turnover of phosphatidylinositol, producing diacylglycerol that activates protein kinase C, which as discussed above phosphorylates pp60^{c-src} at Ser-12. The nature of the protein-serine kinase(s) phosphorylating the other two serine residues is unclear. Many serine phosphorylation events are induced by PDGF treatment, including the phosphorylation of the small ribosomal subunit protein S6. Some of these may be mediated directly by protein kinase C, but others are probably due to protein-serine kinases activated as part of a cascade, initiated either by the protein-tyrosine kinase activity of the liganded PDGF receptor or protein kinase C. For instance, there is evidence that S6 is phosphorylated by a unique protein-serine kinase distinct from protein kinase C.

The N-terminal PDGF-induced tyrosine phosphorylation event is poorly understood. So far we have been unable to identify a discrete phosphotyrosine-containing peptide. N-terminal tyrosine phosphorylation occurs in pp60^{c-src} under other conditions, but the tyrosines involved seem to be distinct from that phosphorylated in PDGF-treated cells. For instance, incubation of pp60^{v-src} *in vitro* at high ATP concentrations leads to tyrosine phosphorylation at multiple sites, apparently by an autophosphorylation reaction (39). This is correlated with an increase in protein kinase activity. Similar N-terminal tyrosine phosphorylations have been observed following treatment of intact cells with the phosphotyrosine phosphatase inhibitor Na_3VO_4 (40,41). Thus, it is possible that the PDGF-induced tyrosine phosphorylation is due to an autophosphorylation. Alternatively pp60^{c-src} could be phosphorylated by another protein-tyrosine kinase or even the PDGF receptor itself following activation by PDGF. We are currently testing the latter possibility using purified PDGF receptor.

With regard to the increased protein kinase activity of pp60^{c-src} isolated from PDGF-treated cells, it is unclear whether this is an attribute of p60$^+$ or both species. The additional serine phosphorylations and the tyrosine phosphorylation are only detected in p60$^+$, and, since Ser-12 phosphorylation does not alter protein kinase activity, it seems likely that the increased phosphotransferase activity is due to p60$^+$. Whether this is a consequence of one of the phosphorylation events unique to p60$^+$ or a combinatorial effect of several phosphorylations is unclear. The elevated protein kinase activity of p60$^+$ may be important in the mitogenic response to PDGF. In this regard it is worth noting that we have not found any other mitogen for NIH 3T3 cells which induces the formation of p60$^+$ or an increase in pp60^{c-src} protein kinase activity (59).

NEURONAL pp60^{c-src}

There are several reports that neuronal cells contain a high level of pp60^{c-src} protein kinase activity, as well as a unique neuronal form of pp60^{c-src} (42–47). We have investigated whether the expression and phosphorylation of pp60^{c-src} are regulated at either the protein or enzymatic level during the embryonic development of the rat striatum, a brain region chosen because it has a well-defined neuroanatomy and ontogeny (47). The level of pp60^{c-src} protein as measured by immunoblotting rises about 2-fold from E15 to E20 days of embryonic (E) life and then falls slowly postnatally (P). Protein kinase activity toward an exogenous substrate increases about 5-fold, peaking again at E20 (60). Thus, during the principal period of neuronogenesis in the striatum there is a 2- to 3-fold increase in specific activity. Comparative studies with the hippocampus show a 50% increase in pp60^{c-src} protein and a 2- to 3-fold increase protein kinase activity between E18 and P2. In contrast, in the cerebellum pp60^{c-src} protein and protein kinase activities stay essentially constant from E20 (the earliest time at which the cerebellum is a definable structure in the rat) to P12, which is the major period for cerebellar development.

Similar changes occur when neurons from E15 striatum are placed in culture and then induced to differentiate by withdrawal of serum at day 4 (47). By day 11 most of the cells have neuronal character by morphological criteria, and all the cells appear to be neurons by virtue of their binding tetanus toxin and the absence of staining with anti-GFAP, which detects glial cells. Comparison of the level of pp60^{c-src} in the undifferentiated and differentiated cultures shows a 50% increase in pp60^{c-src} protein and a 7-fold increase in protein kinase activity. Thus, there is a 4-fold elevation in specific protein kinase activity during differentiation, and this increase in both protein and specific activity is gratifyingly similar to that observed during differentiation of striatal neurons in the intact tissue. These changes in expression and activity of pp60^{c-src}, as well as the existence of the neuron-specific form of pp60^{c-src}, suggest that pp60^{c-src} plays a role in either the differentiation and/or function of neurons.

The phosphorylation state of pp60^{c-src} and the expression of the neuronal form of pp60^{c-src} during differentiation have been examined by immunoprecipitating pp60^{c-src} from [^{35}S]methionine- and ^{32}P-labeled cultures. As noted by others, the neuronal form of pp60^{c-src} migrates about 1 kDa more slowly than the fibroblast form upon SDS–polyacrylamide gel electrophoresis (44,46). Undifferentiated cells display about equal levels of the neuronal and fibroblast forms of pp60^{c-src}, whereas the neuronal form predominates in the differentiated cells. Examination of fragments generated by partial proteolysis indicates that the difference between the fibroblast form of pp60^{c-src} and the neuronal species lies in the N-terminal 18 kDa (44,46,47). The N-terminal fragment from the neuronal form contains only phosphoserine, and we have never detected phosphotyrosine. Phosphopeptide mapping indicates that both forms in the differentiated cells are phosphorylated at Ser-17 and Tyr-527. However, the neuronal form is additionally phosphorylated at Ser-12 and at a second serine site which is not detected in the fibroblast form from the same cells. Phosphorylation at this site may not be totally specific for the neuronal form, however, since the same peptide is apparently present in both forms of pp60^{c-src} in other types of neuronally derived cells (46). This serine site is distinct from that found in p60$^+$ from PDGF-treated cells and could be phosphorylated by a protein-serine kinase which is prevalent in neurons.

What is the nature of the N-terminal difference between the neuronal and fibroblast forms of pp60^{c-src}? Dephosphorylation of neuronal pp60^{c-src} with potato acid phosphatase does not result in a mobility shift, suggesting that phosphorylation is not responsible for its slower mobility (46). This in turn suggests that the neuronal pp60^{c-src} might differ in primary structure. Two-dimensional gel analysis of [^{35}S]methionine-labeled immunoprecipitates from cultured neurons shows that the charge isoforms of the neuronal pp60^{c-src} are slightly more basic than the corresponding fibroblast forms, which is inconsistent with their being derived from the fibroblast forms by any known posttranslational modification (61).

From the cloning work of others it is now apparent that neuronal pp60$^{c\text{-}src}$ differs from the normal form of pp60$^{c\text{-}src}$ by an additional exon of 6 amino acids inserted at residue 117 (48,62). These 6 amino acids have a net charge of +2, and thus the expected pI of neuronal pp60$^{c\text{-}src}$ should be more basic than that of the fibroblast form, exactly as we have observed. The two-dimensional gel analysis also indicates that both forms of pp60$^{c\text{-}src}$ are less highly phosphorylated overall than pp60$^{c\text{-}src}$ from rat fibroblasts, with there being some molecules lacking phosphate altogether. This hypophosphorylation may explain the increased specific activity of pp60$^{c\text{-}src}$ from neurons, since some molecules will lack phosphate at Tyr-527 and thus would be expected to have unrestricted kinase activity. Whatever underlies the increase in specific activity, it is interesting to note that there is no increase in Tyr-416 phosphorylation between pp60$^{c\text{-}src}$ from differentiated and undifferentiated neurons or fibroblasts. This is in distinction to transforming mutants of pp60$^{c\text{-}src}$, such as the Ile-338 mutant, which have high protein kinase activity and show Tyr-416 phosphorylation rather than Tyr-527 phosphorylation.

We have also detected the neuronal form of pp60$^{c\text{-}src}$ in two cell lines. The LAN-5 human neuroblastoma line contains nearly equal amounts of the two forms of pp60$^{c\text{-}src}$ and shows a 2- to 3-fold higher specific protein kinase activity than pp60$^{c\text{-}src}$ from a human fibroblast (D. Middlemas and T. Hunter, unpublished results). Two-dimensional gel analysis shows that the LAN-5 pp60$^{c\text{-}src}$ is relatively hypophosphorylated. Neuronal pp60$^{c\text{-}src}$ was also found in the AtT20 mouse pituitary corticotrope cell line at about 5% the level of the fibroblast form (61). pp60$^{c\text{-}src}$ from these cells also had a 2- to 3-fold increased specific activity.

pp60$^{c\text{-}src}$ AND POLYOMAVIRUS MIDDLE T ANTIGEN

In cells infected or transformed with polyomavirus, middle T (mT) antigen forms a stable complex with at least two of the members of the *src* gene family, pp60$^{c\text{-}src}$ and pp62$^{c\text{-}yes}$ (49,50). In both cases this complex only involves a small fraction of either mT antigen or the protein kinase. The complexed forms of these protein kinases show significant increases in specific protein kinase activity over the free forms. For bound pp60$^{c\text{-}src}$ the increase in activity has been estimated at between 10- and 100-fold (51), whereas for pp62$^{c\text{-}yes}$ the elevation is apparently less dramatic, being only 2- to 3-fold (50). The phosphorylation state of pp60$^{c\text{-}src}$ associated with mT antigen has been determined by peptide mapping. Instead of the normal pattern of phosphorylation at Ser-17 and Tyr-527, pp60$^{c\text{-}src}$ is phosphorylated at Ser-17 and Tyr-416, the site of autophosphorylation (52). Both the absence of Tyr-527 phosphorylation and the Tyr-416 phosphorylation are correlated with the activation of the kinase. It seems likely that the interaction with mT antigen blocks the ability of a second protein-tyrosine kinase to phosphorylate Tyr-527. Alternatively the association with mT antigen in some way improves pp60$^{c\text{-}src}$ as a substrate for a phosphotyrosine phosphatase. It is possible that the lower degree of activation of pp62$^{c\text{-}yes}$ could be due to a constitutively lower occupancy of the "Tyr-527" site.

The requirements for the formation of an activated complex between pp60$^{c\text{-}src}$ and mT antigen have been probed by using site-directed mutations in both mT antigen and pp60$^{c\text{-}src}$. For mT antigen it is essential that the C-terminal hydrophobic region is retained, since this is necessary for membrane association. Deletions at the N terminus also prevent the formation of the complex, as do points mutations near the middle of the protein. The fact that pp60$^{v\text{-}src}$ does not associate with mT antigen had been interpreted to mean that the C terminus of pp60$^{c\text{-}src}$ was essential for this interaction. By analyzing a series of C-terminal truncations of pp60$^{c\text{-}src}$, it has now become clear that residues downstream of 519 are dispensable for this interaction (35;63). In some ways this is surprising given that the phosphorylation of Tyr-527 is affected by the association with mT antigen and yet apparently is not required for the binding.

In vitro phosphorylation of the mT antigen/pp60$^{c\text{-}src}$ complex in immunoprecipitates made with a monoclonal antibody against pp60$^{c\text{-}src}$ or polyomavirus antitumor serum leads to the phosphorylation of mT antigen and pp60$^{c\text{-}src}$. However in immunoprecipitates made with polyomavirus antitumor serum the phosphorylated pp60$^{c\text{-}src}$ migrates more slowly than normal (53,54). The difference has been mapped to the N terminus of the altered pp60$^{c\text{-}src}$, and it has been shown to be due to a novel tyrosine phosphorylation. This tyrosine phosphorylation is not readily detected when the complex is isolated from ^{32}P-labeled cells but can be observed if polyomavirus-transformed cells are treated with Na$_3$VO$_4$ (54). The tyrosine phosphorylation site has been mapped to Tyr-90 or Tyr-92 (W. Yonemoto and J. S. Brugge, personal communication). We find that this tyrosine phosphorylation site differs from that in pp60$^{c\text{-}src}$ isolated from PDGF-treated cells. Whether this tyrosine phosphorylation plays any role in the activation of pp60$^{c\text{-}src}$ is unclear, since the bulk of pp60$^{c\text{-}src}$ associated with mT antigen is not phosphorylated at this site. Nevertheless the mutation of Tyr-90 to Phe converts pp60$^{c\text{-}src}$ to a transforming protein, suggesting that this region of pp60$^{c\text{-}src}$ may have a regulatory function.

VANADATE TREATMENT

Treatment of intact RSV-transformed cells with a phosphotyrosine phosphatase inhibitor, Na_3VO_4, leads to an increase of pp60^{v-src} protein kinase activity when assayed *in vitro*, in concert with an increase in tyrosine phosphorylation in the N-terminal half of the molecule (40,41). Increased protein kinase specific activity and N-terminal tyrosine phosphorylation events are also observed when pp60^{v-src} is allowed to autophosphorylate *in vitro* at high ATP concentrations (39). This implies that certain N-terminal tyrosine phosphorylations may increase activity. In contrast, when pp60^{c-src} is isolated from Na_3VO_4-treated cells there is a decrease in protein kinase activity (55). This is probably due to increased phosphorylation of pp60^{c-src} at Tyr-527. Presumably the negative effect of Tyr-527 phosphorylation is dominant and suppresses any regulatory N-terminal tyrosine phosphorylations which might otherwise occur.

ELEVATION OF pp60^{c-src} PROTEIN KINASE ACTIVITY IN TRANSFORMED CELLS

Elevated pp60^{c-src} protein kinase activity has been observed in a number of tumor cell lines compared to their normal counterparts (56,57). In particular, high levels of pp60^{c-src} activity have been detected in neuroblastoma and colon carcinoma lines. In neuroblastomas this activity is in part accounted for by an elevated level of pp60^{c-src}, but this may be augmented by a higher specific activity in some cases (e.g., LAN-5). In colon carcinoma cells there is no difference in the level of pp60^{c-src} compared to a normal colonic mucosal cell strain, yet the increase in protein kinase activity is as great as 50-fold (58). The phosphorylation state of pp60^{c-src} appears to be normal in these cells, with Ser-17 and Tyr-527 being phosphorylated and no evidence for tyrosine phosphorylation in the N-terminal half of the protein. The only difference detected between pp60^{c-src} from the carcinoma cells and that from mucosal cells is an apparent difference in the rate of turnover at Tyr-527 (58). The half-life of phosphate at this site is 3–4 hr in the normal pp60^{c-src} but less than 1 hr in pp60^{c-src} from the carcinoma cells. It is not known whether this diminished half-life is responsible for the increased protein kinase activity, nor how it is brought about. The relative level of Tyr-527 phosphorylation compared to Ser-17 phosphorylation is the same in the two cell types, although it is not clear that the absolute stoichiometry of phosphorylation at Tyr-527 is the same in both cases. If it is then one possible explanation is that the active conformation which pp60^{c-src} adopts following dephosphorylation is not immediately reversed upon rephosphorylation, allowing the protein to remain in the active state for a greater fraction of the time.

SUMMARY

The phosphorylation of pp60$^{c\text{-}src}$ is clearly a complex process. There are at least six serine phosphorylation sites: Ser-12, Ser-17, the "neuronal specific" serine site, two PDGF-inducible serine sites, and two minor constitutive serine phosphorylation sites (Figs. 1 and 2). All these sites lie in the N-terminal half of the molecule outside the catalytic domain. There are two major tyrosine phosphorylation sites (Tyr-416 and Tyr-527) in the C-terminal half of the protein and two further tyrosine phosphorylation sites in the N-terminal half (PDGF-inducible tyrosine site, mT antigen-associated Tyr-90/92 site). Thus, a minimum of seven different protein kinases can phosphorylate pp60$^{c\text{-}src}$.

There are clear-cut effects of phosphorylation of these sites on pp60$^{c\text{-}src}$ protein kinase activity for only Tyr-416 and Tyr-527. Phosphorylation at the other sites could also affect activity, but our current assays do not reveal this. It should be borne in mind that phosphorylation could well alter other properties of pp60$^{c\text{-}src}$, such as precise subcellular location through interaction with other proteins and substrate specificity. Based on the deleterious effects of deletion mutations in the N-terminal region of pp60$^{v\text{-}src}$, it has been suggested that the N-terminal noncatalytic region regulates the function of pp60$^{v\text{-}src}$, in addition to the C-terminal Tyr-527 site. This notion is supported by the fact that there are point mutations in the region between residues 90 and 96 which lead to oncogenic activation. It seems likely, then, that phosphorylation of pp60$^{c\text{-}src}$ in this domain will be involved in its regulation.

In summarizing the phosphorylation of pp60$^{c\text{-}src}$ it is interesting to note that many of the phosphorylation sites in pp60$^{c\text{-}src}$ lie in the unique N-terminal region of the molecule. As far as we know the only sites that pp60$^{c\text{-}src}$ shares with the other members of the src gene family are the "Tyr-416" autophosphorylation site and the C-terminal "Tyr-527" negative regulatory site (4). Thus, although other members of the family show increased phosphorylation following treatment of cells with TPA, and are therefore presumptive substrates for protein kinase C, none of these enzymes has a site equivalent to Ser-12 in pp60$^{c\text{-}src}$.

Given the multiplicity of both phosphorylation sites in pp60$^{c\text{-}src}$ and the protein kinases which can phosphorylate these sites, it is clear that the regulation of pp60$^{c\text{-}src}$ function is sophisticated. The effects of phosphorylation at the different sites will almost certainly be interactive and may work synergistically or counteractively. There are clear precedents for interactions between phosphorylation sites with other proteins, such as glycogen synthase, which have multiple sites. Such a mechanism will allow pp60^{c-src} to act as a node in a regulatory circuit which receives inputs through the different protein kinases that can phosphorylate it, and which provides an integrated output by phosphorylating its substrates.

REFERENCES

1. Hunter, T., and Cooper, J. A. (1986). *In* "The Enzymes" (P. D. Boyer and E. G. Krebs, eds.), 3rd Ed., Vol. 17, pp. 191–246. Academic Press, Orlando, Florida.
2. Sefton, B. M. (1986). *Curr. Top. Microbiol. Immunol.* **123,** 40–72.
3. Sefton, B. M., and Hunter, T. (1986). *Cancer Surv.* **5,** 160–172.
4. Cooper, J. A. (1990). *In* "Peptides and Protein Phosphorylation" (B. Kemp, ed.), pp. 85–113, CRC Press, Inc.
5. Hunter, T. (1987). *Cell (Cambridge, Mass.)* **49,** 1–4.
6. Shalloway, D., Coussens, P. M., and Yaciuk, P. (1984). *Proc. Natl. Acad. Sci. U.S.A.* **81,** 7071–7075.
7. Parker, R. C., Varmus, H. E., and Bishop, J. M. (1984). *Cell (Cambridge, Mass.)* **37,** 131–139.
8. Iba, H., Takeya, T., Cross, F. R., Hanafusa, T., and Hanafusa, H. (1984). *Proc. Natl. Acad. Sci. U.S.A.* **81,** 4424–4428.
9. Iba, H., Cross, F. R., Garber, E. A., and Hanafusa, H. (1985). *Mol. Cell. Biol.* **5,** 1058–1066.
10. Coussens, P. M., Cooper, J. A., Hunter, T., and Shalloway, D. (1985). *Mol. Cell. Biol.* **5** 2753–2763.
11. Gould, K. L., Woodgett, J. R., Cooper, J. A., Buss, J. E., Shalloway, D., and Hunter, T. (1985). *Cell (Cambridge, Mass.)* **42,** 849–857.
12. Gentry, L. E., Chaffin, K. E., Shoyab, M., and Purchio, A. F. (1986). *Mol. Cell. Biol.* **6,** 735–738.
13. Hunter, T., Ling, N. C., and Cooper, J. A. (1984). *Nature (London)* **311,** 480–483.
14. Woodgett, J. R., Gould, K. L., and Hunter, T. (1986). *Eur. J. Biochem.* **161,** 177–184.
15. Buss, J. E., Kamps, M. P., Gould, K., and Sefton, B. M. (1986). *J. Virol.* **58,** 468–474.
16. Sefton, B. M., Patschinsky, T., Berdot, C., Hunter, T., and Elliot, T. (1982). *J. Virol* **41,** 813–820.
17. Cross, F. R., and Hanafusa, H. (1983). *Cell (Cambridge, Mass.)* **34,** 597–607.
18. Patschinsky, T., Hunter, T., and Sefton, B. M. (1986). *J. Virol.* **59,** 73–81.
19. Collett, M. S., Erikson, E., and Erikson, R. L. (1979). *J. Virol.* **29,** 770–781.
20. Roth, C. W., Richert, N. D., Pastan, I., and Gottesman, M. M. (1983). *J. Biol. Chem.* **258,** 10768–10773.
21. Patschinsky, T., Hunter, T., Esch, F. S., Cooper, J. A., and Sefton, B. M. (1982). *Proc. Natl. Acad. Sci. U.S.A.* **79,** 973–977.
22. Smart, J. E., Opperman, H., Czernilofsky, A. P., Purchio, A. F., Erikson, R. L., and Bishop, J. M. (1981). *Proc. Natl. Acad. Sci. U.S.A.* **78,** 6013–6017.
23. Sugimoto, Y., Erikson, E., Graziani, Y., and Erikson, R. L. (1985). *J. Biol. Chem.* **260,** 13838–13843.

24. Snyder, M. A., Bishop, J. M., Colby, W. W., and Levinson, A. D. (1983). *Cell (Cambridge, Mass.)* **32**, 891–901.
25. Snyder, M. A., and Bishop, J. M. (1984). *J. Virol.* **136**, 375–386.
26. Weinmaster, G., Zoller, M. J., Smith, M., Hinze, E., and Pawson, T. (1984). *Cell (Cambridge, Mass.)* **37**, 559–568.
27. Kmiecik, T. E., and Shalloway, D. (1987). *Cell (Cambridge, Mass.)* **49**, 65–73.
28. Piwnica-Worms, H., Saunders, K. B., Roberts, T. M., Smith, A. E., and Cheng, S. H. (1987). *Cell (Cambridge, Mass.)* **49**, 75–82.
29. Weinmaster, G., and Pawson, T. (1986). *J. Biol. Chem.* **261**, 328–333.
30. Bramson, H. N., Thomas, N., Matsueda, R., Nelson, N. C., Taylor, S. S., and Kaiser, E. T. (1982). *J. Biol. Chem.* **257**, 10575–10581.
31. Courtneidge, S. A. (1985). *EMBO J.* **4**, 1471–1477.
32. Cooper, J. A., Gould, K. L., Cartwright, C. A., and Hunter, T. (1986). *Science* **231**, 1431–1434.
33. Laudano, A. P., and Buchanan, J. M. (1986). *Proc. Natl. Acad. Sci. U.S.A.* **83**, 892–896.
34. Cooper, J. A., and King, C. S. (1986). *Mol. Cell. Biol.* **6**, 4467–4477.
35. Cartwright, C. A., Eckhart, W., Simon, S., and Kaplan, P. L. (1987). *Cell (Cambridge, Mass.)* **49**, 83–91.
36. Kato, J., Takeya, T., Grandori, C., Iba, H., Levy, J. B., and Hanafusa, H. (1986). *Mol. Cell. Biol.* **6**, 4155–4160.
37. Levy, J. B., Iba, H., and Hanafusa H. (1986). *Proc. Natl. Acad. Sci. U.S.A.* **83**, 4228–4232.
38. Ralston, R., and Bishop, J. M. (1985). *Proc. Natl. Acad. Sci. U.S.A.* **82**, 7845–7849.
39. Purchio, A. F., Wells, S. K., and Collett, M. S. (1983). *Mol. Cell. Biol.* **3**, 1589–1597.
40. Collett, M. S., Belzer, S. K., and Purchio, A. F. (1984). *Mol. Cell. Biol.* **4**, 1213–1220.
41. Brown, D. J., and Gordon, J. A. (1984). *J. Biol. Chem.* **259**, 9580–9586.
42. Cotton, P. C., and Brugge, J. S. (1983). *Mol. Cell. Biol.* **3**, 1157–1162.
43. Sorge, L. K., Levy, B. T., and Maness, P. F. (1984). *Cell (Cambridge, Mass.)* **36**, 249–257.
44. Brugge, J. S., Cotton, P. C., Queral, A. E., Barrett, J. N., Nonner, D., and Keane, R. W. (1985). *Nature (London)* **316**, 554–557.
45. Lynch, S. A., Brugge, J. S., and Levine, J. M. (1986). *Science* **234**, 873–876.
46. Brugge, J., Cotton, P., Lustig, A., Yonemoto, W., Lipsich, L., Coussens, P., Barrett, J. N., Nonner, D., and Keane, R. W. (1987). *Genes & Dev.* **1**, 287–296.
47. Cartwright, C. A., Simantov, R., Kaplan, P. L., Hunter, T., and Eckhart, W. (1987). *Mol. Cell. Biol.* **7**, 1830–1840.
48. Martinez, R., Mathey-Prevot, B., Bernards, A., and Baltimore, D. (1987). *Science* **237**, 411–415.
49. Courtneidge, S. A., and Smith, A. E. (1983). *Nature (London)* **303**, 435–439.
50. Kornbluth, S., Sudol, M., and Hanafusa, H. (1987). *Nature (London)* **325**, 171–173.
51. Bolen, J. B., Thiele, C. J., Israel, M. A., Yonemoto, W., Lipsich, L. A., and Brugge, J. S. (1984). *Cell (Cambridge, Mass.)* **38**, 767–777.
52. Cartwright, C. A., Kaplan, P. L., Cooper, J. A., Hunter, T., and Eckhart, W. (1986). *Mol. Cell. Biol.* **6**, 1562–1570.
53. Yonemoto, W., Jarvis-Morar, M., Brugge, J. S., Bolen, J. B., and Israel, M. A. (1985). *Proc. Natl. Acad. Sci. U.S.A.* **82**, 4568–4572.
54. Cartwright, C. A., Hutchinson, M., and Eckhart, W. (1985). *Mol. Cell. Biol.* **5**, 2647–2652.
55. Ryder, J. W., and Gordon, J. A. (1987). *Cell (Cambridge, Mass.)* **7**, 1139–1147.
56. Rosen, N., Bolen, J. B., Schwartz, A. M., Cohen, P., DeSeau, V., and Israel, M. A. (1986). *J. Biol. Chem.* **261**, 13754–13759.

57. Bolen, J. B., Veillette, A., Schwartz, A. M., DeSeau, V., and Rosen, N. (1987). *Proc. Natl. Acad. Sci. U.S.A.* **84**, 2251–2255.
58. Bolen, J. B., Veillette, A., Schwartz, A. M., DeSeau, V., and Rosen, N. (1987). *Oncogene Res.* **1**, 149–168.
59. Gould, K. L., and Hunter T. (1988). *Mol. Cell. Biol.* **8**, 3345–3356.
60. Cartwright, C. A., Simanto, R., Cowan, W. M., Hunter, T., and Eckhart, W. (1988). *Proc. Natl. Acad. Sci. U.S.A.* **85**, 3348–3352.
61. Gould, K. L., Bilezikjian, L. M., and Hunter, T. (1989). *Mol. Endocrinol.* **3**, 79–88.
62. Levy, J. B., Dorai, T., Wang, L. -H., and Brugge, J. S. (1987). *Mol. Cell. Biol.* **7**, 4142–4145.
63. Cheng, S. H., Piwinica-Worms, H., Harvey, R. W., Roberts, T. M., and Smith, A. E. (1988). *Mol. Cell. Biol.* **8**, 1736–1747.
64. Cooper, J. A., and Runge, K. (1987). *Oncogene Res.* **1**, 297–306.
65. Kornbluth, S., Jove, R., and Hanafusa, H. (1987). *Proc. Natl. Acad. Sci. USA* **84**, 4455–4459.
66. Jove, R., Kornbluth, S., and Hanafusa, H. (1987). *Cell (Cambridge, Mass.)* **50**, 937–948.
67. Cooper, J. A., and Macauley, A. (1988). *Proc. Natl. Acad. Sci. USA* **85**, 4232–4236.

5

G Proteins May Exist as Polymeric Proteins: A Basis for the Disaggregation Theory of Hormone Action

MARTIN RODBELL
National Institute of Environmental Health Sciences
Research Triangle Park, North Carolina 27709

INTRODUCTION

The actions of hormones at the surface of cells can be likened to a propagated wave. When initiated, action spreads temporally and spatially throughout the immediate environment. In this manner all elements that control the ability of a cell to adapt to environmental changes are affected. As originally postulated by Erlich at the turn of the twentieth century, initiation begins at recognition elements or receptors on the cell surface. This hypothesis has been proven even to the extent of the fine structure of receptors. However, how the information gained by the transaction between hormone and receptor is conveyed or transduced and the basis of the multitude of changes (pleiotropism) are major unresolved issues.

The "second messenger" hypothesis remains the most popular model for explaining hormone action at cell surfaces. Simply stated, this hypothesis suggests that hormones regulate the levels of cyclic nucleotides, ions, inositol phosphates, and other organic materials that potentially affect the activities of numerous processes involved in cellular adaptation. The original model for this hypothesis was the adenylate cyclase system coupled to β-adrenergic and α-adrenergic receptors. Binding of catecholamines to these receptors stimulate and inhibit, respectively, the enzyme with resultant changes in the levels of cyclic AMP, the so-called second messenger. Pleiotropic responses derive from the fact that the nucleotide activates specific protein kinases. The latter, in cascade fashion, alter through phosphorylation a number of enzyme systems involved in carbohydrate, fat, and protein synthesis as well as those affecting cell growth, differentiation, and secretion of ions and proteins. In more recent times, other enzyme systems involving phospholipases have been invoked as alternate pathways for hormone action. In fact, multiple parallel and/or interactive networks seem now to be the rule for cellular responses to external signals. Thus, it would appear that the parallel neuronal network theory of informational processing may have its counterpart at the cellular and molecular level.

Despite the abundant evidence in support of the second messenger hypothesis, there are arguments against the idea that cyclic nucleotides, ions, etc., are the primary signals formed in response to hormone action and that pleiotropism is solely explained by the actions of these chemicals. For example, it is clearly established that catecholamines acting through β-adrenergic receptors affect several intracellular systems independent of activating adenylate cyclase. Such systems include Mg^{2+} transport, calcium channels, potassium channels, and glucose transport, each of which are altered in response to the hormone independently of cyclic AMP production and activation of protein kinase A. One explanation is that the same receptor independently interacts with each of these processes. If so, the implication is that the receptor and these "effectors" have common structural interacting sites. A more likely explanation is that the GTP-binding proteins (G proteins), which are now known to interact directly with a large number of receptors, are the candidates for mediating hormone action and transmitting this information in pleiotropic fashion. I have given arguments in favor of this hypothesis in the form of the "programmable messenger" theory of hormone action (1). In this theory, the GTP-binding proteins are released from their "moorings" with the receptors and thereby become accessible to specific effector systems as well as to a variety of protein modifying enzymes. Pleiotropism derives in this model from the malleable structures of the GTP-binding proteins and to the differing activities when these proteins are modified by an array of modifying enzymes programmed by each cell to alter the structure and function of the released GTP-binding protein.

The programmable messenger theory was predicated on the assumption that the G proteins are stoichometrically linked to receptors. However, there is increasing evidence that G proteins are likely present in great abundance compared to receptors and effector systems. Another assumption is that the released GTP-binding proteins can diffuse rapidly in transmitting information from the receptor to other parts of the cell. In considering this problem and other issues, the thought occurred that perhaps the GTP-binding proteins are constructed as polymers which associate through "linkage" proteins to the membranes and/or receptors embedded in the membranes such that binding of the hormones initiates instability in the polymers, resulting in either partial or total depolymerization. This type of change, particularly if the polymeric forms of the GTP-binding proteins are part of the cytoskeletal network, could provide rapid, propagated wavelike changes in cellular architecture and function. Moreover, as a consequence, multiple cascades can be introduced rather than the single cascade of events currently in vogue for explaining hormone action. In fact, I had presented in 1980 (2) a theory that GTP-binding proteins and receptors may exist in oligomeric or polymeric form and that hormones may initiate their actions to disaggregate these structures. Here I extend the theory in light of new information on the structure and function of G proteins that has evolved during the past several years.

DEPOLYMERIZATION THEORY

The idea that hormones initiate their actions by inducing a depolymerization of polymeric GTP-binding proteins (N or G) stemmed from target size analysis of adenylate cyclase systems in liver and adipocyte membranes (3,4). It was found that the target sizes of these systems (both in their stimulatory and inhibitory forms) were greater than 10^6 Da when analyzed prior to applying stimuli (hormone plus GTP). The possibility that the target included receptor and G proteins was tested by also examining the binding of glucagon to its receptor when the latter is coupled to N (a high affinity state of receptor that is susceptible to GTP conversion to a lower affinity state). Target size analysis again showed a high molecular size (0.8×10^6). Hormones and guanine nucleotides or fluoride ion decreased the ground state target sizes to approximately one-fourth or less.

The theory (2) developed from these findings is that the ground state of adenylate cyclase systems exists as polymeric structures of complexes between receptors and G units; hormones and GTP act concertedly to induce breakdown of the polymers to monomers that act as the primary "messengers" of hormone action. At that time, little was known of the composition and structure of receptors and G proteins. Hence, the idea that receptors and G proteins were copolymers could only be conjectural.

CURRENT THOUGHTS ON G PROTEINS

Comprehensive reviews on the subject of G proteins discuss various aspects of their structure and function (4–7). G proteins have been isolated as heterotrimeric complexes composed of three distinct protein subunits, termed α, β, and γ, of which the α-subunit uniquely binds and degrades GTP to GDP. Turnover of GTP to GDP represents part of the dynamic control by the G proteins on their effector systems. Based on cDNA analyses, the α-subunits comprise a small family of proteins (perhaps eight or nine) each of which displays regions of sequence identity throughout the animal kingdom. Regions of marked sequence divergence are important for interactions of the heterotrimeric G proteins with receptors, others for the effector systems upon which G proteins act. The β-subunits consist of two types that differ in size but not apparently in function, and they are the products of separate genes. The γ-subunits are thought to differ depending on the type of G protein; however, only the sequence of the γ-subunit of transducin is known. The β/γ-subunits form a tightly interactive complex which is necessary for the transmission of information between receptor and G proteins. Hence, by definition, G proteins are transduction elements requiring, minimally, coordinate interactions of three distinct protein subunits with surface receptors. This definition excludes other GTP-binding proteins, such as RAS proteins (8), which do not have the requisite heterotrimeric structure and which apparently do not interact with or require surface receptors for their actions.

The effector systems with which G proteins interact include, in addition to the classic regulation of adenylate cyclase activity, ion pumps, enzymes such as cyclic nucleotide phosphodiesterases and various types of phospholipases, and secretory processes. The powerful new tool of "patch-clamping" for monitoring single channels in whole cells or isolated patches of surface membrane has revealed regulatory functions of G proteins on various ion channels. Although formerly restricted to events occurring at the plasma membrane, G proteins now appear to interact directly with and regulate the functions of processes within intracellular organelles such as the Golgi complex (9). It is certain that enzymes which are not transmembrane proteins are regulated by direct action of the α-subunits. A prime example is "transducin," in which the α-subunit (Tα) directly interacts with a phosphodiesterase (PDE) that cleaves cyclic GMP to 5'-GMP. Visual excitation in vertebrate retinal rod cells occurs through light activation of rhodopsin receptors linked to transducin (T) (10). Utilizing purified components and various chemical cross-linking agents, it has been shown recently (11) that Tα, the GTP-binding protein that functions as a signal carrier, is linked to a complex of the β/γ-subunits of T. Based on various structural modifications of T and cross-linking studies, it would appear that all three subunits interact with the receptor at the inner face of the membrane. When examined in solution, the heterotrimer can be cross-linked as an oligomer minimally as large as $(T\alpha\beta\gamma)_3$. This structure is observed when T is in its latent, dark-adapted state; in this state GDP is tightly bound to the α-subunit. Upon light excitation of rhodopsin in intact rods, GDP becomes exchangeable with GTP, the physiological activating nucleotide. In the case of T in solution, replacement of GDP with Gpp(NH)p, a nonhydrolyzable GTP analog, diminishes the amount of cross-linked structures. Based on these findings, it was suggested that T in solution exists as an oligomeric protein that dissociates upon activation into free Tα. Using similar cross-linking studies with purified PDE before and after activation of T, the conclusion is drawn that the product of disaggregation, Tα, interacts with the enzyme to displace an inhibitory protein, thus resulting in activation of the enzyme. This cascade of events is thought to be responsible for all the changes induced by light in vertebrate rods.

It should be emphasized that the structure of transducin associated with the rhodopsin-rich plates remains unknown. There is evidence that T exists in both membrane-bound (T_m) and free forms (T_f) after light excitation in the presence of GTP (11). T_f is the structure that, in solution, exists as oligomers. Kinetic studies with reconstituted receptor and T_f suggest that a single activated rhodopsin molecule activates four T molecules in an allosteric fashion (12). However, although more abundant than T_m, T_f is considerably less active than T_m in simulating the activity of PDE (11). As an explanation of these differences, it is conceivable that T_m is associated with the receptors as polymers which, upon activation with light and GTP, yield more active products at lower levels of rhodopsin excitation (i.e., high levels of amplification, as observed with intact rods when excited).

SOME STRUCTURAL ASPECTS OF G PROTEINS

Sequences of α-subunits have been gleaned from analyses of their cDNAs by several laboratories (4). One of the most notable features of these proteins is the abundance of cyst(e)ine residues and the similarity in alignment of these residues among the various types of α-proteins, as illustrated in Fig. 1. Note that with the exception of $α_s$, each protein analyzed contains, as indicated in the enclosed bars, cysteine residues at the same relative positions. Such structural conservation suggests that these cysteine residues are essential for the structure/function of the α-subunits. Although the precise functional role of the conserved cysteine residues is unknown, there is evidence that $α_o$ contains two cysteinyl residues (circled) which are dependent on activation by guanine nucleotides for their reactivity with alkylating chemicals such as N-ethylmaleimide (14). It seems that these particular cysteine residues (as well as those situated identically in the other α-subunits) are fundamental to the transduction process. In this regard, it is of interest that

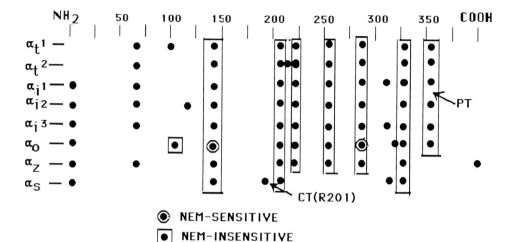

Fig. 1. The various cysteine residues in the α-subunits of G proteins were determined from the cDNA structures (reviewed in Refs. 4 and 10). The positions were determined by arranging residues that gave maximal alignment (the residue numbers are not to be taken as accurate). These cysteine residues are in the enclosed rectangular boxes. Pertussis-sensitive (PT) cysteine residues are four residues in from the carboxy terminus. Note that α_z and α_s are the only two α-proteins lacking cysteine at this position. Cholera toxin (CT) catalyzes ADP-ribosylation of the arginine residue R201 in α_s. Residues enclosed in squares and circles were found in the α_o protein to be either sensitive or insensitive to the covalent modification by N-ethylmaleimide when Go is activated by guanine nucleotides (reviewed in Ref. 4).

the cross-linking studies with transducin noted above revealed that some SH residues are considerably more susceptible to cross-linking than are the abundant lysine residues and that activation of T by Gpp(NH)p reduces cross-linking of the α-subunit with the other two subunits. These observations suggest that there may be a cascade of changes in cyst(e)ine residues induced by activation.

Recently we have discovered that the α-proteins behave during electrophoresis in buffers containing dodecyl sulfate and mercaptide reducing agents as if they are highly folded molecules.[1] For example, using antibodies as the means of detection, α-subunits extracted from brain and human red blood cell (rbc) membranes travel as single bands when electrophoresed in normal Laemmli buffer. By contrast, when electrophoresed with 8 M urea in this buffer, single bands became unstacked. Utilizing two antibodies that distinguish α_o- from α_i-like proteins and one that detects practically all classes of G protein α-subunits, along with labeling the extracts with per-

1. F. Ribeiro-Neto and M. Rodbell, (1989). *Proc. Natl. Acad. Sci. U.S.A.* **86**, 2571–2581.

tussis toxin and ^{32}P-NAD, it was possible to detect several unique bands. In every case, electrophoresis with urea caused some bands to display decreased mobility (apparent increases in molecular weight) suggestive of unfolding of the molecules by urea. When applied to the β-subunits, the same phenomenon was seen. Interestingly, the β-subunits contain even more cysteine residues than the α-subunits (13,14).

We interpret these findings as evidence that disulfide linkages govern the secondary structure of G subunits in a manner that prevents their accessibility to SH-reducing agents. Urea-induced unfolding allows reduction and subsequent separation of the very similarly structured molecules. As noted above with Tα, these findings also indicate that cysteine residues are particularly important in the secondary and probably tertiary structures of the α-subunits and thus provide another clue to the relationship between structure and function in these transduction elements.

PERTUSSIS TOXIN ACTIONS

Numerous studies have shown that pertussis toxin diminishes or abolishes the ability of many hormones and neurotransmitters to activate G proteins. The A-monomer of the toxin catalyzes ADP-ribosylation of a critical SH group four residues from the carboxy terminus. Pertussis toxin blocks receptor function. The same cysteine residue is found in six of the eight α-proteins deciphered from cDNA analysis (Fig. 1). From this, one might expect that all G proteins having this cysteinyl group would be equally susceptible to ADP-ribosylation.

During evaluation of the effects of urea on the electrophoretic properties of α-proteins, we have found that, despite antigenic properties suggestive of α_o/α_i-like proteins, the levels of ADP-ribosylation varied considerably. This unexpected phenomenon caused us to reevaluate the conditions for pertussis toxin action. Out of such studies came the realization that the toxin causes two effects on various α-proteins that are entirely independent of ADP-ribosylation.[2] (1) Enhanced antigenicity on Western blots was observed with all reactive antibodies toward α-proteins that are either poorly or highly susceptible to ADP-ribosylation by the toxin. (2) The electrophoretic mobility of the proteins is diminished (higher apparent molecular weights) by pertussis toxin. The β-subunits of G proteins are not similarly affected, nor are other nonrelated proteins. Both effects of pertussis toxin are not mimicked by cholera toxin (in the presence or absence of NAD) and are dependent on the integrity of SH groups in both the toxin and the various α-

2. F. Ribeiro-Neto and M. Rodbell, (1989). *Proc. Natl. Acad. Sci. U.S.A.* **86**, 2571–2581.

protein substrates (i.e., alkylation leads to loss of all effects of the toxin). As expected, ADP-ribosylation is equally abolished by alkylation of toxin and substrate SH residues. The two effects noted above are observed in the complete absence of NAD but require both an SH-reducing agent and ATP for toxin action without NAD.

The precise mechanism responsible for the NAD-independent actions of pertussis toxin remains unknown. We speculate based on our current knowledge that the toxin initiates its primary actions on α-proteins by catalyzing a thiol–disulfide exchange reaction. This reaction may ultimately induce alterations in the structure/function of the affected molecules, including susceptibility of the cysteine residue in the carboxy-terminal region to ADP-ribosylation. This hypothesis can explain the differing susceptibilities of α-proteins to ADP-ribosylation by pertussis toxin. It is conceivable that the changes induced in the structure other than by ADP-ribosylation are primary and fundamental to the actions of the toxin. These and the previous studies emphasize the importance of critical cyst(e)inyl residues in the structure/function of α-proteins. Clearly, α-proteins have rather specialized properties which likely account for their behavior both when linked with their partners in the G-protein heterotrimers and when liberated from the heterotrimers.

EFFECTS OF DETERGENTS ON G PROTEINS

G proteins have been extracted and purified from cell membranes generally with the use of ionic detergents such as cholate or deoxycholate. Yields of purified proteins range from 3 to 10%. When purified G proteins are subjected in detergent solutions to the activating effects of nonhydrolyzable analogs of GTP such as GTPγS or Gpp(NH)p, the sedimentation properties change from about 4 Svedberg (S) units down to 2 S. These changes suggest disaggregation of the heterotrimeric proteins with liberation of free α-subunits and β/γ-complexes. When G proteins are reconstituted in lipid vesicles and subjected to similar activating conditions, some of the α-subunits are released to the medium as water-soluble proteins (15). The latter studies indicate that the α-subunits are converted in their irreversibly activated [GTPγS or Gpp(NH)p] forms to structures that behave as globular proteins.

We have examined the actions of GTPγS or Gpp(NH)p on the release of α-subunits from brain vesicles, human erythrocyte membranes, kidney nephrons, and thyroid membranes.[3] Generally α-subunits were prelabeled by incubating the membranes with pertussis toxin and ^{32}P-NAD in the absence of detergents. Incubation was followed by centrifuging the membranes at 200,000 g for 2 hr and determining the levels of radioactivity in the supernatant and pellet. From these studies we concluded that, in all cases, the activating nucleotides induced the release of between 40 and 60% of the prelabeled material in a form not sedimentable at 200,000 g. These studies are in accord with the properties of the purified G proteins. However, when determinations were made of the total levels of α-subunits susceptible to pertussis toxin-induced ADP-ribosylation [carried out with Lubrol as activating detergent (16)], it became clear that about 95% of the labeled proteins were not released to the incubation medium. It was subsequently found that release only occurred from the small percentage of vesicles that were inside-out, that is, where the inner membrane surface containing G proteins was directly exposed to the medium.

These findings prompted a search for detergents that might minimally affect the native structure of G proteins but dissolve the membranes sufficiently to allow accessibility of the vesicular contents to probes such as pertussis toxin and antibodies against the various types of G-protein subunits. A most important criterion for a satisfactory detergent was that it exhibit a high critical micelle concentration, thus permitting its bulk removal by dilution or dialysis. A second criterion was to demonstrate whether release of the α-subunits was quantitative and dependent on activation by guanine nucleotides. Of several ionic and neutral detergents examined, the one that proved both interesting in its effects and which satisfied both criteria was octylglucoside.[4] Its effects are summarized as follows.

1. When detected with antibodies against α_o- and various α_i-type subunits, unactivated α-subunits sedimented with cytoskeletal elements (tubulin, various types of microfilaments, actin, etc.) following extraction of brain membrane vesicles with octylglucoside. CHAPS, a zwitterionic detergent also with high critical micelle properties, tends to prevent sedimentation with these elements.

2. Association of the unactivated α-subunits and the cytoskeletal elements in their sedimentable form was temperature dependent. Temperatures above 15°C tended to stabilize the sedimented α-subunits.

3. Mg^{2+} (5mM and above) enhanced sedimentation of α-subunits in presence of octylglucoside; activating guanine nucleotides act in an opposing fashion.

4. Activation of brain G proteins with GTPγS or Gpp(NH)p induced all of the α-subunits detectable with antibodies or pertussis toxin-catalyzed ADP-ribosylation to become soluble in octylglucoside in a non-temperature-sensitive fashion. Following removal of the cytoskeletal fraction and dialysis to remove octylglucoside, about 50% of the activated α-subunits remained water-soluble; the other half sedimented in a membrane-bound form.

5. The β/γ subunits of G proteins were completely extractable with octylglucoside at any temperature and fully reassociated with the lipid vesicles formed after removal of octylglucoside by dialysis.

The properties exhibited by the α-subunits prior to activation (presumably in their heterotrimeric structure) are very similar to those of tubulin. Its polymerized state is also temperature dependent, affected by Mg^{2+} and other divalent cations, and affected by the binding of GTP and GDP (17). Drugs such as colchicine affect tubulin polymerization but do not alter the sedimenting properties of the α-subunits. Hence, although both tubulin and α-subunits are GTP-binding proteins (and contains GTPase activities), they clearly are different in this property as well as in their primary structures.

The properties of the α-subunits in their unactivated form suggest that they preexist as polymers associated with cytoskeletal elements underlying or associated with the plasma membrane. In a recent study, we have found that treatment of octylglucoside or CHAPS extracts of brain membranes with bifunctional cross-linking agents prevents the α-proteins from entering gels during electrophoresis, suggesting cross-linkage and presumed polymerization. However, when extracts were prepared from membranes that had been activated by GTPγS, for example, all of the α-subunits previously in the membrane were detected immunochemically in the electrophoretic bands corresponding to freshly prepared α-proteins. We interpret these findings to suggest that the α-proteins were initially present in oligomeric or polymeric structures such that they were readily linked by the cross-linking agents to form nonpenetrating polymers. The activated α-proteins, by contrast, were liberated from their postulated polymeric structures as monomers since they were no longer susceptible to cross-linkage. Interestingly, immunoblots of the β-subunits in the same extracts of brain membranes treated (or not) with cross-linking reagents revealed that the β-subunits readily penetrated the gels and migrated with the same mobility as native β-subunits. Thus, under no conditions (with several types of cross-linking reagents) did we observe cross-linkage of β-subunits. These findings raise the possibility that, in native membranes, the β-subunits (presumably linked with the γ-subunit) are not associated with the α-subunits in a manner which yields cross-linked polymers either between themselves or with the α-subunits. These findings contrast with cross-linking studies with purified transducin heterotrimers in which it was found that all three subunits of the heterotrimer became cross-linked with bifunctional cross-linking agents (17). It should be emphasized that the transducin preparations employed in those studies were extracted from membranes as water-soluble heterotrimeric structures and may not be representative of the active forms of α_t linked to the rhodopsin receptors.

SIGNAL TRANSDUCTION AND CYTOSKELETAL ELEMENTS

Since the mid-1970s several reports have provided suggestive evidence that hormone action now known to be mediated through G proteins may be associated with cytoskeletal elements (18,19). Other studies have shown in more direct fashion that components of G proteins are associated with cytoskeletal fractions. These include findings that cholera toxin-labeled material (presumably α_s) associates with cytoskeletal fractions isolated in Triton X-100 (20,21); in the same vein, treatment of synaptic vesicles with colchicine or vinblastine releases soluble material that, when incubated with Gpp(NH)p, stimulates adenylate cyclase (20). In S49 murine lymphoma cells, 60% of the β-subunits of G proteins cofractionate with actin and tubulin (21). These and the other cited findings are consistent with the view that G proteins represent a class of cytoskeletal proteins linked to surface membranes through membrane-spanning proteins such as hormone receptors. In this regard, actin, spectrin, and similar proteins represent a well-described class of membrane-linked polymeric proteins that behave as transduction elements in the same sense as G proteins. Upon binding of ligands to surface lectin receptors, for example, the underlying polymeric proteins comigrate on the surface as if the entire network of elements in the cell has been altered in concert.

SUMMARY AND SPECULATIONS

The studies reported here on the properties of G proteins, though in several respects still in their embryonic phase, are consistent with the view, early predicted from target analysis of adenylate cyclase systems, that these transduction proteins are minimally oligomeric and probably are higher order polymeric structures. Their properties resemble those of certain cytoskeletal proteins. Since many studies indicate that these proteins are linked directly to several classes of membrane receptors, it is likely, considering the small number of receptors (rhodopsin is a special case), that the polymeric structures are confined to localized, receptor-rich regions of the surface membrane. It remains unknown how G proteins are linked to receptors and are activated by external signals. Hormones seem generally to increase the exchange of GTP for tightly bound GDP. Since activating guanine nucleotides increase the chemical reactivity of particular cyst(e)ine residues in the α-subunits to the actions of activating guanine nucleotides, it is likely that the hormone-induced exchange reaction is fundamental to the changes in these critical cyst(e)ine residues and the resultant changes in secondary, tertiary, and quaternary structures. As revealed here, the compactness of these structures suggested from the relative inertness of the α- and β-subunits to sulfhydryl reagents in the absence of urea argues for highly coiled structures that, springlike, may rapidly transmit information from subtly changed receptor structure with resultant unfurling and dissipation of interactive energy. The release of the α-subunits as water-soluble proteins observed in our studies with membranes is consistent with this view. The finding that pertussis toxin acts through an SH-dependent process is also consistent with the idea of a highly compact structure dependent on disulfide bridges to control secondary and tertiary structure. ADP-ribosylation seems but one effect of the toxin that relates to the loss of receptor function so often observed after pertussis toxin treatment of cells. It is conceivable that its other structural effects are more fundamental. Clearly, our findings raise questions concerning the correlations to be found between toxin action on cells and the degree of ADP-ribosylation on selective α-proteins.

Finally, there is the essential question of how the suggested polymeric structures are involved in the amplification and signal transduction processes on the numerous activities affected by receptors and G proteins. It is likely that α-subunits have multiple roles in signal transduction; in other words,

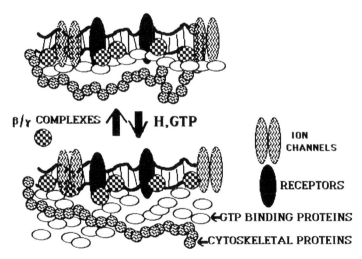

Fig. 2. Possible polymeric structures of the GTP-binding proteins and their linkage through β/γ-complexes to membrane lipids, receptors, and ion channels present in the plasma membrane. Cytoskeletal proteins are suggested to interact with the GTP-binding proteins such that the latter are an intimate part of the cytoskeletal network. The actions of hormones (H) and GTP on their receptors and GTP-binding proteins, respectively, cause in concerted fashion disaggregation of the polymeric structures into monomers, dimers, etc., of the α-proteins. In this fashion it is suggested that the cytoskeletal network is altered in structure and function by hormones and other external sensory signals.

each type may have several actions either in unmodified or modified form, as postulated in the so-called programmable messenger theory (1). Also to be considered is the finding that the G polymers resemble sliding filaments which "move" along the membrane as each α-subunit is released in response to changes induced by hormones and GTP. If the polymers are displaced with respect to membrane processes such as ion channels, one can imagine rapid opening and closing of these channels depending on their relative location to G polymers. In this regard, recently it has been shown that ankyrin and spectrin associate with voltage-dependent sodium channels in brain (22). Breakdown of the polymer (and its possible association with other cytoskeletal elements) by injection of GTPγS or Gpp(NH)p into cells might bypass the specific, localized actions of hormones/neurotransmitters and grossly change the positioning and actions of the polymers on channels. By the same reasoning, addition of α- or β/γ-complexes to isolated "patches" of membranes containing channels might also disturb the structures of associated G proteins or other cytoskeletal proteins that may form linkages with G proteins. Such considerations might apply to the controversial effects of G-protein subunits on ion channels (24,25). These ideas are illustrated in Fig. 2.

REFERENCES

1. Rodbell, M. (1985). *Trends Biochem. Sci.* **10**, 461–464.
2. Rodbell, M. (1980). *Nature (London)* **284**, 17–22.
3. Schlegel, W., Kempner, E. S., and Rodbell, M. (1979). *J. Biol. Chem.* **254**, 5168–5176.
4. Schlegel, W., Cooper, D. M. F., and Rodbell, M. (1980). *Arch. Biochem. Biophys.* **2011**, 678–682.
5. Stryer, L., and Bourne, H. R. A. (1986). *Annu. Rev. Cell Biol.* **2**, 391–419.
6. Gilman, A. G. (1987). *Annu. Rev. Biochem.* **56**, 615–650.
7. Neer, E. J., and Clapham, D. E. (1988). *Nature (London)* **333**, 129–134.
8. Barbacid, M. (1987). *Annu. Rev. Biochem.* **56**, 779–827.
9. Melancon, P., Glick, B. S., Malhotra, V., Weidman, P. J., Serafini, T., Gleason, M. L., Orci, L., and Rothman, J. E. (1987). *Cell (Cambridge, Mass.)* **51**, 1053–1062.
10. Hurley, J. B. (1987). *Annu. Rev. Physiol.* **49**, 793–812.
11. Hingorani, V. N., Tobias, D. T., Henderson, J. T., and Ho, Y.-K. (1988). *J. Biol. Chem.* **263**, 6916–6926.
12. Wensel, T. G., and Stryer, L. (1985). *J. Cell Biol.* **101**, 221.
13. Wessling-Resnick, M., and Johnson, G. L. (1987). *J. Biol. Chem.* **262**, 3696–3706.
14. Winslow, J. W., Bradley, J. D., Smith, J. A., and Neer, E. J. (1987). *J. Biol. Chem.* **262**, 4501–4507.
15. Florio, V. A., and Sternweiss, P. C. (1985). *J. Biol. Chem.* **260**, 3477–3483.
16. Ribeiro-Neto, F., Matera, R., Sekura, R., Birnbaumer, L., and Fields, J. (1987). *Mol. Endocrinol.* **1**, 472–481.
17. Cassimeris, L. U., Wadsworth, P., and Salmon, E. D. (1986). *J. Cell Biol.* **102**, 2023–2032.
18. Kennedy, M. S., and Insel, P. A. (1979). *Mol. Pharmacol.* **16**, 215–223.

19. Rasenick, M. M., Stein, P. J., and Bitensky, M. W. (1981). *Nature (London)* **294**, 560–562.
20. Sayhoun, N. E., LeVine III, H., Davis, J., Hebdon, G. M., and Cuatrecasas, P. (1981). *Proc. Natl. Acad. Sci. U.S.A.* **78**, 6158–6162.
21. Carlson, K. E., Woolkalis, M. J., Newhouse, M. G., and Manning, D. R. (1986). *Mol. Pharmacol.* **30**, 463–468.
22. Srinivasan, Y., Elmer, L., Davis, J., Bennett, V., and Angelides, K. (1988). *Nature (London)* **333**, 177–180.
23. Logothetis, D. E., Kurachi, Y., Galper, J., Neer, E. J., and Clapham, D. E. (1987). *Nature (London)* **325**, 321–326.
24. Codina, J., Yatani, A., Grenet, D., Brown, A. M., and Birnbaumer, L. (1987). *Science* **236**, 442–445.

6
Pleiotropic Functions of the Insulin Receptor

ALAN R. SALTIEL
The Rockefeller University
New York, New York 10021

INTRODUCTION

Despite considerable attention over the past twenty years, the molecular mechanisms involved in signal transduction in insulin action remain unexplained. This problem has been complicated by the diversity of actions of insulin which can be distinguished on the basis of time course. Certain events occur within minutes, such as enhancement of glucose transport, regulation of cyclic nucleotide levels, and activation or deactivation of key intracellular enzymes that regulate carbohydrate and lipid metabolism. Others occur over the course of hours, such as stimulation of protein synthesis and cellular growth (Fig. 1). These distinct time frames suggest different mechanisms of signal transduction of bifurcation of transduction pathways.

A number of molecules have been proposed to act as intracellular second messengers for insulin, including cAMP, cGMP, Ca^{2+}, H_2O_2, and insulin itself (see Ref. 1 for review); however, each has proved inadequate to explain insulin action (see Refs. 2 and 3 for review). In the last few years, a general consensus has developed concerning the role of protein phosphorylation in mediating at least the acute effects of the hormone. Many of the rate-limiting enzymes involved in the major metabolic pathways are regulated by control of their phosphorylation state, and insulin is known to modulate several of these activities through regulation of phosphorylation/dephosphorylation processes (1). Certain enzymes are activated by an insulin-stimulated dephosphorylation reaction. Other proteins are phosphorylated in response to insulin, although the biological significance of enhanced phosphorylation is unclear in many cases.

Two predominant hypotheses have emerged to explain the coupling of the

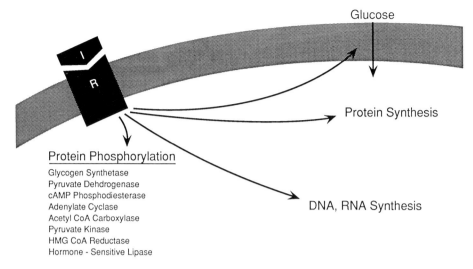

Fig. 1. Cellular actions of insulin. This general model depicts some of the major cellular actions of insulin. These actions can be categorized into four groups: (a) enhancement of transport phenomena at the cell surface, resulting in increased transport of glucose in some cells; (b) acute modulation of metabolic enzymes, many of which are regulated by insulin-induced changes in their state of phosphorylation; (c) stimulation of protein synthesis, resulting in the appearance of certain enzymes such as tyrosine aminotransferase and lipoprotein lipase; and (d) regulation of DNA and RNA synthesis, accounting for the mitogenic activities of insulin and its regulation of specific gene expression.

insulin–receptor complex to the intracellular control of protein phosphorylation: (i) a phosphorylation cascade, perhaps initiated by the insulin-dependent tyrosine kinase activity of the receptor (2); and (ii) the generation of a chemically undefined second messenger (3) (Fig. 2). These two pathways are not mutually exclusive and, in fact, may work synergistically to provide a cellular response to the hormone. Evaluation of the phosphorylation cascade hypothesis has centered on the identification and functional characterization of cellular substrates for the receptor kinase. It appears that the insulin receptor kinase may be necessary for insulin action (4,5), and several substrates have been identified that are phosphorylated on tyrosine residues upon exposure of cells to insulin (6–8). At the present time, however, none of these proteins have been functionally characterized, and the precise role of the tyrosine kinase activity of the receptor remains undetermined. The search for a second messenger has been underway since the last 1970s. Larner et al. (9) first reported the detection of an insulin-sensitive, enzyme-modulating activity from selected muscle. Similar substances were later identified that regulated the activities of several insulin-sensitive enzymes, including glycogen synthetase, pyruvate dehydrogenase, cAMP phosphodi-

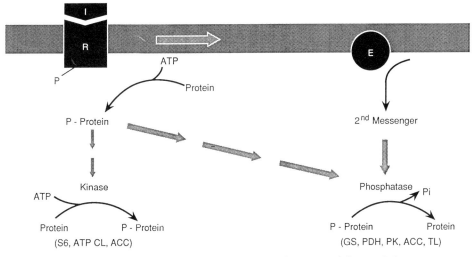

Fig. 2. Regulation of protein phosphorylation by insulin. Many of the metabolic activities under the control of insulin are regulated by changes in protein phosphorylation. Certain proteins are phosphorylated in response to insulin, including ribosomal S6, ATP citrate lyase, and acetyl-CoA carboxylase. Dephosphorylation of other proteins is observed in response to the hormone, including glycogen synthetase, pyruvate dehydrogenase, hormone-sensitive triglyceride lipase, and pyruvate kinase. Two possible pathways which may mediate these changes are illustrated: (left) phosphorylation cascade, initiated by the tyrosine kinase activity of the insulin receptor; (right) generation of a second messenger produced by a receptor-coupled effector system.

esterase, acetyl-CoA carboxylase, adenylate cyclase, glucose-6-phosphatase, and phospholipid methyltransferase (see Ref. 3 for review), although there were some inconsistencies regarding the chemical properties of these enzyme-modulating activities. Despite the detection of these substances in several laboratories, however, definitive information concerning the chemical structure(s) responsible for these activities has not been forthcoming. Thus, our work in this area was initiated with the goal of uncovering the precise chemical identity of these insulin-sensitive, enzyme-modulating substances.

PURIFICATION AND BIOLOGICAL CHARACTERIZATION OF THE INSULIN-SENSITIVE ENZYME MODULATORS

Two structurally similar enzyme-modulating substances were purified that were released from hepatic plasma membranes in response to insulin (10). These substances were assayed by the ability to regulate certain insulin-

TABLE I
Chemical Properties of the Insulin-Sensitive
Enzyme Modulators

Net negative charge
Acid stable
Alkaline labile
Not adsorbed to charcoal
Not adsorbed to reversed phase resins
Polar
Molecular weight 800–1000
Inactivated by methylation, acetylation, periodate oxidation,
  nitrous acid deamination
Contains myo-inositol

---

sensitive enzymes. Initial purification studies focused on modulation of the low K_m cAMP phosphodiesterase in fat cell membranes and relied mainly on ion exchange, molecular sizing, and phase partitioning procedures. The chemical properties of these substances are summarized in Table I. They exhibited dissimilar net negative charges, even at pH 2.0, but were similar in all other chemical properties. The relative hydrophilicity, negative charge, apparent molecular weight of 800–1000, and sensitivity to periodate and nitrous acid suggested an oligosaccharide phosphate structure.

The purified enzyme-modulating substances regulate several insulin-sensitive enzymes assayed in subcellular fractions, including low K_m cAMP phosphodiesterase (10), adenylate cyclase (11), and pyruvate dehydrogenase (11). In some cases, the modulation of enzyme activity was biphasic with respect to concentration, reminiscent of the paradoxical dose responses observed for some of the metabolic activities of insulin. Although the precise biochemical mechanisms by which this substance elicits its actions are unclear, the regulation of each of these enzymes might be explained by control of the state of phosphorylation of proteins (11). In preliminary experiments, we have observed the specific activation of a Mg^{2+}-dependent protein phosphatase which may mediate some of the dephosphorylating effects of insulin. The acute regulation of key rate-limiting enzymes suggests a role for these modulators in the regulation of carbohydrate and lipid synthesis and degradation.

THE ENZYME MODULATORS ARE INOSITOL PHOSPHATE GLYCANS

Preliminary compositional analyses of the carbohydrate phosphate enzyme-modulating substances suggested inositol as a component. Several of the known inositol phosphates were evaluated, yet none exhibited enzyme-modulating activity, nor shared the chemical properties, chromatographic or electrophoretic behavior, or insulin sensitivity of the enzyme modulators. These results suggested the possibility that these carbohydrate substances might be inositol phosphate derivatives. A glycosylated derivative of inositol phosphate was recently described as a product of a covalent bond between certain proteins and phosphatidylinositol (PI) (see Refs. 12–14 for review). This unusual linkage serves as an anchor for these proteins to the plasma membrane, in which the protein is coupled to an oligosaccharide exhibiting a terminal non-N-acetylated hexosamine covalently linked to PI. The membrane-bound protein can be converted to soluble form containing an inositol phosphate glycan by digestion with a *Staphylococcus aureus* PI-specific phospholipase C (PLC) (15). This enzyme specifically hydrolyzes PI and glycosyl-PI, but it is inactive in the hydrolysis of the polyphosphoinositides or other phospholipids.

To evaluate the possibility that the enzyme modulators might arise from the phosphodiesteratic hydrolysis of a structurally similar, PI-containing glycolipid, this specific bacterial phospholipase was added to liver plasma membranes, and the release of the enzyme modulators was assayed. In a series of experiments, PI-PLC was found to reproduce the effect of insulin in facilitating the generation of the enzyme modulators (10,11). These PI-specific PLC generated activities were chromatographically, electrophoretically, and chemically identical to those produced by insulin. Moreover, a precursor for the PLC-generated enzyme modulator was extracted from liver membranes and resolved by thin-layer chromatography (TLC). This PI-specific PLC substrate exhibited the properties of a glycophospholipid and was tentatively identified as a PI glycan (10).

Studies on the chemical structure of the inositol glycan enzyme modulator and the precursor PI glycan have centered on evaluation of the biological activity and chromatographic behavior after chemical treatments. The presence of inositol was validated by production with a PI-specific PLC (10), insulin-sensitive [^{3}H]inositol labeling (16), and mass spectral analysis after hydrolysis (17). Glycosidically linked, non N-substituted hexosamine was suggested by insulin-sensitive glucosamine incorporation (16) and nitrous acid sensitivity (10). The composition and orientation of remaining monosaccharides distal to glucosamine remain to be determined.

The generation of the inositol glycans in response to insulin was also evaluated in a cultured myocyte line, BC$_3$H1, by following the radioisotopic labeling of the PI glycan precursor and carbohydrate products. The PI glycan was labeled with inositol, glucosamine, and myristic acid and purified by both TLC and high-performance liquid chromatography (HPLC) (16,18). This glycolipid was hydrolyzed in response to insulin in intact BC$_3$H1 myocytes (16,18) and H-4 hepatoma (19) cells and in response to PI-specific PLC in a reconstituted system. Following digestion with the PI-specific PLC, the

TABLE II
Properties of the Phosphatidylinositol Glycan

Metabolically labeled with inositol, glucosamine, fatty acids, phosphate
Hydrolysis with PI-specific PLC produces diacylglycerol
  and inositol phosphate glycan
Hydrolysis with PLA_2 produces saturated fatty acids
Nitrous acid deamination produces PI

---

water-soluble carbohydrate products, labeled both with [^{3}H]inositol and [^{3}H]glucosamine, were evaluated with respect to chemical stability and chromatographic and electrophoretic behavior. They appeared to be identical to the insulin-sensitive enzyme modulators derived from the same cells, as well as to those from liver or fat cell plasma membranes (16).

Mato et al. (19) found that in cultured hepatoma cells the PI glycan was labeled with [^{3}H]glucosamine or [^{3}H]palmitic acid, but not with myristate or inositol. However, the isolated glycolipid was a substrate for hydrolysis by a PI-specific PLC, and deamination with nitrous acid under mildly acidic conditions resulted in the production of PI, providing substantial evidence for glycosidically linked PI in the molecule. The properties of the PI glycan are summarized in Table II. These structural studies on the enzyme modulators and their glycolipid precursor suggest an inositol phosphate glycan (IPG) structure for the enzyme modulators, consisting of inositol phosphate glycosidically linked to non-N-acetylated hexosamine, which in turn is glycosidically linked to an oligosaccharide. The difference between the two species of enzyme modulators may be due to the presence of a 1,2-cyclic versus 1- or 2-monophosphate inositol glycan, based on their differential sensitivity to alkaline phosphatase (16). Thus far, these differences appear unrelated to the enzyme-modulating activities. This insulin-sensitive hydrolysis reaction is summarized in Fig. 3.

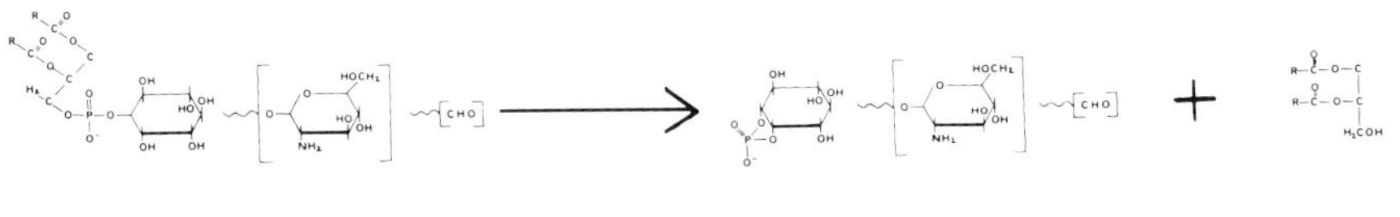

PHOSPHATIDYLINOSITOL-GLYCAN   INOSITOL PHOSPHATE GLYCAN   DIACYLGLYCEROL
(PI-GLYCAN)   (IPG)   (DAG)

Fig. 3. Phosphodiesteratic hydrolysis of the phosphatidylinositol glycan. A summary of the insulin-sensitive hydrolysis of the PI glycan is represented. Insulin stimulates the activity of a specific phospholipase C that catalyzes the cleavage of this glycolipid, resulting in the generation of two potential signals: (1) an inositol phosphate glycan and (2) diacylglycerol.

THE INOSITOL PHOSPHATE GLYCAN ENZYME MODULATORS ARE PRODUCED BY AN INSULIN-SENSITIVE PHOSPHATIDYLINOSITOL GLYCAN-SPECIFIC PHOSPHOLIPASE C ACTIVITY

The insulin-sensitive hydrolysis of the [^{3}H]inositol-labeled PI glycan was accompanied by simultaneous production of the labeled IPG (16,18). In cells labeled with [^{3}H]myristic acid, the transient decrease of the labeled PI glycan was also observed, along with the production of [^{3}H]myristate-labeled diacylglycerol over the same time course. The rapid production of this specifically labeled diacylglycerol was not observed with agonists known to stimulate the hydrolysis of the phosphoinositides, although these agonists did specifically cause the production of [^{3}H]arachidonate-labeled diacylglycerol (18). These studies indicate that the specifically labeled, insulin-sensitive diacylglycerol and the IPGs arise from this distinct PI glycan substrate, suggesting the possible importance of this glycolipid as a rate-limiting component in insulin action. Recent reports (20,21) have suggested a multiplicity of genes for protein kinase C, the assumed target enzyme for diacylglycerol, resulting perhaps in distinct isozymes of this enzyme (22). The insulin-sensitive species of diacylglycerol with a distinct fatty acid composition may be directed to specific isozymes of kinase C. This possibility along with the absence of increased intracellular calcium, may explain, in part, the perplexing relationship between insulin action and protein kinase C.

In addition to descriptive studies on the insulin-sensitive hydrolysis of the PI glycan, we have also recently examined the enzymology of this reaction. While insulin is known to cause the increased labeling of several phospholipids, it has not been found to stimulate the hydrolysis of the phosphoinositides and does not share the activities of those hormones which induce calcium mobilization through the generation of inositol trisphosphate. Thus, the insulin-dependent generation of IPG and diacylglycerol is

TABLE III
Properties of the Phosphatidylinositol Glycan Phospholipase C

Localized in the plasma membrane in liver
Solubilized with neutral, nonionic detergent
Apparent molecular weight of 52,000
Active as a monomer
Catalyzes the phosphodiesteratic hydrolysis of glycosyl-PI
Inactive in the hydrolysis of PI, PIP$_2$, or other phopholipids
No requirement for Ca^{2+}
Inhibited by mercurial compounds

likely to be due to the stimulation of a PLC which selectively hydrolyzes glycosyl-PI. Such an enzyme has been purified to near homogeneity from rat liver plasma membranes (23). Its properties are summarized in Table III. This catalytic activity appears to reside in a single polypeptide with an apparent molecular weight of 52,000. The enzyme is specific for glycosyl-PI; no hydrolysis of PI, phosphatidylinositol bisphosphate (PIP_2), or other phospholipids is observed under a variety of conditions. In addition to its unique substrate specificity, the PI glycan PLC is calcium independent. These data suggested a role for this novel enzyme as an effector in mediating some of the actions of insulin.

TURNOVER OF GLYCOSYLPHOSPHATIDYLINOSITOL IN INSULIN ACTION

The precise role of the insulin-sensitive phosphodiesteratic hydrolysis of glycosylated phosphatidylinositol in the action of the hormone remains unclear. It appears that insulin–receptor interaction leads to the rapid, dose-dependent activation of a selective PLC that hydrolyzes the PI glycan. Preliminary studies suggest that this glycolipid is predominantly localized on the cytoplasmic face of the plasma membrane, since extensive treatment of cells with PI-specific PLC does not alter the ability of insulin to stimulate the intracellular accumulation of inositol glycan (24). Another potential function of this reaction may be the release of proteins covalently linked to glycosyl-PI at the cell surface. One such example may be the glycosyl-PI-anchored heparan sulfate proteoglycan, which was released from cultured hepatocytes in response to exogenously added PI-specific PLC or insulin (25). Although the concentrations of insulin used were probably not physiologically relevant, this action was presumed to be mediated by the activation of a PI glycan-specific PLC. It was further suggested that the released form of heparan sulfate proteoglycan behaved as an autocrine growth factor, owing to its specific internalization that occurs at binding sites which recognize the inositol phosphate moiety of the released proteoglycan (25). These observations suggest that, in addition to activating the hydrolysis of the PI glycan on the cytoplasmic side of the plasma membrane, insulin binding may also lead to the hydrolysis of the structurally analogous glycosyl-PI anchor, resulting in the release of certain proteins from the cell surface. This action of insulin may be advantageous in both the regulation of the concentration of a protein at the cell surface as well as its regulated secretion and subsequent activity (14). The molecular characterization of the insulin-sensitive cleavage of this novel class of glycolipids as it relates to the control of protein phosphorylation (through the generation of inositol glycans and diacylglycerol) as well as the regulation of proteins anchored to biological membranes may provide further insights into the pleiotropic nature of insulin action.

REFERENCES

1. Czech, M. P. (1985). *Annu. Rev. Physiol.* **47,** 357–381.
2. Kasuga, M., Karlsson, F. A., Kahn, C. R. (1982). *Science* **215,** 185–186.
3. Larner, J. (1982). *J. Cyclic Nucleotide Res.* **8,** 289–296.
4. Ch, C. K., Dull, T. J., Russell, D. S., Gherzi, R., Lebwohl, D., Ullrich, A., and Rosen, O. M. (1987). *J. Biol. Chem.* **262,** 1842–1849.
5. Ebin, Y., Areki, E., Taira, M., Shimada, F., Mori, M., Craik, C. S., Siddle, K., Pierce, S. B., Roth, R. A., and Rutter, W. J. (1987). *Proc. Natl. Acad. Sci. U.S.A.* **84,** 704–708.
6. Rees-Jones, R. W., and Taylor, S. I. (1985). *J. Biol. Chem.* **260,** 4461–4467.
7. White, M. F., Maron, R., and Kahn, C. R. (1985). *Nature (London)* **318,** 182–186.
8. Bernier, M., Laird, D. M., and Lane, M. D. (1987). *Proc. Natl. Acad. Sci. U.S.A.* **84,** 1844–1848.
9. Larner, J., Galasko, G., Chang, E. K., DePaoli-Roach, A. A., Huang, L., Daggy, D., and Kellog, J. (1979). *Science* **206,** 1408–1409.
10. Saltiel, A. R., and Cuatrecasas, P. (1986). *Proc. Natl. Acad. Sci. U.S.A.* **83,** 5793–5797.
11. Saltiel, A. R. (1987). *Endocrinology* **120,** 967–972.
12. Low, M. G., Ferguson, M. A. J., Futerman, A. H., and Silman, I. (1986). *Trends Biochem. Sci.* **11,** 212–213.
13. Low, M. G. (1987). *Biochem. J.* (in press).
14. Low, M. G., and Saltiel, A. R. (1987). *Science* (in press).
15. Low, M. G., and Finean, J. B. (1977). *Biochem. J.* **167,** 281–284.
16. Saltiel, A. R., Fox, J. A., Sherline, P., and Cuatrecasas, P. (1986). *Science* **233,** 967–972.
17. Sherman, W. R., and Saltiel, A. R. Unpublished observations.
18. Saltiel, A. R., Sherline, P., and Fox, J. A. (1987). *J. Biol. Chem.* **262,** 1116–1121.
19. Mato, J. M., Kelly, K. C., and Jarett, L. (1987). *J. Biol. Chem.* **262,** 2131–2137.
20. Coussens, L., Parker, P., Chee, L., Yang-Feng, T. L., Chem, E., Waterield, M. D., Francke, V., and Ullrich, A. (1986). *Science* **233,** 859–866.
21. Knopf, J. L., Lee, M.-H., Sultzman, L. A., Kriz, R. W., Loomis, C. R., Hewick, R. M., and Bell, R. M. (1986). *Cell (Cambridge, Mass.)* **46,** 491–502.
22. Huang, K.-P., Nakabayashi, H., and Huang, F. L. (1986). *Proc. Natl. Acad. Sci. U.S.A.* **83,** 8535–8539.
23. Fox, J. A., Soliz, N. M., and Saltiel, A. R. (1987). *Proc. Natl. Acad. Sci. U.S.A.* **84,** 2663–2667.
24. Saltiel, A. R. Unpublished observations.
25. Ishihara, M., Fedarko, N. S., and Conrad, H. E. (1987). *J. Biol. Chem.* **262,** 4708–4716.

7

Cooperative Regulation of Gene Expression in Liver Cells by Hormones and Extracellular Matrix

LOLA M. REID, ISABEL ZVIBVEL, TOHRU WATANABE,
YASU FUJITA, MARIA AGELLI, KATE MONTGOMERY,
ANDREA KRAFT, AND ANDREAS OCHS

Departments of Molecular Pharmacology and Microbiology and Immunology
Albert Einstein College of Medicine
Bronx, New York 10461

INTRODUCTION

Differentiation[1] *in vivo* and *in vitro* in metazoan cells is dependent on signals derived from various cellular interactions (1–11). These signals serve to coordinate the activities of the many specialized cell types in the organism. During embryogenesis, complex cellular interactions result in the "commitment" of cells, that is, cause cells to restrict their genetic potential to a particular developmental pathway leading to a specific cell type (1,4,5). The commitment process is, as far as we know, unidirectional and irreversible. Thus, a cell committed to be a liver cell cannot be changed to another cell type by any known condition.

In committed or adult cells, the signals from cell–cell interactions serve to maintain and to regulate gene expression, but, obviously, only of those genes

---

1. Abbreviations: SSM, serum-supplemented medium; HDM, serum-free, hormonally defined medium; EGF, epidermal growth factor; TGF-α, tumor growth factor-α; IGF-I, insulin-like growth factor-I; IGF-II, insulin-like growth factor-II; GAGs, glycosaminoglycans; PGs, proteoglycans.

within the repertoire of the committed cell (1–11). The same signal can elicit quite distinct responses from different cell types. Thus, the specificity of the response is dictated by the cell type (6–9).

The known signals from cell–cell interactions include a set that are soluble and a set that are insoluble, the two sets having effects that are synergistic. The set of soluble signals includes autocrine factors, factors produced by a cell and active on the same cell; paracrine factors, factors secreted by a cell and active on its neighbor; and endocrine factors, factors produced by a cell and active on a target cell at some distance from the source cell and requiring that the factor be delivered through the bloodstream or lymphatic fluid (12–18). The set of insoluble signals includes the components in the extracellular matrix, an insoluble material produced by all cells and found between cells and around the outside of the cellular membrane (19).

All metazoan tissues are organized into distinctive architectural patterns that indicate some of the critical cell–cell interactions (1,10,20). The basic cellular relationship governing the architecture of all tissues is the epithelial–mesenchymal relationship: sheets of epithelial cells, hooked to each other by an array of intercellular junctions, are glued by a thin layer of extracellular matrix (the basement membrane) to a layer of mesenchymal cells, either fibroblasts or endothelial cells (1,11,20). Multiple epithelial–mesenchymal dyads are grouped together near a blood vessel (also an epithelial–mesenchymal dyad). In tissues, such as the liver, where there must be rapid transport of products into or out of the epithelium, the epithelial or epithelioid cell is bound directly onto the endothelial cell forming the blood vessel. Thus, in the liver, the epithelial–mesenchymal dyad is formed by the hepatocyte and the sinusoidal endothelial cell. The cellular interactions between the hepatocyte and the endothelial cells are the source of primary regulatory mechanisms governing hepatocellular functions.

Some anomalies and mysteries revealed in studies of liver cells in culture have been the findings that one can observe only limited growth of hepatocytes at relatively high density in culture; one cannot subculture hepatocytes, and one cannot clone them despite the fact that the liver is the most renowned regenerative organ in the body (21). The work in several laboratories (22–51) over the past decade may have solved these paradoxical findings. If the hypotheses of these investigators are correct, the liver may prove to be a stem cell system in which putative stem cells, called "oval" cells, differentiate through a series of discrete stages into hepatocytes, which are assumed to be the nearly terminally differentiated cells in the lineage. If true, the liver would be analogous to its embryological cousin, the gastrointestinal system. In both systems, the stem cells (in the crypts of the gut versus putative stem cells around the portal triads in the liver) would proliferate and give rise to daughter cells that differentiate (as they slide away from the crypts in the gut versus slide along the sinusoids in the liver) and then die. A validation of this theory will revolutionize investigations on the liver and will offer totally new approaches to molecular biologists who want to do gene transfection studies in liver. A brief review on what is known about oval cells is available in the reviews by Sell (22–24). [See also p. 96, Endnote 1.]

CELL CULTURE

Utility and Validity of Cell Culture for Studies on Gene Expression

The use of cell cultures permits one to control the numerous variables that affect the phenomena under investigation. However, in the past, cell culture, as an experimental tool, has been empirically of limited value, since only certain cell types (fibroblasts, various other mesenchymal cells, and tumor cells) survive under classic culture conditions (52). Those that survive and grow under these conditions exhibit patterns of gene expression that are more similar to pathological states *in vivo*. The limitations in cell culture technology have been largely eliminated with the realization that culture conditions must incorporate the signals (hormones and extracellular matrix components) derived from cell–cell interactions and normally present *in vivo* (52–59). With the use of defined and purified hormones, growth factors, and other soluble signals and with defined and purified matrix components, normal hepatocytes and hepatomas can be cultured and will retain most of their usual repertoire of gene expression and their usual pharmacological and endocrinological responses. In summary, these new approaches to cell culture have produced results which reveal that gene expression in cultured cells can accurately reflect the gene expression of those cells *in vivo*. For example, the data derived from cells under the classic cell culture conditions (serum-supplemented medium and tissue culture plastic) are thought to reflect how cells respond under wounding conditions, that is, when scar tissue is formed. This is the *in vivo* state that results in the formation of serum.

Types of Cell Culture

When tissues are disrupted and first plated into culture, the cultures are referred to as "primary cultures" (52). Under classic cell culture conditions, primary liver cultures last with retention of liver-specific functions for approximately 1 week (20,53). By contrast, under defined hormonal and matrix conditions, the cultures can last for varying lengths of time depending on which hormones and matrix components are included. Cultures have been reported to survive for months (20).

If one is successful in isolating a specific cell type by cloning the cells (deriving an entire cell culture from one cell) and by passaging the cells from dish to dish, the cultures are referred to as a "cell strain." By definition, cell strains have a limited life span, usually 50–60 passages. If a permanent, clonal cell population is established, the cultures are referred to as a "cell line." Cell lines can occur, but do so infrequently, as a result of spontaneous "transformation" of a cell strain. However, the frequency of conversion of a cell strain to a cell line is greatly augmented by exposing the cells to some oncogenic event: treatment with radiation, chemicals, or viruses. Thus, a standard approach to the establishment of a specific cell type in culture is to establish a cell strain and then intentionally transform it. This approach entails significant risks: usually the differentiated state of the cells can be partially or completely lost (52). An alternative has been to resort to minimally deviant neoplastic cells carried as transplantable tumor lines in syngeneic or immunosuppressed hosts (reviewed in Ref. 20). Such tumor lines can be adapted more readily to culture than can normal cells. Using this approach, many partially differentiated cell cultures have been established (reviewed in Refs. 20 and 52). However, even in highly differentiated cell lines that demonstrate high cytoplasmic abundance of tissue-specific mRNAs and that show expression which can be modulated by appropriate hormonal or pharmacological agents, the transcription rates for these mRNA species are always a small fraction of the known rates *in vivo* (54). Furthermore, the hormonal or pharmacological modulations of the tissue-specific mRNA abundance are almost always due to posttranscriptional regulatory mechanisms (53–55).

Classic Cell Culture

All methods of preparing cells for culture start with the disruption of the tissue and dispersal into chunks or into single cell suspensions. In classic cell culture, the dispersed tissue or cells are then plated onto an inflexible plastic substratum and suspended in or covered with a liquid medium (52). The plastic substratum, in the form of culture dishes, consists of polystyrene dishes or containers that are exposed to ionizing conditions (ionizing plasma gas) to induce a negatively charged surface that will permit the cells to attach by charge binding. The cells attach and spread via the negative charges and then are fed with a liquid medium that typically consists of a basal medium containing a specific mixture of salts, amino acids, trace elements, carbohydrates, etc., and which is supplemented from 5 to 25% with a biological fluid, usually serum. The basal media that are commonly used (e.g., RPMI, DME, BME, Waymouth's) were originally developed for cultures of fibroblasts (52,56–59). Although some investigators utilize serum autologous to the cell types to be cultured, it is more common that the serum derives from animals which are routinely slaughtered for commercial usage such as cows, horses, sheep, or pigs.

Defined Conditions based on Signals for Cell–Cell Interactions

Hormonally Defined Media for Hepatocytes and Hepatomas: Defining the Soluble Signals from Cell–Cell Interactions

An approach to defining the soluble signals from cell–cell interactions has been to replace the serum supplements in medium with known and purified hormones and growth factors and to test the defined conditions for their effects on the appropriate biological responses (57–59). This results in the development of serum-free, hormonally defined media (HDM) (57–59). Use of HDM in cell culture has greatly improved our ability to isolate certain cell types, especially many epithelial cell types, in culture and to maintain a variety of differentiated functions (52,55,57–59). The composition of an HDM has been found to be dependent on the cell type (57–59). In general, the development of an HDM for a cell type begins by defining the hormonal and growth factor requirements for growth of a minimally deviant neoplastic cell line. Since a cell line is already adapted to culture, it can readily be used in clonal growth assays to define growth requirements under serum-free conditions. The HDM for the tumor cells usually contains a subset of the requirements and is, therefore, a starting point in defining the requirements for the normal cellular counterparts to these tumor cells (57–63).

This approach has been used to develop an HDM for hepatocytes. Since the mid-1970s, an increasing number of studies have utilized serum-free media to which one or more hormones or growth factors were added to test for their influence on growth or on some liver-specific function (64–73). In general, serum-free conditions result in substantial improvement in the expression of tissue-specific functions (53,55,63). Furthermore, some hormonal effects on liver cells can be observed only under serum-free conditions (55). Thus, it has become apparent that serum supplements to the medium must be eliminated for studies on tissue-specific gene expression (53,55).

A rigorously defined HDM for hepatomas was developed by Gatmaitan *et al.* (62), who systematically screened over 70 factors to determine the ideal combination for clonal growth of a Reuber rat hepatoma cell line. It was found that the composition of HDM which resulted in optimal clonal growth was dependent on the type of substratum used for the cells (see Table I). Thus, on collagenous substrata as opposed to tissue culture plastic, the cells require fewer factors for clonal growth. The same medium was found effective with hepatoma cell lines derived from mice, rats, humans, and other species (53,62,74). However, for some hepatoma cell lines (e.g., HepG2), which show deficient production of one or more of the lipoproteins, more complex lipids are required than linoleic acid. By adding high density lipoprotein (HDL), these hepatoma cell lines were able to grow and be cloned under serum-free conditions (75).

TABLE I
Constituents of Hormonally Defined Media for Normal and Neoplastic Rat Hepatocytes[a,b]

Factor	Hepatocytes on tissue culture plastic	Hepatomas on tissue culture plastic	Hepatomas on type I collagen
Epidermal growth factor	50 ng/ml	Not required	Not required
Insulin	5–10 µg/ml	10–100 µg/ml[e]	1–5 µg/ml
Glucagon	10 µg/ml	10 µg/ml	5 µg/ml
Hydrocortisone	Nor required	1×10^{-8} M	Not required
Transferrin	Not required	10 µg/ml	Not required
Linoleic acid[c]	5 µg/ml	10 µg/ml	5 µg/ml
HDL[d]	10 µg/ml	10 µg/ml	n.t.
Triiodothyronine	Not required	1×10^{-9} M	Not required
Prolactin	20 Mu/ml	2 mU/ml	2 mU/ml
Growth hormone	10 µU/ml	10 µU/ml	10 µU/ml
Trace elements			
Zinc	5×10^{-11} M	5×10^{-11} M	5×10^{-11} M
Copper	1×10^{-7} M	1×10^{-7} M	1×10^{-7} M
Selenium	3×10^{-10} M	3×10^{-10} M	3×10^{-10} M

[a] These culture conditions permit clonal growth of many hepatoma cell lines (human, rat, and mouse cell lines have been tried). However, normal hepatocytes will grow at high densities (~10^5 cells/60 mm dishes) but not at clonal seeding densities (10^2–10^3 cells/60 mm dishes) when maintained in the above hormonally defined medium. [See also p. 97, 2.] Furthermore, unless protease inhibitors and/or specific matrix substrata are used, the cells must be seeded in a mixture of 5–10% serum plus the hormones for a few hours (4–6 hr) and then transferred into the serum-free hormonally defined medium. n.t., Not tested.

[b] Pituitary hormones can be purchased from commercial sources or alternatively obtained through the NIH hormone distribution service.

[c] Linoleic acid must be present in combination with fatty acid-free bovine serum albumin (Pentax, Inc.) at a 1:1 molar ratio. If added alone, it is quite toxic to the cells. It is important to obtain the purest form of BSA possible, since serum-free conditions do not buffer the cells against any toxin bound to BSA.

[d] Some hepatoma cell lines have more complex lipid requirements, especially those that are poor in making some of the lipoproteins forming high density lipoprotein (HDL) or low density lipoprotein (LDL). These cell lines must be given a more complex lipid source such as HDL (75). Long-term primary liver cultures do better with HDL supplements.

[e] The amount of insulin needed varies with the cell line and correlates inversely with the amount of proteases produced by the tumor cells. One can reduce the amount of insulin needed by adding protease inhibitors or by plating the cells on collagenous substrata; collagenous substrata are known to inhibit protease secretion.

Utilizing the HDM developed by Gatmaitan et al. (62) as a starting point, Enat et al. (63) developed an HDM for primary cultures of normal rat liver cells. The major difference between the HDM used for the hepatomas and that for the hepatocytes was the requirement by the normal hepatocytes for epidermal growth factor (EGF). In recent studies, an explanation for this has become apparent: many hepatoma cell lines produce transforming growth factor-α (TGF-α), a tumor-derived factor that is an analog to EGF (76; Zvibel, I., unpublished data) (see Table V). Thus, at least some hepatoma cells are independent of EGF by making an equivalent factor. It is unknown at this time whether the hepatoma cells that are independent of EGF in their growth and yet do not produce TGF-α have in some other way activated an EGF-driven metabolic pathway.

HDM has been found to select for parenchymal cells even when the cells are on tissue culture plastic (52,55,63). This results, within a few days, in cultures that are predominantly hepatocytes. However, if the cultures are plated onto tissue culture plastic and in HDM, the life span of the primary cultures has been found to be approximately 1 week, at which time the cells peel off the plates in sheets. Achievement of longer culture life spans has been found to be dependent on using collagenous or matrix substrata in combination with the defined media (reviewed in Ref. 52).

Extracellular Matrix Molecules: Insoluble Regulatory Signals, Derived from Cell–Cell Interactions and Needed for Normal Liver Gene Expression

The extracellular matrix is an insoluble mixture of molecules in between and around cells. Although extracellular matrix has been known for many years to be relevant to differentiation (reviewed in Refs. 20, 53, and 77), molecular approaches have greatly facilitated our abilities to both isolate matrix components and analyze their effects (77–91). Thus, there has been a great surge in recent years in the number of studies detailing the role of matrix in the regulation of growth and differentiation of cells (reviewed in Refs. 19, 20, 53, 59, and 77).

General Rules Regarding Extracellular Matrix. There are many chemically distinct types of extracellular matrix produced by cells (19,77,92,93). Although no type of matrix has been studied sufficiently to know all of its components, all matrices consist of collagen scaffoldings (93,94) to which cells are bound by adhesion proteins (95,96). In association with the collagens, adhesion proteins, and the cell surface are polymers of sulfated saccharides, glycosaminoglycans (GAGs), and proteoglycans (PGs) (97). All cells produce extracellular matrix, and the extracellular matrix in between any given set of cells contains components derived from all the cell types in contact with the matrix.

The different extracellular matrices are readily grouped into types dictated by the type of collagen used as the scaffolding (92,93). Each collagen type is associated with specific adhesion proteins and proteoglycans that are typically synthesized and assembled with it (19,77,92–97). For example, a "type I" matrix consists of type I collagen, fibronectin (plus other anchorage proteins), and chondroitin sulfate proteoglycan or dermatan sulfate proteoglycan. By contrast, a "type IV" matrix consists of type IV collagen, laminin, and heparan sulfate proteoglycan.

Every cell that is bound into a tissue structure secretes and is associated with one or more types of matrix. When the cell undergoes a change in its physiological status (e.g., from quiescence to growth), it produces either an altogether distinct matrix (e.g., it shifts from a type III to a type IV matrix), or it chemically alters one or more components in the matrix (e.g., produces a proteoglycan with less sulfate content) (19,77,92–97). [See also Endnote 3.]

Matrix Types Produced by Liver Cells. Although more detailed studies on the matrix composition of the liver are given elsewhere (98) [see also Endnotes 4 and 5], a brief overview of the liver matrix chemistry is presented here to identify the insoluble regulatory signals pertinent to gene expression studies. Analyses of the extracellular matrix of the liver have indicated that it contains multiple matrix types (types I, III, IV, V, and VI) which are susceptible to change during regeneration or during pathological conditions (98–114). (See Tables II, IV.) In the quiescent liver, the hepatocytes contribute a type III and type VI matrix; the endothelial cells and Ito cells contribute a type IV matrix; and fibroblasts and other nonparenchymal cells contribute a type I matrix (107). Over most of the hepatocyte–sinusoidal surface, the matrix is a mixture of type III/type VI and type IV matrices (104,107). However, in regions of architectural stress, type I matrix is present (104,107).

In the regenerating liver, the chemistry of the matrix changes: the hepatocytes convert to synthesizing a type IV matrix. Thus, the hepatocyte–sinusoidal area becomes a basal lamina produced by the hepatocytes, endothelial cells, and Ito cells (107,110,114).

Under pathological conditions in which the liver endures chronic insults and becomes cirrhotic, the matrix chemistry again changes: now the hepatocytes as well as certain nonparenchymal components synthesize a type I matrix (98,104,107,113). These results are particularly useful for interpreting some of the findings of studies in cell culture. In Tables II and III are listed the known matrix types produced by liver cells.

Adhesion Proteins and Collagens. Adhesion protein is a generic term for molecules that attach cells to collagens (or to a substratum) (95,96). There are only two anchorage proteins that have been identified and well characterized in liver: fibronectin and laminin (100,104,110,114–119). Several other putative anchorage proteins have been identified, but very little is known about them (120,121). Similarly, a thorough presentation on the chemistry of collagen types in the liver (99,101–109,111–121), their cellular sources, and the regulation of their synthesis under various hormonal and physiological conditions has been extensively reviewed elsewhere and will not be presented here.

TABLE II
Extracellular Matrix of the Liver: Parenchymal Cells[a]

Cell type	Collagen type	Associated adhesion protein	Cell surface receptor	Proteoglycan (PG)
Hepatocytes (quiescent)	Types III and IV	Fibronectin	Integrin	Heparin PG
Hepatocytes (regenerating)	Type IV	Laminin	Laminin receptor	Heparan sulfate PG
Hepatocytes (cirrhotic)	Type I	Fibronectin	Integrin	Chondroitin sulfate PG and dermatan sulfate PG
Liver stem cells ("oval cells")	Types IV and V	Laminin	Laminin receptor	Heparan sulfate PG

[a] The collagen scaffolding, along with associated proteoglycans and adhesion proteins, is present on the basal, sinusoidal side of the hepatocytes (facing the endothelial cells). Proteoglycans and cell adhesion molecules (CAMs), without a collagen scaffolding, are present on lateral sides of the hepatocytes. [See also Endnotes 4 and 5.]

TABLE III
Extracellular Matrix of the Liver: Nonparenchymal Cells[a]

Cell type	Collagen type	Associated adhesion protein	Cell surface receptor	Proteoglycan (PG)
Fibroblasts Sinusoids	Type III	Fibronectin	Integrin	Heparan sulfate PG
Fibroblasts (areas other than sinusoids)	Type I	Fibronectin	Integrin	Chondroitin sulfate PG and dermatan sulfate PG
Epithelia: Ito cells and bile duct cells (quiescent)	Type IV	Laminin	Laminin receptor	Heparin PG
Epithelia: Ito cells and bile duct (regenerating)	Type IV	Laminin	Laminin receptor	Heparan sulfate PG
Endothelia (quiescent)	Types IV and VI	Laminin fibronectin	Laminin receptor integrin	Heparin PG
Endothelia (regenerating)	Type III	Fibronectin	Integrin	Heparan sulfate PG

[a] See also Endnote 5.

Proteoglycans and Glycosaminoglycans in Liver. Since the proteoglycans have been less studied in general as well as to a very limited extent in specific tissues, a brief background is given. [See also Endnote 6.] Proteoglycans are macromolecules which consist of variable numbers of glycosaminoglycan chains covalently bound to a protein core (77,97). Glycosaminoglycans are sugar polymers composed of alternating glucuronic acid and sulfated hexosamine residues, which carry closely spaced negatively charged groups. Thus, the GAG chains are composed of disaccharide repeating units. The major form of proteoglycans present in the plasma membranes of hepatocytes and hepatomas are forms of heparan sulfate PG and heparin PG (122–129), both being polymers of glucuronic acid and glucosamine. An epimerase and a sulfatransferase, located in the Golgi, dictate the amount of epimerization of glucuronic acid to iduronic acid and the degree of sulfation of the polymer (130,131). The heparin PG results from increased epimerase and sulfatransferase activity, enzymatic activities found to be induced by high cell densities.

The rat liver-derived heparan sulfate PG has been purified and chemically characterized. It is about 80,000 in molecular weight and contains four heparan sulfate chains bound to a core protein. The core protein is known to be closely associated with the cell surface (126–129). However, the data are inconclusive on the exact nature of its association. There is some evidence that the core protein of heparan sulfate PG is a transmembrane molecule connected intracellularly with the cytoskeleton (126,127,129,132,133). There are other data to suggest that it can also be bound to a putative heparan sulfate PG receptor (129). The heparan sulfate PG present in plasma membranes of quiescent hepatocytes forms complexes with fibronectin, and these complexes mediate cell attachment to an appropriate form of interstitial, fibrillar collagen (134). During regeneration, hepatocytes are associated with type IV collagen by fibrils containing laminin associated with a heparan sulfate PG (114). The synthesis of heparan sulfate PG in cultured liver cells shows density dependence (135–140).

Recent studies by Conrad and associates have shown that hepatocytes produce another form of heparan sulfate PG that is bound to the cell surface by inositol phosphate (141–143). This heparan sulfate proteoglycan is subsequently internalized, the protein core eliminated, and the GAG chains (heparan sulfates) translocated to the nucleus (144). Conrad and associates have also shown that cells at confluence have nuclear GAG chains that are much more sulfated, that is, are more like heparin GAG chains (143). The relevance of this phenomenon to growth regulation and differentiation is assumed but not yet fully understood.

Neoplastically transformed cells usually do not assemble a normal extracellular matrix (145–149). Robinson *et al.* (149) demonstrated that hepatoma cells synthesize aberrant forms of heparan sulfate PGs that are undersulfated and have lowered affinity for fibronectin. Sulfation decreased by 20–40%, particularly in $GlcNH_2$ residues bearing 6-O-sulfate groups (149). Uneven distributions of sulfate groups in oligosaccharide chains were also observed (149). The reduced negative charge of the PGs due to undersulfation was shown to affect not only the formation of complexes with collagen, anchorage proteins, and other matrix components but also the permeability of matrix around the cells.

Effects of Culture Conditions on Matrix Synthesis in Liver Cells. Specific culture conditions dictate both quantitatively and qualitatively the matrix

TABLE IV
Known Matrix Chemistry in Cultured Liver Cells and in Hepatoma Cell Lines

Cell or culture type	Culture substratum	Medium[a]	Predominant matrix chemistry (localization)
Primary liver cultures	Tissue culture plastic	SSM	Type I matrix (mostly secreted into medium)
	Tissue culture plastic	SSM plus glucocorticoids	Type III and small amounts of type I (some to medium, some to basal surface of cell)
	Tissue culture plastic	HDM	Type IV matrix and some type I matrix (mostly secreted into medium)
	Tissue culture plastic	HDM plus glucocorticoids	Type III matrix; greatly elevated fibronectin levels (some into medium but some increased localization at basal surface of cell)
	Type I collagen	SSM	Types III and I matrix (mostly localized on basal cell surface)
	Type I collagen	HDM	Type III matrix (predominantly on basal cell surface)
	Type IV collagen	SSM	Types I, III, and IV matrix (predominantly on basal cell surface)
	Type IV collagen	HDM	Type IV matrix (predominantly on basal cell surface)
Hepatoma cell lines	Tissue culture plastic	SSM	Types I and IV matrix with a poorly sulfated heparan sulfate PG (mostly secreted into medium)
	Tissue culture plastic	HDM	Type IV matrix (mostly secreted into medium)
	Type I collagen	SSM	Types IV and I (mostly secreted into medium, although some is localized on the basal cell surface)

[a] SSM, Serum-supplemented medium; HDM, serum-free, hormonally defined medium.

chemistry synthesized by liver cells and determine whether the matrix components will be secreted into the medium or will be found in close association with the cells (53,137,145–149). For example, collagen gel substrata remarkably reduced the degradation of GAGs synthesized by hepatocytes and caused those GAGs synthesized to remain in the substratum of the cells rather than being secreted into the medium (reviewed in Ref. 53).

Once again, tumor cells respond differently in culture when compared to normal cells. On tissue culture plastic both transformed cells and the normal, parental counterparts have identical rates of GAG synthesis and secretion into the medium (147). On collagen substrata, normal cells incorporate the GAGs into a basal lamina, whereas transformed cells continue to secrete the GAGs into the medium. This impairment, thought to be due to production of matrix-degrading enzymes, may prevent the neoplastic cells from forming an appropriate matrix and is known to dictate invasive and metastatic potential. The ability of tumor cells to be invasive and metastatic has been shown to be strictly correlated with production of enzymes such as type IV collagenase and/or endoheparinase that degrade components of the extracellular matrix (reviewed in Ref. 53). A summary of some of the known and important affects of culture conditions on matrix synthesis by hepatocytes are summarized in Table IV.

GENE EXPRESSION IN NORMAL LIVER

Gene Expression *in Vivo*

It is now established that there are many steps in which regulation of gene expression in eukaryotic cells can occur. [See also Endnote 7.] The potential regulatory levels include the site and rate of initiation of transcription, RNA polyadenylation, splicing, transport, mRNA stability, and translation (150–156). Early experiments (157) showed that the primary level of regulation for tissue-specific genes *in vivo* was transcriptional. In that study, analysis of nuclear RNA showed that the genes classified as "tissue-specific" were transcribed only in appropriate tissues, or were transcribed at highest rates in particular tissues, while they were transcribed at very low rates or were not detectable in other tissues. Thus, it would seem that tissue-specific genes must have a special set of transcriptional regulatory signals, which differ from those of the "common" genes expressed in all cells. However, it is becoming increasingly apparent that posttranscriptional regulatory mechanisms, such as mRNA stabilization and translational regulation, are not only very common, but they can even be the dominant form of regulation depending on the gene, the cell type, and the developmental stage [150–156; see also recent textbooks (162, 163)]. Therefore, it is necessary now to reevaluate some of the previous data regarding tissue-specific gene regulation *in vivo*.

There are two classes of cis-regulatory elements which are now generally recognized to affect the site and rate of initiation of transcription for all protein coding genes. These are promoter elements, located within about 100 bases 5' to the transcription start site, and enhancer elements, found in a variety of positions more distant from the transcriptional start site. Cellular protein factors, some of which have been identified, interact with these cis elements in a positive or negative way. The two commonly identified promoter elements are the TATA box, found about 20–30 bases upstream from the cap site, and the CCAAT box, somewhat further upstream (158–160). Enhancers are more diverse. The first clear example of a sequence capable of enhancing transcription was the 72-base pair repeat in the DNA virus, SV40 (161). In the past several years, many other enhancers have been identified in various systems (150–156). Enhancers can be tissue or cell type specific, and they can be inducible or constitutive. Presumably, tissue specificity reflects the presence or absence of cellular factors which must interact with enhancers to mediate their effect. Unlike promter elements, enhancers are, by definition, active in either orientation and in different locations, sometimes several thousand bases away from the transcription unit affected, sometimes within the unit, and sometimes 3' to the gene itself. Furthermore, enhancing activity may be observed in heterologous systems when enhancers are linked to unrelated genes. [See also Endnote 8.] The mechanisms for posttranscriptional regulation are not yet defined.

Although studies on gene regulation to date have proved most informative, they have been constrained by the fact that, until now, they have had to be done either *in vivo* by isolation and analysis of RNA from intact tissues, under which conditions there are numerous, undefined variables, or in cells cultured under classic conditions, in which tissue-specific genes are transcribed at undetectable or extremely low rates. Therefore, new regulatory mechanisms and/or new understandings of currently identified mechanisms are likely to emerge in studies using cells cultured under defined conditions.

Gene Expression in Primary Liver Cultures under Classic Cell Culture Conditions

Hepatocytes under standard, classic cell culture conditions spread to a squamouslike shape, appear increasingly translucent and agranular, and rapidly lose their granular endoplasmic reticulum. Rapid morphological deterioration is accompanied by decreased biochemical functions, resulting in a loss, within 3–5 days, of all tissue-specific functions and, within 1–2 weeks, of viable cells (164–165, reviewed in Refs. 10, 20, and 53).

The morphological and biochemical findings have been extended in recent years by analyses using recombinant DNA technologies and examining specific gene expression. Within a few hours of tissue dissociation, the transcription of most tissue-specific mRNAs in liver is significantly reduced and by 24 hr is less than 1% of normal (55,166). By contrast, the transcription of mRNAs for common genes such as actin and tubulin increases manyfold within a few hours but returns to normal by 24 hr (55,166). Over the life span of these primary cultures, usually 1–2 weeks, the tissue-specific mRNA transcription remains low or not detectable, where the common gene mRNA transcription remains at levels comparable to those *in vivo*. Tissue-specific mRNAs already transcribed at the time of dissociation of the liver degrade at an exponential rate such that, by 5 days of culture, the cytoplasmic mRNA levels for even the most abundant tissue-specific mRNA species are no longer detectable by Northern blot analysis. Thus, the classic phenomenon of dedifferentiation of primary cultures is, on a molecular level, a combination of loss of transcription and rapid degradation of tissue-specific mRNAs (55, 166).

Classic cell culture conditions do not always result in the demise of tissue-specific functions. The pioneering efforts of Michalopoulos and Pitot (167–169) have shown that plating liver cells on floating, type I collagen gels results in cultures with prolonged viability (up to 1 month) and with sustained liver-specific gene expression for 2–3 weeks. The cultures, however, undergo a fetalization process (169) in which they gradually revert to expression of fetal antigens and fetal proteins. Clayton and Darnell showed that if liver is put into culture, even if under classic cell culture conditions, in the form of tissue slices or small chunks of tissue, the transcription of tissue-specific mRNAs is sustained for as long as the slices or chunks are intact, a matter of days (170). Further evidence for the relevance of cell–cell interactions and tissue structure in the regulation of gene expression derives from the extensive studies by the Guillouzos and co-workers on the relevance of cell–cell interactions in gene expression of primary liver cultures (171–183). They have shown, using classic cell culture conditions, that coculture of primary cultures of liver cells with a liver-derived epithelial cell results in maintenance of many if not most of the liver-specific functions for 2–4 weeks (171–183). Under the coculture conditions, transcription rates for various liver-specific genes can be near normal levels (176). The influence of the coculture has been shown to be, in part, due to the production of extracellular matrix between the hepatocytes and the liver-derived epithelial cells (181).

Influence of Serum-Free, Hormonally Defined Media on Gene Expression in Cultured Liver Cells Plated onto Tissue Culture Plastic

Serum-free medium and/or HDM improves the differentiation of liver cultures, as evidenced in studies in which differentiation was assayed at the protein level for liver-specific functions (64–73). Furthermore, in studies using molecular probes, the cultures in HDM exhibited liver-specific gene transcription that was detectable, albeit low (3–12% of normal) (55,59,62,63). Of considerable interest, the cytoplasmic mRNA concentrations for liver-specific genes were almost equal to *in vivo* levels during the life span of the cultures on tissue culture plastic (55). The discrepancy between low transcriptional activity and yet near normal cytoplasmic levels of specific mRNA species was found to be due to mRNA stabilization in the serum-free, hormonally defined medium (55). In hepatocytes plated onto tissue culture plastic, the predominant form of regulation of gene expression, whether of common genes or of liver-specific genes, is, thus, shown to be by a posttranscriptional mechanism(s) (55).

When primary cultures are maintained under serum-free conditions and on tissue culture plastic (i.e., in the absence of matrix components), addition or deletion of individual protein hormones or growth factors to the medium can result in dramatic changes in the abundance of individual tissue-specific mRNAs. These changes, where tested, have always been due to changes in the mRNA half-life rather than changes in synthesis of the mRNAs (184). Thus, individual hormones have been found to regulate the stability of individual mRNA species. Although transcriptional regulation is known to be the dominate form of regulation *in vivo* (157), it is unknown to what extent such posttranscriptional regulatory mechanisms are also present *in vivo*. Certainly, regulation of mRNA stability is now recognized as a frequent regulatory mechanism (and sometimes the dominant form) in gene expression in both invertebrates and vertebrates (185, 186). Thus, only further studies will determine the extent of posttranscriptional regulation of gene expression in liver cells as a complementary regulatory mode of action to that of transcriptional regulation.

Of the hormones tested, only glucocorticoids have been shown to affect gene expression by transcriptional mechanisms when the cultures are on tissue culture plastic (the papers referenced are representative of those in the field: 187–207). Three classes of regulation have been described. Some of the genes are induced by dexamethasone through an increased transcription rate and by means of an identified glucocorticoid regulatory element. For example, Marceau *et al.* (199,200,207) showed that dexamethasone induced a fibronectin-rich matrix in primary liver cultures. Weiner *et al.* (197) analyzed this further and showed that the induction of fibronectin was by a transcriptional mechanism and identified the glucocorticoid element. Similarly, γ-glutamyl transpeptidase (188) and tyrosine aminotransferase (194) are also induced by dexamethsone. A different glucocorticoid response has been found for the collagen genes (189,196,197). These groups have documented that dexamethasone suppresses the synthesis of type I collagen (189,196) and also of type IV collagen (197). This suppressive effect of dexamethasone involves partly transcriptional regulation but mostly posttranscriptional regulation (189,197). As for the genes positively influenced by glucocorticoids, a glucocorticoid-responsive element has been found in the upstream regulatory region of the collagen genes (197).

More puzzling and contradictory findings derive from studies on various other tissue-specific genes such as albumin and α_1-acid glycoprotein (189,192–194,201). Potentially there is a third and distinct class of glucocorticoid-responsive genes. Although the addition of dexamethasone to primary liver cultures has long been known to result in augmented expression of albumin and α_1-acid glycoprotein (190), analysis of transcription rates of these genes *in vivo* and in primary liver cultures suggests that it has little or no direct affect (189,201). The effects that do occur, occur only when the cells are cultured under classic culture conditions and are largely posttranscriptional (201). Similarly, studies by Ringold and associates (192,193,202,203) on α_1-acid glycoprotein first suggested that dexamethasone acts posttranscriptionally. However, subsequent studies using a transfected construct containing the regulatory regions of the gene along with a chloramphenicol acetyltransferase (CAT) recorder gene strongly indicated that dexamethasone was affecting the gene transcriptionally (203). These investigators speculate that there may be an upstream promoter which is independent of dexamethasone regulation and which results in mRNAs that do not appear in the cytoplasm. A distinct, dexamethasone-regulated promoter is hypothesized to give rise to the elevated cytoplasmic mRNA levels. [See also Endnote 9.]

Although this may be true for α_1-acid glycoprotein, it cannot be true for the other genes in this class (e.g., albumin), since the cells can express these mRNAs cytoplasmically even in untreated cultures as long as the cells are under serum-free conditions. In other words, such tissue-specific mRNAs are not stable in serum, are more stable in serum containing dexamethasone, and are very stable in serum-free media. Under serum-free conditions, dexamethasone has little effect either on transcription rates or on mRNA abundance. Similarly, several careful studies have documented the effect of dexamethasone on albumin transcription but only *in vivo* in adrenalectomized animals (193) or in primary liver cultures in a serum-supplemented medium within the first day or two of culture (194). If the cultures are analyzed thereafter, the posttranscriptional effects but not the transcriptional effects are observed (189,201). These paradoxical findings have yet to be resolved. Perhaps the effects of dexamethasone on these genes is indirect: it may be stimulating one gene that in turn induces or enhances the transcription of albumin and α_1-acid glycoprotein; the expression of the first gene is lost after several days of culture. Empirically it appears to increase the stability of these tissue-specific mRNAs in cultures in serum-supplemented medium. [See also Endnote 10.]

The pattern of expression of common genes, represented by actin and tubulin, is quite distinct from that of the tissue-specific genes (55,167,170,184). *In vivo* in quiescent liver, the common genes are strongly transcribed but are expressed at very low abundance in the cytoplasm, implicating some mechanism for rapid turnover of mRNA molecules. When the liver is dissociated into a single cell suspension, the transcriptional signals for common genes increase manyfold within a few hours but return to normal by 24 hr (55,167,184). Thereafter, in primary liver cultures, the rate of transcription for these genes remains at levels comparable to that seen in quiescent liver *in vivo* (55,167). However, measures of cytoplasmic mRNA levels by Northern blots for the common genes stay at high levels under all medium conditions tested: with and without serum and in all the combinations of hormones (55,184). Thus, as was found for tissue-specific genes, there are changes observed in the cytoplasmic mRNA levels that do not correlate with a comparable change in the transcription rates for those genes. This implies that the expression of common genes in cultured liver cells is also regulated by a posttranscriptional mechanism, presumably mRNA stabilization. However, the mRNAs for the common genes are stabilized in serum-supplemented media, in serum-free media, and in HDM.

In later studies, it has been found that, even in serum-free medium with no hormonal additives, liver-specific mRNAs achieve some measure of stability such that cytoplasmic levels are at least 25–50% of levels seen *in vivo* (184). With the addition of the hormones and growth factors found active on liver cultures, the cytoplasmic concentrations of tissue-specific mRNAs increase to near normal levels even though the transcription rates for these same mRNAs remain at levels similar to that in serum-free medium with no hormonal additives. Individual hormones or growth factors could preferentially stabilize or destabilize specific mRNA species (184,204). For example, epidermal growth factor stabilized albumin mRNA (184). Glucagon stabilized gap junction mRNA but destabilized ligandin (glutathione S-transferase) mRNAs (184,204). Insulin stabilized ligandin but destabilized gap junction mRNA (204).

Clearly, in all the investigations to date on liver cultures on tissue culture plastic and in any serum-free HDM, the differentiation of the cells has been primarily posttranscriptionally regulated, most commonly by mRNA half-life (55,184). These findings are in sharp distinction to the results on liver *in vivo* or in cultures in which the liver tissue structure remains intact; under these conditions, differentiation of liver is well documented to include predominantly (although now known to be not exclusively) transcriptional control (157,170). Empirically, normal transcription rates are dependent on various aspects of tissue structure. To date, normal transcription rates for most tissue-specific genes are not observed in cultures maintained on tissue culture plastic whether in serum-supplemented media or in serum-free, hormonally defined media (reviewed in Ref. 53).

Influence of Adhesion Proteins and Collagenous Substrata on Gene Expression in Primary Liver Cell Cultures

Hepatocytes, like most epithelial cells, attach better and maintain differentiated properties for longer periods of time when maintained on substrata coated with adhesion proteins or collagens (167–169,205–210; reviewed in Refs. 20 and 53). Fibronectin and laminin substrata have been well documented to improve attachment and culture life span. However, their effects on gene expression appear, to date, to be relatively minor. Only a few genes (e.g., α_1-antitrypsin) respond transcriptionally in cultures maintained on either fibronectin or laminin (208). Few or no posttranscriptional effects have been noted that can be distinguished from the effects of hormones or growth factors. By contrast, these adhesion proteins have been shown to facilitate growth (205,206,209,210).

On the other hand, collagens can have quite striking, albeit mostly posttranscriptional, effects (208). In most studies, an acid extract of rat tail tendon has been used due to its ease of isolation and preparation. Such an extract consists mostly of type I collagen with variable amounts of fibronectin, dermatan sulfate proteoglycan, and other matrix components. The collagen is used to coat dishes to be used as substrata (168,169), but several variations in how it is presented to the cell have proved useful for hepatocyte cultures.

Primary liver cultures plated onto floating collagen gels or on nylon meshes coated with collagen (168,169) have shown far more efficacy in extending the life span and, even more dramatically, in increasing differentiated functions. For example, the life span of primary liver cultures plated onto dishes coated with collagens is typically 2–3 weeks. If the collagenous substrata are detached and permitted to float, the hepatocytes remain viable and functional for 4–5 weeks (168,169). Permitting the collagen substratum to float results in cells that assume polygonal shapes which become cuboidal over time. Histology and ultrastructure indicate formation of cellular junctions, development of polarity, and well-developed endoplasmic reticulum and Golgi bodies. The cytolasmic abundance of liver-specific mRNAs can increase 5- to 10-fold or more in cultures on floating gels as opposed to attached gels. Some functions, such as cytochrome P-450 and glucocorticoid-inducible tyrosine aminotransferase, survive for 10–14 days (168,169). Other functions (e.g., albumin) are expressed throughout the life span of the culture (168, 169). However, as the cultures age, there is gradual fetalization in which adult-specific functions, such as albumin synthesis, are replaced by fetal-specific ones, such as α-fetoprotein synthesis (169). Thus, albumin levels, which peak in a few days and remain on a plateau for about 2 weeks, gradually decrease over the next 2–4 weeks; in a parallel fashion, α-fetoprotein is not expressed for 2–3 weeks, and then increases, peaking by the end of the life of the culture. By 4–6 weeks, the cultures, although viable, express few or very low levels of their tissue-specific functions and express high levels of embryonic proteins.

Characterization of the cultures on collagens using molecular probes has revealed that liver-specific gene expression can be near normal for several weeks (208), although this expression is due, once again, to stabilization phenomena. The mRNA stability of some genes (e.g., the cytochromes P-450) was found to be quite dependent on collagen substrata (208). Thus, the stability of some mRNA species such as albumin or gap junction is dependent on hormones or growth factors (184, 204), and the stability of others such as the cytochromes P-450 are dependent on matrix molecules (208). Furthermore, the dramatic augmentation of liver-specific mRNAs in liver cultures on floating gels has been found, surprisingly, to be entirely posttranscriptional (208), implying that shape changes also contribute to stabilization mechanisms. The mechanisms of the shape change to mRNA stability relationship remain obscure.

The type of collagen has also proved to be important and dictates the general profile of the physiological response of the cultures, that is, whether the cells grow or do not grow and whether they express adult versus fetal functions. Type III collagen substrata result in cultures that do not grow well and that have a pattern of gene expression quite similar to quiescent liver (208). Type IV collagen substrata result in cultures that appear to be the analogs of regenerating liver: the cultures grow rapidly and have a pattern of gene expression similar to that in regenerating liver (208,210,211). Type I collagen substrata result in cultures that are biphasic: they show limited growth, transiently express adult functions, and then gradually increase their expression of fetal proteins such as α-fetoprotein (169). It is thought that cultures on type I collagen might reflect the cirrhotic liver condition. There are no studies to date that explain the mechanism(s) by which the collagen types dictate these responses.

Influence of Glycosaminoglycans and Proteoglycans on Liver Gene Expression

The highly sulfated proteoglycans and glycosaminoglycans have long been known to inhibit the attachment of cells to dishes, to cause striking cell shape changes in cells already attached (211–216), and to slow or stop the growth of hepatocytes (213) and hepatomas (211,212). However, recent studies have documented that these matrix molecules are potent regulators of gene expression in hepatocytes and hepatomas (204,214–216). Proteoglycans and glycosaminoglycans have been found essential for the synthesis and expression of gap junctions and thus regulate cell–cell communication in primary liver cultures and in hepatomas (214–216). They also participate synergistically with hormones or growth factors to regulate transcription rates for tissue-specific genes in primary rat liver cultures. The effects of the proteoglycans plus hormones on transcription rates can be observed as long as the cultures are in a serum-free medium (215,216). Of the proteoglycans and glycosaminoglycans tested to date, the most active forms on primary liver cultures are heparin and heparan sulfate proteoglycan derived from liver (214–216). The tissue specificity of the glycosaminoglycans and proteoglycans was evidenced by their relative potencies and by the time necessary to induce the biological response. The addition of biologically active species to primary liver cultures results in restoration of transcription and in augmentation of the liver-specific mRNAs and, simultaneously, in suppression, by posttranscriptional mechanisms, of common gene mRNAs such as actin and tubulin. The proteoglycans and glycosaminoglycans are the first factors known which simultaneously increase tissue-specific mRNAs and decrease common gene mRNAs in liver cultures (214–216). In summary, protein hormones and growth factors in the absence of proteoglycans (or glycosaminoglycans) have been found to regulate gene expression by posttranscriptional mechanisms and in the presence of these matrix molecules to regulate gene expression by both transcriptional and posttranscriptional mechanisms.

Studies with Matrix Extracts

The longest culture survival and the most nearly normal gene expression observed in primary liver cultures have been observed in studies using tissue extracts enriched in multiple extracellular matrix components. Most of the protocols for isolating extracts enriched in extracellular matrix involve negative selection: the tissue is repeatedly extracted with solutions that solubilize cellular components other than the extracellular matrix. Since the highly cross-linked extracellular matrix is largely insoluble, this approach can be a highly successful one.

Using such a protocol, Reid and associates prepared NaCl extracts, referred to as "biomatrices," from liver and other tissues (217,218). The biomatrices were pulverized in liquid nitrogen and then painted onto culture surfaces. Normal liver cells plated onto liver biomatrices attached and spread within hours and then, in a medium containing 5% serum and supplemented with the hormones from the HDM (Table I), could survive for more than 6 months (217). Such long-term survivals have never been seen in cultures under any medium conditions with substrata of individual, purified matrix components. The cultures maintained on biomatrix were initially differentiated to the same degree as those on type I collagen gels. However, within 5–7 days, differentiated functions were restored to normal and remained at near normal levels throughout the culture life span. Culturing cells on such biomatrices, especially if in combination with a serum-free, hormonally defined medium, can result in cells with greatly elevated cytoplasmic levels of tissue-specific mRNAs that show increased sensitivity to other regulators (218). Furthermore, these defined matrix and hormonal conditions can be permissive for posttranscriptional regulatory mechanisms such as adenylation mechanisms that affect translation efficiency of the mRNAs (218).

In more recent studies, Kleinman and associates have developed a protocol for a matrix extract derived from urea-extracted EHS tumors, transplantable mouse embryonal carcinomas that constitutively secrete basement membrane proteins (219–224). The extract, referred to as "matrigel," is prepared by extracting the EHS tumor tissue with urea and then allowing the urea-solubilized components to gel after dialyzing out the urea and bringing the solution to room temperature. The gel consists largely of laminin (~90%) along with some type IV collagen, entactin, and a form of heparan sulfate PG (219). In the limited studies to date using matrigel (209, 210, 221), cells attached very poorly, did not grow, and remained in clusters of highly aggregated, three-dimensional cells that survived for approximately 2–4 weeks. The transcription rates and the mRNA abundance for the liver-specific genes that have been tested in cells cultured on matrigel are near normal or are normal. Thus, there is the exciting potential for having short-term liver cultures which are nearly that of normal, quiescent liver. The future studies utilizing matrigel or other matrix conditions along with appropriate hormones or growth factors should revolutionize our abilities to analyze tissue-specific gene expression under completely defined conditions.

Synergies in the Effects of Matrix and Hormones

As described above, plating liver cells on particular collagen types can dictate their responsiveness to specific hormones. Thus, primary liver cultures plated onto type I or III collagen show limited growth even in an HDM that supports growth of cells on tissue culture plastic or on type IV collagen (208). Similarly, hepatoma cells plated onto collagens demonstrate distinct hormone requirements for clonal growth compared to cells plated onto plastic substrata (62). The mechanism(s) by which collagen or a specific collagen type is permissive or nonpermissive for responsiveness to a specific hormone is unknown. It is assumed to involve, at least in part, the stabilization by a particular collagen scaffolding of other matrix molecules (e.g., proteoglycans) which are thought to be the critical regulatory signals. It has long been known that different collagen types are synthesized along with and are assembled with distinct matrix components.

The synergies of proteoglycans and glycosaminoglycans with hormones have been clarified by more readily understandable mechanisms. The binding of cationic proteins to heparins has been utilized routinely in the purification of such molecules using heparin columns (225–236). Recently, it has been shown that heparins have low affinity binding sites (dissociation constants of 10^{-4}–10^{-5}) for many cationic factors and high affinity binding sites (dissociation constants of 10^{-9}–10^{-10}) for a few specific growth factors and hormones (234). In structure–function analysis of the effects of heparin on anticoagulation, a pentasaccharide sequence was found responsible (237–241). In studies on chondroitin sulfate proteoglycans, the sulfation patterns are especially relevant (242). Furthermore, heparins from different tissue sources show a differential binding affinity for specific hormones and growth factors (234). Last, heparin-binding growth factors are dependent on the presence of small amounts of heparin for their activity; so long as they are bound to heparin, they are less sensitive to proteolysis (234). This is especially interesting in the context of the findings that heparins are required along with hormones/growth factors for tissue-specific gene transcription (214–216). In the absence of the hormones, the heparins lose their effects on gene expression (204). These data support the hypothesis that complexes of hormones/growth factors and proteoglycans or glycosaminoglycans may be either a more potent stimulus or a totally distinct stimulus from that of hormones alone on induction of gene expression.

GENE EXPRESSION STUDIES IN HEPATOMA CELLS

Studies Using Classic Cell Culture Conditions

Hepatoma cell lines can but do not necessarily express many of the known liver-specific genes, and expression can sometimes be modulated by the appropriate hormonal or pharmacological agents (54,62,191,192,202,203, 243–254). Thus, hepatoma cell lines have, with some selectivity, been used as model systems for studying the regulation of gene expression in normal liver (54,191,192,202,203,246,247,250,253). Although the ability of hepatoma cultures to express liver-specific functions is directly correlated with the degree of differentiation of the original tumor and inversely correlated with the extent of tumor progression, there is, in addition, extensive heterogeneity in differentiated expression in the cultured tumor cells, even in cloned culture populations (54,62,243). However, from studies on the same tumor cell lines under defined culture conditions, it has become apparent that some of the heterogeneity is a reflection of the specific culture conditions employed for particular studies (62,218). [See also Endnotes 1b and 1c.]

Under classic cell culture conditions, it has been found that posttranscriptional regulatory mechanisms, most commonly stabilization mechanisms, are the primary cause of the high cytoplasmic abundance of tissue-specific mRNAs in the well-differentiated hepatomas. Furthermore, hormonal or pharmacological regulation of expression of the liver-specific functions has also proved to be almost entirely by posttranscriptional mechanisms (54). Tumor cell lines have been shown to have undetectable or exceptionally low transcription rates (at most 10–12% of normal) for adult tissue-specific genes when measured by nuclear run-on assays (54). Of even greater concern, the transcription of tissue-specific genes is often insensitive to the customary regulatory factors. Therefore, hepatoma cell lines, at least when grown under classic culture conditions, are of limited use in studying regulation of transcription of tissue-specific mRNAs. [Endnotes 1b and 1c.]

Gene Expression in Hepatoma Cells in Hormonally Defined Media

Culturing hepatoma cells in serum-free, hormonally defined media results in an increase in mRNA levels of those tissue-specific genes that the cells are capable of expressing but has no effect on silent genes (62,218,245). However, the increased mRNA abundance of these tissue-specific genes in hepatoma cells has been found to be due mostly if not entirely to posttranscriptional mechanisms (218).

Effects of Matrix Substrata

Use of matrix substrata in combination with specific hormonally defined media enables tumor cells to express morphological, growth, and differentiative properties that resemble those of their normal counterparts *in vivo* (62,218). On collagenous substrata, tumor cells attach and grow more efficiently, with a reduced requirement for serum supplementation (62). Collagenous and matrix substrata reduce the hormone requirements for clonal growth both qualitatively and quantitatively for hepatoma cells (62).

Whereas normal epithelial cells respond stably to a matrix, tumor cells show responses that are transient. Hepatomas plated onto biomatrix and in HDM go into growth arrest after one or two divisions and remain in growth arrest for 10–12 days (218). During the growth arrest, tissue-specific functions were augmented 5- to 50-fold above those observed under any other culture conditions. The phase of growth arrest accompanied by augmented tissue-specific functions is transient for all tumor cell lines studied. At the end of growth arrest, regions of the dishes contain cells growing in piles on top of one another. The cells in the piles eventually detach and float into the

TABLE V
Distinctions in Regulation of Normal versus Tumor Cells

Variable	Hepatocytes	Hepatoma cells
Serum	Qualitative effect: Completely blocks transcription of tissue-specific mRNAs	Quantitative effect: Reduces transcription rate of tissue-specific mRNAs
Serum	Qualitative effect: Rapid loss (hours) of tissue-specific mRNAs due to short half-life	Quantitative effect: Slow loss in half-life of tissue-specific mRNAs
Critical requirements for growth	Matrix: Type IV collagen and heparan sulfate PG	Matrix: None (hepatomas rapidly synthesize type IV collagen and heparan sulfate PG)
	Hormones: EGF and insulin are strict mitogens; other secondary requirements are growth hormone, prolactin, certain lipids	Hormones: Usually insulin (or IGF-II) and lipids; varies with the cell line; dependent on which autocrine growth factors (e.g. TGF-α, IGF-II) the cells are making, which must be supplied exogenously as EGF, insulin, or growth hormone if the cells do not make them
Requirements for differentiation	Matrix: Fibrillar collagen and heparin PG	Matrix: Fibrillar collagen and heparin PG
	Hormones: Dependent on which gene	Hormones: Dependent on which gene
Stability of responses	Quite stable	Transient: due, in part, to the fact that tumors produce enzymes which degrade hormones and matrix

medium. If the floating cells or cells in the piles are transferred to dishes with serum-free, hormonally defined medium and biomatrix, they again go into growth arrest. The response of tumor cells to the matrix and defined medium conditions indicate both that tumor cells are capable of responding to regulatory signals and that the cells are able to escape, most probably through the release of degradative enzymes (218). See Table V for a summary of the distinctions in regulation of normal hepatocytes for hepatomas by matrix and hormones.

Transfection Studies into Hepatoma Cells or Hepatocytes

A great deal of progress has been made recently in the identification and characterization of cis- and trans-regulatory signals for tissue- and cell type-specific transcription. [See also Endnotes 7 and 8.] The most common procedure for this type of analysis is to introduce DNA constructs, including the presumed regulatory regions linked to a marker sequence, into some appropriate cultured cell. Then expression of the marker sequence can be assayed within a few days, either at the RNA level or at the protein level if the regulatory sequences are linked to a marker gene which has an assayable activity such as CAT (255). Different deletions and rearrangements of the sequences to be tested can be transfected in parallel, and the relative efficiencies of transcription quantitated. For most studies, the cultured cells selected have been well-differentiated but neoplastically transformed tumor cell lines, because these are the easiest to work with. Since these cells are known to express some or all differentiated genes, it has been assumed that they are appropriate vehicles for such studies. Much of what we know about tissue-specific gene regulation has been derived from such systems.

For instance, particular DNA sequence elements which seem to play a role in promoting or enhancing tissue- or cell type-specific transcription have been found associated with a number of genes (for detailed reviews of specific genes, see Refs. 256–258), including several from the liver: for example, α-fetoprotein (259,260), albumin (261), and α_1-antitrypsin (262). In at least one case, an element which represses tissue-specific expression in an inappropriate cell type has been identified (263). The regulatory elements are most frequently found upstream of the transcription initiation site of the gene in question; however, immunoglobulin genes often have regulatory elements which lie within the gene, and in the globin system an enhancer has been found in the 3' flanking region.

Despite the fruitful results of gene transfection studies, one must be aware of the potential limitations of such experiments. Neoplastic cells are clearly not subject to many of the normal regulatory signals—certainly not for growth—and frequently they do not express all of the very small percentage of genes we can assay. One should not, therefore, assume that they contain all of the as yet undefined trans-acting protein factors which cooperate *in vivo* to modulate gene expression. As stated earlier, it is certain that the rate of transcription of tissue-specific genes in tumor cells is far below that of the same genes in normal cells *in vivo*. Thus, a part of the story has been and may continue to be revealed by transfection studies into neoplastic cells, but the results must be recognized as probably incomplete at the very least. Complementary but much more complicated studies can be done by inserting these constructs into transgenic animals (264), and *in vitro* transcriptional studies using nuclear extracts derived from different tissues can give other information.

An alternative to using neoplastic cell lines for transfection studies is to use primary cell cultures for transient expression assays (265, 266). This can be done by modifying transfection techniques slightly, depending on the fragility and special sensitivities of the cell type to be studied, and by using the modern cell culture systems to ensure that the cells are expressing their differentiated functions as fully as possible. The usefulness of primary cultures in gene expression studies is just beginning to be understood. Because the culture methods for hepatocytes have been so fully defined, this is a particularly good place to begin to exploit transfections in primary cultures.

HYPOTHESES ON THE MECHANISMS OF MATRIX AND HORMONAL REGULATION OF GENE EXPRESSION: DELIVERY OF PROTEIN HORMONES TO THE NUCLEUS BY PROTEOGLYCAN/GLYCOSAMINOGLYCAN CARRIERS

Of possible relevance to the proteoglycan–hormone interactions are the particularly startling findings recently reported from Conrad and associates (144) that a specific form of heparan sulfate PG, synthesized by hepatocytes, is bound to the cell surface by inositol phosphate and is subsequently translocated to the nucleus. The biological role(s) for this heparan sulfate PG is unknown. However, it is possible that the biologically active glycosaminoglycans and proteoglycans added to primary liver cultures could be acting either at the cell surface, within the nucleus, or both. Thus, the finding of heparan sulfate chains being translocated to the nucleus in combination with

TABLE VI
Levels of Regulation of Expression of Liver-Specific Messenger RNAs

Level	Process	External signals found to affect expression at this level
DNA	Transcription	A. *Matrix-regulated genes* (e.g. α-1-antitrypsin). Heparin-PGs, heparins, and laminin have been found active B. *Matrix/hormone-regulated genes* (e.g. albumin, connexins, IGF II). Peptide hormones in the presence of heparin PGs, heparins. Each mRNA species is regulated by different hormones and growth factors in combination with heparin PGs or heparins. C. *Steroid-regulated genes* (e.g. tyrosine aminotransferase). Although steroids can regulate genes on their own, GAGs and PGs can modify that regulation quantitatively
mRNA	Processing	Steroids
	mRNA stability (half-life)	Collagens, adhesion proteins, all PGs, heparins and dermatan sulfates, and peptide hormones in the absence of PGs or GAGs. Some of these posttranscriptional effects can be mimicked by membrane permeant analogs of cAMP.
Protein	Translation efficiency	Hormones + matrix (specific components not yet identified)
	Insertion into	Lipids; factors that affect lipid or lipoprotein synthesis
	Secretion	PGs and GAGs through influence on pH and calcium; certain peptide hormones whose effects can be mimicked by second messengers; collagens; adhesion proteins
Removal	Internalization and degradation	PGs and GAGs affect turnover of proteins via affects on cytoskeleton; hormones via phosphorylation mechanisms; cAMP derivatives (phosphorylation;); calcium; factors that disrupt the cytoskeleton

the findings that heparins can bind hormones and growth factors suggest an even more startling concept: that heparins (and other glycosaminoglycans or proteoglycans?) could serve as vehicles for hormones or growth factors to reach the nucleus of cells and/or that heparin–hormone complexes are the direct trigger for an intranuclear target. A more detailed presentation of this hypothetical model is given elsewhere (267).

SUMMARY

The potential for rigorous studies of gene expression is now greatly expanded by the availability of defined hormone and matrix conditions that can be used for maintaining normal aspects of gene expression in cultures of normal and neoplastic liver cells from any species. These defined hormone and matrix conditions reflect the complicated signaling from cell–cell interactions now recognized to effect and maintain differentiation of cells *in vivo*. Recognition of the roles of cell–cell interactions and of the hormones and matrix components that are the vehicles of those interactions is providing a rigorous scientific foundation for cell culture technology, a field that for so long has been empirically defined. Furthermore, the development of these defined conditions has dramatically indicated the multiple levels at which gene expression is regulated (see Table VI).

As in other scientific fields, the advent of sophisticated technologies paves the way for scientific discoveries. Even though the development of many of the new cell culture technologies has only recently taken place, important scientific discoveries have already been made: the recognition of the central role for mRNA stabilization mechanisms, the recognition that serum components destabilize tissue-specific mRNAs, the recognition of the synergies in the effects of hormones and matrix components on gene expression. The coming years should prove very exciting as such discoveries continue to be made.

ACKNOWLEDGMENTS

This research was supported by a grant from the American Cancer Society (BC-439) and by grants from the National Institutes of Health (CA30117, P30-CA13330, AM17702-12). Lola Reid received salary support through a Career Development Award (NIH CA00783). Ms. Rosina Passela has contributed through excellent secretarial assistance. Technical assistance has been provided by Mrs. Dinish Williams, Ms. Elaine Halay, Mr. Luis de la Vega, and Mr. Errol Thompson. The studies described from the laboratory of L.R. were done by many excellent students and research associates, whose individual contributions are acknowledged through authorship on various publications. However, special tribute is due to several of them whose dedicated efforts proved invaluable in clarifying some of the complicated phenomena. These include Zenaida Gatmaitan, who developed the original serum-free, hormonally defined medium for liver cells; Douglas Jefferson, who initiated the gene expression studies on liver cultures under defined conditions; Michiyasu Fujita, who screened the effects on gene expression of numerous proteoglycans and glycosaminoglycans; Tohru Watanabe, who began the efforts to define the complicated synergies of the hormones and the glycosaminoglyans/proteoglycans in regulating transcription; Kate Montgomery, who has worked to define cis-regulatory elements in tissue-specific genes and regulated by matrix molecules; Isabel Zvibel and Andrea Kraft, who analyzed matrix/hormonal regulation of autocrine growth factor genes; and Maria Agelli and Andreas Ochs, who have been working to define liver stem cells and lineages.

Endnotes

1. The consensus of opinion in research has become that the liver is, indeed, a stem cell and lineage system. Two recent reviews discuss this in more detail. [Reid, L. (1990). *Current Opinions in Cell Biology* **2**, 121–130, Sell, S. (1990). *Cancer Research* **50**, 3811–3815.]

Some of the implications of stem cell concepts emerging with respect to liver are providing clarification of a number of paradoxes:

a) The limited growth of hepatocytes (intermediates in the lineage) in culture is due to molecular mechanisms that are lineage-position dependent. Only the stem cells and perhaps the early precursor cells should be capable of clonal growth in culture, reflecting their inherent, greater capacity to divide. The number of divisions observed in hepatocytes in culture is a reflection of their limited *in vivo* proliferation ability.

b) The heterogeneity of gene expression over the liver is also a reflection of the lineage and of molecular changes in gene expression that are dependent upon the number of divisions undergone by the cell or influenced by a position-dependent gradient of signals (matrix or soluble signals) that vary over the lineage. We hypothesize that changes in certain of the transcription factors should correlate with lineage-dependent changes in gene expression, that is that there will be a lineage-dependent gradient of transcription factors. Two stages of transcriptional activation, both lineage-position dependent, have been observed: (1) transcriptional activation with mRNA synthesized constitutively at low levels (posttranscriptional regulation dominates) and (2) activation of transcriptional inducibility by exogenous stimuli, matrix components and hormones.

c) Some pathogenic events (malignancy, certain viral infections) are proving to be lineage position dependent. For example, certain hepatitis viruses can replicate in immature cells but do not result in lytic processes (or, alternatively, in immune rejection) until the middle of the lineage. Similarly, the Pierce/Potter hypothesis that stem cells are targets for oncogenesis (*Prog. Clin. Biol. Res.* (1986), **226**, 67–77) is supported in studies of the liver. These findings indicate that hepatomas represent transformed stem cells or early precursor cells, not transformed hepatocytes. Therefore, patterns of gene expression in hepatomas are, in part, reflections of the phenotypes of normal liver progenitor cells. For example, the first studies on gene expression in normal liver progenitors (Angelli, Zvibel, Kraft, Ochs and Reid, in preparation) indicates that both normal liver progenitor cells and hepatomas express IGF II, ras, α-fetoprotein. These "early" genes are regulatable in hepatomas (and we predict, in normal liver progenitors) by matrix and hormonal signals. By contrast, minimally deviant tumors expressing both "early" and some "middle" genes show largely or only posttranscriptional regulation of the "middle" genes. Hepatomas have never been observed to express "late" genes such as glutamine synthetase or major urinary protein (MUP) at any significant level. Stem cell hypotheses suggest an explanation: the hepatomas are blocked and have not progressed far enough into the lineage to have activated relevant transcription factors. The cells would have to complete the lineage to activate the lineage to activate these factors, and hepatomas do not have this capacity.

2. Stem cell hypotheses (see #1) make it apparent that hepatocytes can never be cloned or subcultured. They can go through only a few divisions, the number of divisions being defined by their position in the lineage.

3. It is no longer clear whether the shifts in matrix chemistry, observed in tissues undergoing dramatic changes in physiological status (e.g. quiescence to growth), are due to changes in synthesis by specific cells or selection of cells at a stage of the lineage expressing a particular matrix chemistry.

4. The cell adhesion molecules (CAMs) were completely ignored in this review. An excellent recent review of CAMs, especially those in the liver, has recently been written by Stanley Hoffman and Kathryn Crossin (Rockefeller University): "Adhesion Molecules in Embryogenesis and Histogenesis," *In* "Extracellular Matrix: Its Chemistry, Biology and Pathobiology." (M. Zern and L. Reid, eds.) Marcel Dekker, N. Y., 1991 (in press).

5. Antonio Martinez-Hernandez and Peter Amenta (Jefferson Medical School) have written an extensive review detailing the matrix chemistry of the liver ("Morphology, Localization and Origin of the Hepatic Extracellular Matrix" *In:* "Extracellular Matrix: Its Chemistry, Biology and Pathobiology." (M. Zern and L. Reid, eds.) Marcel Dekker, N. Y., 1991 (in press). They have identified the cellular sources for all known matrix molecules in the liver as determined in extensive ultrastructural/immunochemical analyses. Moreover, they have surveyed the changes in this matrix chemistry, including the changes by cellular source, in fetal versus adult liver and in adult liver in quiescence, regeneration, and cirrhosis.

6. In recent years, it has been realized that the plasma membrane-associated proteoglycans connect to the cell surface by one of several mechanisms and can be internalized and relocated to the nucleus: (*a*.) The protein core can be an integral membrane protein (P. Marynen, *et al.* (1989). *J. Biol. Chem.* **264**, 7017–7024; (*b*.) The protein core can be coupled to the membrane by a lipid anchor (phosphotidyl inositol) (Bienkowski and Conrad, E. (1984). *J. Biol. Chem.* **259**, 12989–12996; Ishihara, M. *et al.* (1989). *J. Cell Physiol.* **128**, 467–476). This species can traffic to the nucleus and is suspected of being involved in growth regulation (Fedarko, N. *et al.* (1989). *J. Cell. Physiol.* **139**, 287–294. (*c*.) The glycosaminoglycan chains can bind to proteins at or in the plasma membrane or in the basal lamina (Folkman, J. *et al.* (1988). *Am. J. Pathol.* **130**, 393–400; Vlodavsky, I. *et al.* (1987). *Proc. Natl. Acad. Sci.* **84**, 2292–2296; Bashkin *et al.* (1989). *Biochemistry* **28**, 1737–1743.

In addition, the protein cores of a number of the proteoglycans have now been cloned and sequenced. A sampling of some of these papers is listed below: Bourdon, M. *et al.* (1986). *J. Biol. Chem.* **261**, 12534–12537; Krueger, R. *et al.* (1989). *J. Biol. Chem.* **265**, 12088–12097; Zimmerman, D. and Ruoslahti, E. (1989). *EMBO J.* **8**, 2975–2981; Sasaki, M. *et al.* (1988). *J. Biol. Chem.* **263**, 16536–16544; Noonan, D. M. *et al.* (1988). *J. Biol. Chem.* **263**, 16379–16389; and Tantravahi, R. V. *et al.* (1986). *Proc. Natl. Acad. Sci.* **83**, 9207–9210.

Also, suramin, a heparin-like compound, has proven a potential anti-tumor agent having effects on particular growth factors and their receptors (Rocca, R. *et al.* (1990). *Cancer Cells* **2**, 106–115). Suramin has been found active against human prostatic cancers.

7. The field of enhancers and transcription factors, especially those pertaining to the liver, has flourished in the last several years. There are now a number of the transcription factors cloned and sequenced. For information on this rapidly developing field, see the following articles and reviews: Frain, M. *et al.* (1989). *Cell* **59**, 145–157; Nicosia, A. *et al.* (1990). *Cell* **61**, 1225–1236; Pugh, B. and Tjian, R. (1990). *Cell* **61**, 1187–1197; Ptashne, M. (1988). *Nature* **335**, 683–689; and Mitchel, P. and Tjian, R. (1989). *Science* **245**, 371–378.

8. Although evidence for regulation of mRNA stability has been available for more than 10 years (Guyette, W. A. *et al.* (1979). *Cell* **17**, 1013), investigations into mechanisms governing

mRNA stability have increased dramatically in the last few years. There are now known to be sequences in the 3' coding region, or even in the coding region near the promoter, that affect stability. Recent papers discussing some of the mechanisms include: Shaw, G. and Kamen, R. (1986). *Cell* **46**, 659–667; Yen, T. J. *et al.* (1988). *Mol Cell. Biol.* **8**, 1224–1235; and Mueller, C. *et al.* (1990). *Cell* **61**, 279–281.

9. The confusion in the studies by Ringold and associates on whether or not α1-acid glycoprotein is transcriptionally regulated by dexamethasone proved due, in part, to a technical error: use of a double-stranded probe resulted in a transcriptional signal with and without dexamethasone because of transcription from the negative strand. Use of an M13, single-stranded probe confirmed the transfection studies (203) using constructs of α1-acid glycoprotein that this gene is indeed regulated transcriptionally by dexamethasone. However, even with a single-stranded probe, there was still a high basal trancriptional signal due, it is assumed, to read through from an upstream promoter (studies published as a PhD Thesis by Elliot Klein, 1988, Stanford University).

10. Indeed, there is evidence from other studies that glucocorticoids can regulate some genes posttranscriptionally. A recent example in which it was clearly documented is a study on glucocorticoid regulation of P_{450d} in rat liver cells: Silver G. *et al.* (1990). *J. Biol. Chem.* **256**, 3134–3138. Interestingly, the glucocorticoid effect on P_{450d} was shown to be posttranscriptional but intranuclear and did not involve changes in mRNA stability.

REFERENCES

1. Brachet, J., and Alexander, H. (1986). "Introduction to Molecular Embryology." Springer-Verlag, New York.
2. Kincade, P. W., Lee, G., Paige, C. J., and Scheid, M. P. (1981). *J. Immunol.* **127**, 255–260.
3. Cunha, G. R. (1976). *Int. Rev. Cytol.* **47**, 137–194.
4. Protero, J. (1980). *J. Theor. Biol.* **84**, 725–736.
5. Differentiation of normal and neoplastic hematopoietic cells. (1978). Edited by Bayard Clarkson, Paul A. Marks, James E. Till. [Cold Spring Harbor, N.Y.] : Cold Spring Harbor Laboratory, 1978. 2 v. (xiv, 994 p.). (Cold Spring Harbor conferences on cell proliferation, vol. 5.)
6. Nakamura, T., Nagao, M., and Ichihara, A. (1987). *Exp. Cell Res.* **169**, 1–14.
7. Denis, K. A., Treiman, L. J., St. Claire, J. I., and Witte, O. N. (1984). *J. Exp. Med.* **160**, 1087–1101.
8. Rosenstraus, M. J., Sterman, B., Carr, A., and Brand, L. (1984). *Exp. Cell Res.* **152**, 378–389.
9. Levine, J. F., and Stockdale, F. E. (1984). *Exp. Cell Res.* **151**, 112–122.
10. Reid, L. M., and Jefferson, D. M. (1984). *Hepatology* **4**, 548–559.
11. Fleischmajer, R., and Billingham, R. E., eds. (1968). " Epithelial–Mesenchymal Interactions." Williams & Wilkins, Baltimore, Maryland.
12. Hopkins, C. R., and Hughes, R. C. (1985). *J. Cell Sci.* (Suppl. 3).
13. Growth factors and transformation. (1985). Edited by James Feramisco, Brad Ozanne, and Charles Stiles. Cold Spring Harbor, N.Y. : Cold Spring Harbor Laboratory.
14. Walsh-Reitz, M. M., Gluck, S. L., Waack, S., and Toback, F. G. (1986). *Proc. Natl. Acad. Sci. U.S.A.* **83**, 4764–4768.
15. Cuttitta, F., Carney, D. N., Mulshine, J., Moody, T. W., Fedorko, J., Fischler, A., and Minna, J. D. (1985). *Nature (London)* **316**, 823–826.

16. Danielpour, D., and Sirbasku, D. A. (1984). *In Vitro* **20**, 975–980.
17. Stoker, M., and Gherardi, E. (1987). *Ciba Found. Symp.* **125**, 217–239.
18. Harel, L., Blat, C., and Chatelain, G. (1985). *J. Cell. Physiol.* **123**, 139–143.
19. Mecham, R. P., ed. (1986). "Biology of Extracellular Matrix," Vol. 1. Academic Press, New York.
20. Reid, L., and Jefferson, D. (1984). *In* "Mammalian Cell Culture" (J. Mather, ed.), pp. 239–280. Plenum, New York.
21. Sobczak, J., and Duguet, M. (1986). *Biochimie* **68**, 957–967.
22. Dunsford, H. A., and Sell, S. (1989). 49:4887–4893.
23. Dunsford, H. A., Karnasuta, C., Hunt, J. M., and Sell, S. (1989). 49:4894–4900.
24. Sell, S., Hunt, J. M., Knoll, B. J., and Dunsford, H. A. (1987). Cellular events during hepatocarcinogenesis in rats and the question of premalignancy. *Adv. Cancer Res.* **48**, 37–111.
25. Farber, E. (1956). *Cancer Res.* **16**, 142–148.
26. Grisham, J. W., and Hartroft, W. S. (1961). *Lab. Invest.* **10**, 317–332.
27. Steiner, J. W., and Carruthers, J. S. (1961). *Am. J. Pathol.* **38**, 639–661.
28. Wilson, J. W., Groat, C. S., and Leduc, E. H. (1963). *Ann. N.Y. Acad. Sci.* **11**, 8–24.
29. Rubin, E. (1984). *Exp. Mol. Pathol.* **3**, 279–286.
30. Grisham, J. W., and Porta, E. A. (1964). *Exp. Mol. Pathol.* **3**, 242–261.
31. Inaoka, Y. (1967). *Gann* **58**, 355–366.
32. Kalengayi, M. M. R., and Desmet, V. J. (1975). *Cancer Res.* **35**, 2845–2852.
33. Hadjilov, D. L. (1965). *Z. Krebforsch.* **66**, 473–477.
34. Sneider, T. W., Krawitt, E. L., and Potter, V. R. (1970). *Cancer Res.* **30**, 44–47.
35. Yoshimura, H., Harris, R., Yokoyama, S., Takahashi, S., Sells, M. A., Pan, S. F., and Lombardi, B. (1983). *Am. J. Pathol.* **10**, 322–332.
36. Yaswen, P., Hayner, N. T., and Fausto, N. (1984). *Cancer Res.* **44**, 324–331.
37. Hayner, N. T., Braun, L., Yaswen, P., Brooks, M., and Fausto, N. (1984). *Cancer Res.* **44**, 332–338.
38. Hixson, D. C., and Allison, J. P. (1985). *Cancer Res.* **45**, 3750–3760.
39. Germain, L., Goyette, R., and Marceau, N. (1985). *Cancer Res.* **45**, 673–681.
40. Sells, M. A., Katyal, S. L., Shinozuka, H., Estes, L. W., Sell, S., and Lombardi, B. (1981). *J. Natl. Cancer Inst.* **66**, 355–362.
41. Miller, S. B., Pretlow, T. P., Scott, J. A., and Pretlow II, T. G., (1982). *J. Natl. Cancer Inst.* **68**, 851–857.
42. Evarts, R. P., Marsden, E., Hanna, P., Wirth, P. J., and Thorgeirsson, S. S. (1984). *Cancer Res.* **44**, 5718–5724.
43. Tatematsu, M., Kaku, T., Ekem, J. K., and Farber, E. (1984). *Am. J. Pathol.* **114**, 418–430.
44. Sirica, A. E., and Cihla, H. P. (1984). *Cancer Res.* **44**, 3454–3466.
45. Germain, L., Noel, M., Gourdeau, H., and Marceau, N. (1988). *Cancer Res.* 48:368–378.
46. Grisham, J. W. (1980). *Ann. N.Y. Acad. Sci.* **349**, 128–137.
47. Tsao, H. S., Smith, J. D., Nelson, K. G., and Grisham, J. N. (1984). *Exp. Cell Res.* **154**, 38–52.
48. Sirica, A. E., Sattler, C. A., and Cihla, H. (1985). *Am. J. Pathol.* **120**, 67–78.
49. Ogawa, K., Minase, T., and Onoe, T. (1974). *Cancer Res.* **34**, 3379–3386.
50. Sell, S. (1983). *Cancer Res.* **43**, 1761–1767.
51. Braun, L., Goyette, M., Yaswen, P., Thompson, N. L., and Fausto, N. (1987). *Cancer Res.* 47:4116–4124.

52. Freshney, R. I., ed. (1985). "Animal Cell Culture: A Practical Approach." IRL Press, New York.
53. Reid, L. M., Narita, M., Fujita, M., Murray, Z., Liverpool, C., and Rosenberg, L. (1986). In "Isolated and Cultured Hepatocytes" (A. Guillouzo and C. Guguen-Guillouzo, eds.), pp. 225–258. John Libbey Eurotext Ltd./INSERM, Paris.
54. Clayton, D. F., Weiss, M., and Darnell, J. E., Jr. (1985). *Mol. Cell. Biol.* **5**, 2633–2641.
55. Jefferson, D. M., Clayton, D. F., Darnell, J. E., Jr., and Reid, L. M. (1984). *Mol. Cell. Biol.* **4**, 1929–1934.
56. Ham, R., and McKeehan, W. (1979). In "Methods in Enzymology," (W. B. Jakoby and I. Pastan, eds.), Vol. 58, pp. 44–93. Academic Press, New York.
57. Barnes, D., and Sato, G. (1980). *Cell (Cambridge, Mass.)* **22**, 649–655.
58. Barnes, D., and Sato, G., eds. (1984). "Cell Culture Methods for Molecular and Cellular Biology," Vols. 1–4. Alan R. Liss, New York.
59. Mather, J., ed. (1987) "Mammalian Cell Culture," Plenum Press, Inc., New York.
60. Reid, L. M., Stiles, C., Saier, M., Jr., and Rindler, M. (1979). *Cancer Res.* **39**, 1467–1473.
61. Cherington, P. V., Smith, B. L., and Pardee, A. B. (1980). *Proc. Natl. Acad. Sci. U.S.A.* **76**, 3937–3941.
62. Gatmaitan, Z., Jefferson, D., Ruiz-Opazo, N., Leinwand, L., and Reid, L. M. (1983). *J. Cell Biol.* **97**, 1179–1190.
63. Enat, R., Jefferson, D. M., Ruiz-Opazo, N., Gatmaitan, Z., Leinwand, L. A., and Reid, L. M. (1984). *Proc. Natl. Acad. Sci. U.S.A.* **81**, 1411–1415.
64. Patsch, W., Franz, S., and Schonfeld, G. (1983). *J. Clin. Invest.* **7**, 1161–1174.
65. Plant, P. W., Deeley, R. G., and Grieninger, G. (1983). *J. Biol. Chem.* **258**, 15355–15360.
66. Schwarze, P. E., Solheim, A. E., and Seglen, P. O. (1982). *In Vitro*, **18**, 43–54.
67. Schudt, C. (1979). *Eur. J. Biochem.* **98**, 77–82.
68. Yoshimoto, K., Nakamura, T., and Ichihara, A. (1983). *J. Biol. Chem.* **258**, 12355–12360.
69. Wilson, E. J., and McMurray, W. C. (1983). *Can. J. Biochem. Cell Biol.* **61**, 636–643.
70. Williams, G. M., Bermudez, E., San, R. H. C., Goldblatt, P. J., and Laspia, M. F. (1978). *In Vitro* **14**, 824–837.
71. Vaartjes, W. J., de Haas, C. G., and van den Bergh, S. G. (1986). *Biochem. Biophys. Res. Commun.* **138**, 1328–1333.
72. Georgoff, I., Secott, T., and Isom, H. C. (1984). *J. Biol. Chem.* **259**, 9595–9602.
73. Koch, K. S., Shapiro, P., Skelly, H., and Leffert, H. L. (1982). *Biochem. Biophys. Res. Commun.* **109**, 1054–1063.
74. Sells, M. A., Chernoff, J., Cerda, A., Bowers, C., Shafritz, D. A., Kase, N., Christman, J. K., and Acs, G. (1985). *In Vitro Cell. Dev. Biol.* **21**, 216–220.
75. Gherardi, E., Thomas, K., and Bouyer, D. (submitted; personal communication).
76. Yeh, Y. C., Tasi, J. F., Chuang, L. Y., Yeh, H. W., Tsai, J. H., Florine, D. L., and Tam, J. P. (1987). *Cancer Res.* **47**, 896–901.
77. Piez, K. A., and Reddi, A. H., eds. (1984). "Extracellular Matrix Biochemistry," Elsevier, New York.
78. Kornblihtt, A. R., Vibe-Pederson, K., and Baralle, F. E. (1984). *EMBO J.* **3**, 221–226.
79. Kishimoto, T. K., O'Connor, K., Lee, A., Roberts, T. M., and Springer, T. A. (1987). *Cell (Cambridge, Mass.)* **48**, 681–690.
80. Wewer, U. M., Liotta, L. A., Jaye, M., Ricca, G. A., Drohan, W. N., Claysmith, A. P., Rao, C. N., Wirth, P., Coligan, J. E., Albrechtsen, R., Mudryj, M., and Sobel, M. E. (1985). *Proc. Natl. Acad. Sci. U.S.A.* **83**, 7137–7141.
81. Wang, S. Y., and Gudas, L. J. (1983). *Proc. Natl. Acad. Sci. U.S.A.* **80**, 5880–5884.

82. Dean, D. C., Bowlus, C. L., and Bourgeois, S. (1987). *Proc. Natl. Acad. Sci. U.S.A.* **84**, 1876–1980.
83. Oldberg, A., Franzen, A., and Heinegard, D. (1986). *Proc. Natl. Acad. Sci. U.S.A.* **83**, 8819–8823.
84. Pihlajaniemi, T., Myllyla, R., Seyer, J., Kurkinen, M., and Prockop, D. J. (1987). *Proc. Natl. Acad. Sci. U.S.A.* **84**, 940–944.
85. Jimenez, S. A., Feldman, G., Bashey, R. I., Bienkowski, R., and Rosenbloom, J. (1986). *Biochem. J.* **237**, 837–843.
86. Griffin, C. A., Emanuel, B. S., Hansen, J. R., Cavenee, W. K., and Myers, J. C. (1987). *Proc. Natl. Acad. Sci. U.S.A.* **84**, 512–516.
87. Schwarz-Magdolen, U., Oberbaumer, I., and Kuhn, K. (1986). *FEBS Lett.* **208**, 203–207.
88. Ruoslahti, E., Bourdon, M., and Krusius, T. (1985). *Ciba Found. Symp.* **124**, 260–271.
89. Kosher, R. A., Gay, S. W., Kamanitz, J. R., Kulyk, W. M., Rodgers, B. J., Sai, S., Tanaka, T., and Tanzer, M. L. (1986). *Dev. Biol.* **118**, 112–117.
90. Doege, K., Hassell, J. R., Caterson, B., and Yamada, Y. (1985). *Proc. Natl. Acad. Sci. U.S.A.* **83**, 3761–3765.
91. Durkin, M. E., Carlin, B. E., Vergnes, J., Bartos, B., Merlie, J., and Chung, A. E. (1987). *Proc. Natl. Acad. Sci. U.S.A.* **84**, 1570–1574.
92. Miller, E. J. (1988). *In* "Collagen, Chemistry, Biology and Biotechnology," (Marcel E. Nimni, ed.), CRC Press, Boca Raton, Florida. Vol 1. pp. 139–156.
93. Miller, E. J. (1987). *Methods Enzymol.* **144**, 3–41.
94. Bornstein, P., and Sage, H. (1980). *Annu. Rev. Biochem.* **49**, 957–1003.
95. Mosher, D. F. (1984). *Annu. Rev. Med.* **35**, 561–575.
96. Akiyama, S. K., and Yamada, K. M. (1983). *Monogr. Pathol.* **24**, 55–96.
97. Evered, D., Hascall, V. C., and Whelan, J., eds. (1986). *Ciba Found. Symp.* vol. 124.
98. Rojkind, M., and Ponce-Noyola, P. (1982). *Collagen Relat. Res.* **2**, 151–175.
99. Dardenne, A. J., Burns, J., Sykes, B. C., and Kirkpatrick, P. (1983). *J. Pathol.* **141**, 55–69.
100. Foidart, J. M., Berman, J. J., Paglia, L., Rennard, S., Abe, S., Perantoni, A., and Martin, G. R. (1980). *Lab. Invest.* **42**, 525–532.
101. Diegelmann, R. F., Guzelian, P. S., Gay, R., and Gay, S. (1983). *Science* **219**, 1343–1345.
102. Grimand, J. A., Druquot, M., Peyrol, S., Chevalier, O., Herbage, G. U., and Badrawy, N. (1980). *J. Histochem. Cytochem.* **28**, 1145–1156.
103. Grimand, J. A., and Borojevic, R. (1980). *Cell Mol. Biol.* **26**, 555–562.
104. Hahn, E., Wick, G., Pencer, D., and Timpl, R. (1980). *Gut* **21**, 63–71.
105. Guzelian, P. S., and Diegelmann, R. F. (1979). *Exp. Cell Res.* **123**, 269–279.
106. Hata, R., Ninomiya, Y., Nagai, Y., and Tsukada, Y. (1980). *Biochemistry* **19**, 169–176.
107. Martinez-Hernandez, A. (1984). *Lab. Invest.* **51**, 57–74.
108. Saber, M. A., Zern, M. A., and Shafritz, D. A. (1983). *Proc. Natl. Acad. Sci. U.S.A.* **80**, 4017–4020.
109. Sakakibara, K., Umeda, M., Satio, S., and Nagase, S. (1977). *Exp. Cell Res.* **110**, 159–165.
110. Sell, S., and Ruoslahti, E. (1982). *J. Natl. Cancer Inst.* **69**, 1005–1014.
111. Sano, J., Sato, S., Ishizaki, M., Yajima, G., Konomi, H., Fujiwara, S., and Najai, Y. (1981). *Biomed. Res.* **2**, 546–551.
112. Voss, B., Ranterberg, J., Allam, S., and Pott, G. (1980). *Pathol. Res. Pract.* **170**, 50–60.
113. Wick, G., Brunner, H., Penner, E., and Timpl, R. (1978). *Int. Arch. Allergy Appl. Immunol.* **56**, 316–324.
114. Carlsson, R., Engvall, E., Freeman, A., and Ruoslahti, E. (1981). *Proc. Natl. Acad. Sci. U.S.A.* **78**, 2403–2406.

115. Johansson, S., and Hook, M. (1984). *J. Cell Biol.* **98**, 810–817.
116. Rubin, J., Hook, M., Obrink, B., and Timpl, R. (1981). *Cell (Cambridge, Mass.)* **24**, 463–470.
117. Schwartz, C. E., and Ruoslahti, E. (1983). *Exp. Cell Res.* **143**, 456–461.
118. Tamkun, J. W., and Hynes, R. O. (1983). *J. Biol. Chem.* **258**, 4641–4647.
119. Kleinman, H. K., Cannon, F. B., Laurie, G. W., Hassell, J. R., Aumailley, M., Terranova, V. P., Martin, G. R., and DuBois-Dalcq, M. (1985). *J. Cell. Biochem.* **27**, 317–325 (review).
120. Ocklind, C., Odin, P., and Obrink, B. (1984). *Exp. Cell Res.* **151**, 29–45.
121. Ponce, P., Cordero, J., and Rojkind, M. (1981). *Hepatology* **1**, 204–210.
122. Akasaki, M., Kawasaki, T., and Yamashina, I. (1975). *FEBS Lett.* **59**, 100–104.
123. Prinz, R., Klein, U., Sudhakaran, P. R., Sinn, W., Ullrich, K., and von Figura, K. (1980). *Biochim. Biophys. Acta* **630**, 402–413.
124. Gressner, A. M., Pazen, H., and Greiling, H. (1977). *Hoppe-Seyler's Z. Physiol. Chem.* **358**, 825–833.
125. Glimelius, B., Busch, C., and Hook, M. (1978). *Thromb. Res.* **12**, 773–782.
126. Kjellen, L., Oldberg, A., and Hook, M. (1980). *J. Biol. Chem.* **255**, 10407–10413.
127. Kjellen, L., Pettersson, I., and Hook, M. (1981). *Proc. Natl. Acad. Sci. U.S.A.* **78**, 5371–5375.
128. Oldberg, A., and Hook, M. (1977). *J. Cell Biol.* **164**, 75–81.
129. Oldberg, A., Kjellen, L., and Hook, M. (1979). *J. Biol. Chem.* **254**, 8505–8510.
130. Mischiu, L., Marin, A., and Bostinaru, A. (1980). *Rev. Roum. Morphol., Embryol. Physiol., Morphol. Embryol.* **26**, 231–235.
131. Hollmann, J., Thiel, J., Schmidt, A., and Buddecke, E. (1986). *Exp. Cell Res.* **167**, 484–494.
132. Rapraeger, A. C., and Bernfield, M. (1983). *J. Biol. Chem.* **258**, 3632–3636.
133. Norling, B., Glimelius, B., and Wasteson, A. (1981). *Biochem. Biophys. Res. Commun.* **103**, 1265–1272.
134. Laterra, J., Silbert, J. E., and Culp, L. A. (1983). *J. Cell Biol.* **96**, 112–123.
135. Kraemer, P. M. (1971). *Biochemistry* **10**, 1437–1445.
136. Kraemer, P. M. (1971). *Biochemistry* **10**, 1445–1451.
137. Ninomiya, Y., Hata, R.-I., and Nagai, Y. (1981). *Biochim. Biophys. Acta* **675**, 248–255.
138. Ohnishi, T., Ohshima, E., and Ohtsuka, M. (1975). *Exp. Cell Res.* **93**, 136–142.
139. Mutoh, S., Funakoshi, I., and Yamashina, I. (1978). *J. Biochem.* **84**, 483–489.
140. Mutoh, S., Funakoshi, I., Nobuo, U., and Yamashina, I. (1980). *Arch. Biochem. Biophys.* **202**, 137–143.
141. Bienkowski, M. J., and Conrad, H. E. (1984). *J. Biol. Chem.* **259**, 12989–12996.
142. Fedarko, N. S., and Conrad, H. E. (1986). *J. Cell Biol.* **102**, 587–599.
143. Delaney, S. R., and Conrad, H. E. (1983). *Biochem. J.* **209**, 315–322.
144. Ishihara, M., Fedarko, N. S., and Conrad, H. E. (1986). *J. Biol. Chem.* **261**, 13575–13580.
145. Muto, M., Yoshimura, M., Okayama, M., and Kaji, A. (1977). *Proc. Natl. Acad. Sci. U.S.A.* **74**, 4173–4177.
146. Hayman, E. G., Oldberg, A., Martin, G. R., and Ruoslahti, E. (1982). *J. Cell Biol.* **94**, 28–35.
147. Karasaki, S., and Raymond, J. (1981). *Differentiation* **19**, 21–30.
148. Marsillo, E., Sobel, M. E., and Smith, B. D. (1984). *J. Biol. Chem.* **259**, 1401–1404.
149. Robinson, J., Viti, M., and Hook, M. (1984). *J. Cell Biol.* **98**, 946–953.
150. Nevins, J. R. (1983). *Annu. Rev. Biochem.* **52**, 441–466.
151. Moldave, K. (1985). *Annu. Rev. Biochem.* **54**, 1109–1149.
152. Platt, T. (1986). *Annu. Rev. Biochem.* **55**, 339–372.

153. Leff, S. E., Rosenfeld, M. G., and Evans, R. (1986). *Annu. Rev. Biochem.* **55**, 1091–1117.
154. Padgett, R. A., Grabowski, P. J., Konarska, M. M., Seiler, S., and Sharp, P. A. (1986). *Annu. Rev. Biochem.* **55**, 1119–1150.
155. Birnsteil, M. L., Busslinger, M., and Strub, K. (1985). *Cell (Cambridge, Mass.)* **41**, 349–359.
156. Greene, M. R. (1986). *Annu. Rev. Genet.* **20**, 671–708.
157. Derman, E., Krauter, K., Walling, L., Weinberger, C., Ray, M., and Darnell, J. E., Jr. (1981). *Cell (Cambridge, Mass.)* **23**, 731–739.
158. Breathnach, R., and Chambon, P. (1981). *Annu. Rev. Biochem.* **50**, 349–383.
159. Benoist, C., O'Hare, K., Breathnach, R., and Chambon, P. (1980). *Nucleic Acids Res.* **8**, 127–142.
160. Efstradiatis, A., Posakony, J. W., Maniatis, T., Lawn, R. M., O'Connell, C., Spritz, R. A., Riel, J. K., deForget, B. S., Weissman, S. W., Slighton, J. L., Blechl, A. E., Smithies, O., Baralle, F. E., Shoulders, C. C., and Proudfoot, N. J. (1980). *Cell (Cambridge, Mass.)* **21**, 653.
161. Moreau, P., Hen, R., Wasylyk, B., Everett, R., Gaub, M. P., and Chambon, P. (1981). *Nucleic Acids Res.* **9**, 6047–6068.
162. Darnell, J. E., Lodish, H. F., and Baltimore, D., eds. (1990). "Molecular Cell Biology." (2nd Edition) W. H. Freeman, New York, New York.
163. Watson, J., Hopkins, N., Roberts, J., Steitz, J., and Weiner, A. (1987). *In* "Molecular Biology of the Gene," 4th ed., Benjamin/Cummings, Reading, Massachusetts.
164. Tang, C., Laspia, M. F., Telang, S., and Williams, G. M. (1981). *Environ. Mutagen.* **3**, 477–487.
165. Williams, G. M. (1976). *Am. J. Pathol.* **85**, 739–753.
166. Clayton, D. F., and Darnell, J. E. (1983). *Mol. Cell. Biol.* **3**, 1552–1561.
167. Michalopoulos, G., and Pitot, H. C. (1975). *Exp. Cell Res.* **94**, 70–78.
168. Michalopoulos, G., Sattler, G. L., and Pitot, H. C. (1976). *Life Sci.* **18**, 1139–1144.
169. Sirica, A. E., Richards, W., Tsukada, Y., Sattler, C. A., and Pitot, H. C. (1979). *Proc. Natl. Acad. Sci. U.S.A.* **76**, 283–287.
170. Clayton, D. F., Harrelson, A. L., and Darnell, J. E., Jr. (1985). *Mol. Cell. Biol.* **5**, 2623–2632.
171. Clement, B., Guguen-Guillouzo, C., Campion, J. P., Glaise, D., Bourel, M., and Guillouzo, A. (1984). *Hepatology* **4**, 373–380.
172. Guguen-Guillouzo, C., Clement, B., Baffet, G., Beaumont, C., Morel-Chany, E., Glaise, D., and Guillouzo, A. (1983). *Exp. Cell Res.* **143**, 47–54.
173. Guguen-Guillouzo, C., and Guillouzo, A. (1983). *Mol. Cell. Biochem.* **53/54**, 35–56.
174. Lebreton, J. P., Daveau, M., Hiron, M., Fontaine, M., Biou, D., Gilbert, D., and Guguen-Guillouzo, C. (1986). *Biochem. J.* **235**, 421–427.
175. Guguen-Guillouzo, C., Bourel, M., and Guillouzo, A. (1986). *Prog. Liver Dis.* **8**, 33–50 (review).
176. Fraslin, J. M., Kneip, B., Vaulont, S., Glaise, D., Munnich, A., and Guguen-Guillouzo, C. (1985). *EMBO J.* **4**, 2487–2491.
177. Guillouzo, A., Beaune, P., Gascoin, M. N., Begue, J. M., Campion, J. P., Guengerich, F. P., and Guguen-Guillouzo, C. (1985). *Biochem. Pharmacol.* **34**, 2991–2995.
178. Lescoat, G., Theze, N., Clement, B., Guillouzo, A., and Guguen-Guillouzo, C. (1985). *Cell Differ.* **16**, 259–268.
179. Foliot, A., Glaise, D., Erlinger, S., and Guguen-Guillouzo, C. (1985). *Hepatology* **5**, 215–219.
180. Guguen-Guillouzo, C., Clement, B., Lescoat, G., Glaise, D., and Guillouzo, A. (1984). *Dev. Biol.* **105**, 211–220.

181. Clement, B., Guguen-Guillouzo, C., Campion, J. P., Glaise, D., Bourel, M., and Guillouzo, A. (1984). *Hepatology* **4**, 373–380.
182. Begue, J. M., Le-Bigot, J. F., Guguen-Guillouzo, C., Kiechel, J. R., and Guillouzo, A. (1983). *Biochem. Pharmacol.* **32**, 1643–1646.
183. Guguen-Guillouzo, C., Clement, B., Baffet, G., Beaumont, C., Morel-Chany, E., Glaise, D., and Guillouzo, A. (1983). *Exp. Cell Res.* **143**, 47–54.
184. Jefferson, D. M., Clayton, D. F., Darnell, J. E., Jr., and Reid, L. M. (1984). *Mol. Cell. Biol.* **4**, 1929–1934.
185. Brock, M. L., and Shapiro, D. J. (1983). *Cell (Cambridge, Mass.)* **34**, 207–213.
186. Chung, S., Landfear, S. M., Blumber, D. D., Cohen, N. S., and Lodish, H. F. (1981). *Cell (Cambridge, Mass.)* **24**, 785–797.
187. von-der-Ahe, D., Janich, S., Scheiderit, C., Renkawitz, R., Schutz, G., and Beato, M. (1985). *Nature (London)* **313**, 706–709.
188. Edwards, A. M., and Lucas, C. M. (1985). *Carcinogenesis* **6**, 733–739.
189. Jefferson, D. M., Reid, L. M., Giambrone, M. A., Shafritz, D. A., Zern, M. A. (1985). *Hepatology* **5**, 14–20.
190. Spence, J. T., Haars, L., Edwards, A., Bosch, A., and Pitot, H. C. (1980). *Ann. N.Y. Acad. Sci.* **349**, 99–110.
191. Vannice, J. L., Grove, J. R., and Ringold, G. M. (1983). *Mol. Pharmacol.* **23**, 779–785.
192. Vannice, J. L., Ringold, G. M., McLean, J. W., and Taylor, J. M. (1983). *DNA* **2**, 205–212.
193. Nawa, K., Nakamura, T., Kumatori, A., Noda, C., and Ichihara, A. (1986). *J. Biol. Chem.* **261**, 16883–16888.
194. Schmid, E., Schmid, W., Jantzen, M., Mayer, D., Jastorff, B., and Schutz, G. (1987). *Eur. J. Biochem.* **165**, 499–506.
195. Gomez-Lechon, M. J., Garcia, M. D., and Castell, J. V. (1983). *Hoppe-Seyler's Z. Physiol. Chem.* **364**, 501–508.
196. Guzelian, P. S., Lindblad, W. J., and Diegelmann, R. F. (1984). *Gastroenterology* **86**, 897–904.
197. Weiner, F. R., Czaja, M. J., Jefferson, D. M., Giambrone, M.-A., Tur-Kaspa, R., Reid, L. M., and Zern, M. A. (1987). *J. Biol. Chem.* **262**, 6955–6958.
198. Lin, Q., Blaisdell, J., O'Keefe, E., and Earp, H. S. (1984). *J. Cell. Physiol.* **119**, 267–272.
199. Marceau, N., Goyette, R., Valet, J. P., and Deschenes, J. (1980). *Exp. Cell Res.* **125**, 497–502.
200. Marceau, N., Goyette, R., Guidoin, R., and Antakly, T. (1982). *Scanning Electron. Microsc.*, Part 2, 815–823.
201. Silver, G., Reid, L. M., and Krauter, K. S. (1990). Dexamethasone-mediated regulation of 3-methylcholanthrene induced cytochrome P_{450d} mRNA accumulation in primary rat hepatocyte cultures. *J. Biol. Chem.* **256**: 3134–3138.
202. Vannice, J. L., Taylor, J. M., and Ringold, G. M. (1984). *Proc. Natl. Acad. Sci. U.S.A.* **81**, 4241–4245.
203. Klein, E. S., Reinke, R., Feigelson, P., and Ringold, G. M. (1987). *J. Biol. Chem.* **262**, 520–523.
204. Zvibel, I., Halay, E., and Reid, L. M. (1991). Heparin/hormonal regulation of autocrine growth factor mRNA sysnthesis and abundance: Relevance to clonal growth of tumors. *Mol. & Cell. Biol.* (January issue; in press).
205. Hirata, K., Yoshida, Y., Shiramatsu, K., Freeman, A. E., and Hayasaka, H. (1983). *Exp. Cell Biol.* **51**, 121–129.
206. Hirata, K., Usui, T., Koshiba, H., Maruyama, Y., Oikawa, I., Freeman, A. E., Shiramatsu, K., and Hayasaka, H. (1983). *Gann* **74**, 687–692.

207. Marceau, N., Goyette, R., Pelletier, G., and Antakly, T. (1983). *Cell. Mol. Biol.* **29,** 421–435.
208. Narita, M., Jefferson, D. M., Miller, E. J., Clayton, D. F., Rosenberg, L., and Reid, L. M. (1985). In "Growth and Differentiation of Cells in Defined Environments." (H. Murakami, I. Yamane, J. P. Mather, D. B. Barnes, and G. H. Sato, eds.) Springer Verlag, New York. pp. 89–96.
209. Tomomura, A., Sawada, N., Sattler, G. L., Kleinman, H. K., and Pitot, H. C. (1987). *J. Cell. Physiol.* **130,** 221–227.
210. Sawada, N., Tomomura, A., Sattler, C. A., Sattler, G. L., Kleinman, H. K., and Pitot, H. C. (1986). *Exp. Cell Res.* **167,** 458–470.
211. Kawakami, H., and Terayama, H. (1980). *Biochim. Biophys. Acta.* **599,** 301–314.
212. Kawakami, H., and Terayama, H. (1981). *Biochim. Biophys. Acta* **646,** 161–168.
213. Nakamura, T., Nakayama, Y., and Ichihara, A. (1984). *J. Biol. Chem.* **259,** 8056–8058.
214. Spray, D. C., Fujita, M., Saez, J. C., Choi, H., Watanabe, T., Hertzberg, E., Rosenberg, L. C., and Reid, L. M. (1987). *J. Cell Biol.* **105,** 541–551.
215. Fujita, M., Spray, D. C., Choi, H., Saez, J. C., Watanabe, T., Rosenberg, L. C., Hertzberg, E. L., and Reid, L. M. (1987). *Hepatology* **7,** 1S–9S.
216. Saez, J., Gregory, W., Watanabe, T., Dermictzel, R., Hertzberg, E., Reid, L. M., and Spray, D. (1989). cAMP delays disappearance of gap junctions between pairs of rat hepatocytes in primary culture. *Amer. J. Phys.* **257,** C1–C11.
217. Rojkind, M., Gatmaitan, Z., Mackensen, S., Giambrone, M.-A., Ponce, P., and Reid, L. M. (1980). *J. Cell Biol.* **87,** 255–263.
218. Muschel, R., Khoury, G., and Reid, L. M. (1986). *Mol. Cell. Biol.* **6,** 337–341.
219. Kleinman, H. K., McGarvey, M. L., Hassell, J. R., and Martin, G. R. (1983). *Biochemistry* **22,** 4969–4974.
220. Martin, G. R., Kleinman, H. K., Terranova, V. P., Ledbetter, S., and Hassell, J. R. (1984). *Ciba Found. Symp.* **108,** 97–212 (review).
221. Hadley, M. A., Byers, S. W., Suarez-Quian, C. A., Kleinman, H. K., and Dym, M. (1985). *J. Cell Biol.* **101,** 1511–1522.
222. Grant, D. S., Kleinman, H. K., Leblond, C. P., Inoue, S., Chung, A. E., and Martin, G. R. (1985). *Am. J. Anat.* **174,** 387–398.
223. Inoue, S., and Leblond, C. P. (1985). *Am. J. Anat.* **174,** 373–386.
224. Kleinman, H. K., McGarvey, M. L., Hassell, J. R., Star, V. L., Cannon, F. B., Laurie, G. W., and Martin, G. R. (1986). *Biochemistry* **25,** 312–318.
225. Matuo, Y., Nishi, N., Matsui, S., Sandberg, A. A., Isaacs, J. T., and Wada, F. (1987). *Cancer Res.* **47,** 188–192.
226. Chodak, G. W., Shing, Y., Borge, M., Judge, S. M., and Klagsbrun, M. (1986). *Cancer Res.* **46,** 5507–5510.
227. Hauschka, P. V., Mavrakos, A. E., Isfrati, M. D., Doleman, S. E., and Klagsbrun, M. (1986). *J. Biol. Chem.* **261,** 12665–12674.
228. Harper, J. W., Strydom, D. J., and Lobb, R. R. (1986). *Biochemistry* **25,** 4097–4103.
229. Gautschi, P., Frater-Schroder, M., and Bohlen, P. (1986). *FEBS Lett.* **204,** 203–207.
230. Sakai, A., Ebina, T., and Ishida, N. (1986). *Arch. Virol.* **90,** 73–85.
231. Lagente, O., Diry, M., and Courtois, Y. (1986). *FEBS Lett.* **202,** 207–210.
232. Lobb, R. R., Harper, J. W., and Fett, J. W. (1986). *Anal. Biochem.* **154,** 1–14.
233. Strydom, D. J., Harper, J. W., and Lobb, R. R. (1986). *Biochemistry* **25,** 945–951.
234. Lobb, R. R., Sasse, J., Sullivan, R., Shing, Y., D'Amore, P., Jacobs, J., and Klagsbrun, M. (1986). *J. Biol. Chem.* **261,** 1924–1928.
235. Lobb, R. R., Alderman, E. M., and Fett, J. W. (1985). *Biochemistry* **24,** 4969–4973.
236. Lobb, R. R., Rybak, S. M., St.-Clair, D. K., and Fett, J. W. (1986). *Biochem. Biophys. Res. Commun.* **139,** 861–867.

237. Atha, D. H., Stephens, A. W., Rimon, A., and Rosenberg, R. D. (1984). *Biochemistry* **23**, 5801–5812.
238. Marcum, J. A., and Rosenberg, R. D. (1984). *Biochemistry* **23**, 1730–1737.
239. Marcus, J. A., and Rosenberg, R. D. (1985). *Biochem. Biophys. Res. Commun.* **126**, 365–372.
240. Rosenberg, R. D., and Lam, L. (1979). *Proc. Natl. Acad. Sci. U.S.A.* **76**, 1218–1222.
241. Aiken, M., Ciaglowski, R. E., and Walz, D. A. (1986). *Arch. Biochem. Biophys.* **250**, 257–262.
242. Couchman, J. R., Caterson, B., Christner, J. E., and Baker, J. R. (1984). *Prog. Clin. Biol. Res.* **151**, 31–46.
243. Fougene-Deschatrette, C., Schimke, R. T., Weil, D., and Weiss, M. C. (1984). *J. Cell Biol.* **99**, 497–502.
244. Razzouk, C., McManus, M. E., Hayashi, S., Schwartz, D., and Thorgeirsson, S. S. (1985). *Biochem. Pharmacol.* **34**, 1536–1542.
245. Nakabayashi, H., Taketa, K., Yamane, T., Oda, M., and Sato, J. (1985). *Cancer Res.* **45**, 6379–6383.
246. Chasserot-Golaz, S., Flaig, C., and Beck, G. (1985). *Cancer Biochem. Biophys.* **8**, 95–101.
247. Babiss, L. E., Bennett, A., Friedman, J. M., and Darnell, J. E., Jr. (1986). *Proc. Natl. Acad. Sci. U.S.A.* **83**, 6504–6508.
248. Corral, M., Defer, N., Paris, B., Raymondjean, M., and Corcos, D. (1986). *Cancer Res.* **46**, 5119–5124.
249. Wu, G. Y., and Wu, C. H. (1986). *J. Biol. Chem.* **261**, 16834–16837.
250. Lambert, M. A., Simard, L. R., Ray, P. N., and McInnes, R. R. (1986). *Mol. Cell. Biol.* **6**, 1722–1728.
251. Poli, V., Altruda, F., and Silengo, L. (1986). *J. Biochem.* **35**, 355–360.
252. Sakuma, K., Cook, J. R., Smith, C. L., and Chiu, J. F. (1987). *Biochem. Biophys. Res. Commun.* **143**, 447–453.
253. Zannis, V. I., Breslow, J. L., SanGiacomo, T. R., Aden, D. P., and Knowles, B. B. (1981). *Biochemistry* **20**, 7089–7096.
254. Knowles, B., Howe, C., and Aden, D. (1980). *Science* **209**, 497–498.
255. Gorman, C., Moffat, L., and Howard, B. (1982). *Mol. Cell. Biol.* **2**, 1044–1054.
256. Karlsson, S., and Nienhuis, A. W. (1985). *Annu. Rev. Biochem.* **54**, 1071–1108.
257. Hamer, D. H. (1986). *Annu. Rev. Biochem.* **55**, 913–951.
258. Maxson, R., Cohn, R., and Kedes, L. (1983). *Annu. Rev. Genet.* **17**, 239–277.
259. Hammer, R. E., Krumlauf, R., Camper, S. A., Brinster, R. L., and Tilghman, S. M. (1987). *Science* **235**, 53–58.
260. Widen, S. G., and Papconstantinou, J. (1986). *Proc. Natl. Acad. Sci. U.S.A.* **83**, 8196–8200.
261. Godbout, R., Ingram, R., and Tilghman, S. M. (1986). *Mol. Cell. Biol.* **6**, 477–487.
262. Ciliberto, G., Dente, L., and Cortese, R. (1985). *Cell (Cambridge, Mass.)* **41**, 531–540.
263. Larsen, P. R., Harney, J. W., and Moore, D. D. (1986). *Proc. Natl. Acad. Sci. U.S.A.* **83**, 8283–8287.
264. Palmiter, R. D., and Brinster, R. L. (1986). *Annu. Rev. Genet.* **20**, 465–499.
265. Hesse, J. E., Nickol, J. M., Lieber, M. R., and Gelsenfeld, G. (1986). *Proc. Natl. Acad. Sci. U.S.A.* **83**, 4312–4316.
266. Tur-Kaspa, R., Teicher, L., Levine, B. J., and Skoultchi, A. I., and Shafritz, D. A. (1986). *Mol. Cell. Biol.* **6**, 716–718.
267. Reid, L. (1990). *Current Opinion in Cell Biology* **2**, 121–130.

8

Receptor–Ligand Interactions: Role in Cancer Invasion and Metastasis

MARK E. SOBEL AND LANCE A. LIOTTA
Laboratory of Pathology
National Cancer Institute
Bethesda, Maryland 20892

INTRODUCTION

Metastasis is the major cause of morbidity and death for patients with solid tumors. It is a complex multistep process in which a tumor cell must evade host defenses while it (a) detaches from the primary tumor, (b) invades the primary tumor border and adjacent host tissue barriers, (c) intravasates the vascular wall or lymphatic channel, (d) survives the mechanical stress of the circulatory system, (e) extravasates the vascular wall, (f) enters an organ parenchyma, and (g) colonizes at that distant site. Our long-term goal is to delineate possible genetic changes which are associated with the acquisition of the ability of a tumor cell to metastasize. Elucidation of genetic mechanisms associated with the "metastatic phenotype" should define specific differences between metastatic cells and their more normal counterparts. Identification of genetic changes associated with metastatic behavior may lead to the development of new methodologies (e.g., manipulation of specific proteins, targeted monoclonal antibodies, *in situ* hybridization of cDNA probes) for diagnostic and therapeutic approaches to neoplastic disease.

One of the approaches we are using to study the molecular genetics of metastasis is to clone specific genes which encode proteins thought to be involved in the pathophysiology of metastasis. Our working hypothesis is based on the belief that the interaction of the tumor cell with the extracellular matrix, and in particular with the basement membrane through which it must traverse at various stages of the metastatic process, plays an important role in determining its invasive properties (1). We have placed particular emphasis on tumor cell–receptor interactions with the basement membrane.

BASEMENT MEMBRANE AND THE THREE-STEP HYPOTHESIS OF INVASION

Basement membranes are thin extracellular matrices which separate epithelia and endothelia from their underlying connective tissue (2,3). They are insoluble, continuous, but flexible structures which are impermeable to large proteins (4). Basement membranes can become focally permeable to cell movement during specific processes such as tissue healing and remodeling, inflammation, and neoplasia (5). We have proposed a three-step hypothesis describing the sequence of biochemical events which occur during tumor cell invasion of the basement membrane (6). The first step is tumor cell attachment to the basement membrane via cell surface receptors. The anchored tumor cell then secretes enzymes, or induces host cells to secrete enzymes, which can locally degrade the basement membrane. Finally, the third step is tumor cell locomotion via the production of, and response to, motility factors.

COMPONENTS OF BASEMENT MEMBRANE

Major components of basement membrane include type IV collagen, heparan sulfate proteoglycan, and laminin. Type IV collagen is a basement membrane-specific collagen (7) thought to be arranged in a netlike structure (8) which provides the basement membrane with tensile strength. Heparan sulfate proteoglycan consists of a basement membrane-specific core protein linked to glycosaminoglycan side chains, and it may function to block passage of anionic molecules through the basement membrane (9).

Laminin is a large and complex glycoprotein which consists of three major types of disulfide-bonded glycosylated polypeptide chains. Two "B" chains have an approximate molecular mass of 200 kilodaltons (kDa), and a large "A" chain has an approximate mass of 400 kDa. The molecule has been visualized by rotary shadowing electron microscopy as a cruciform molecule with three short arms and one long arm. All the arms have globular end regions. The specialized structure of laminin may contribute to its multiple biological functions, which include cell attachment, cell spreading, mitogenesis, neurite outgrowth, morphogenesis, and cell movement (1,10). Laminin also plays a role in the architecture of the basement membrane via its ability to bind to multiple membrane components, including type IV collagen, heparan sulfate proteoglycan, and entactin (11).

LAMININ RECEPTOR

The interaction of the tumor cell with laminin has been under intensive investigation in our laboratory. This interaction involves a receptor which plays a role in the first (attachment) step of the basement membrane invasion process described above. In 1983, we identified a cell surface receptor in human breast carcinoma cells which exhibits specific, saturable binding to radiolabeled laminin (12,13). Scatchard analysis of laminin binding was linear, with a relatively high affininty binding constant of 2 nM. The number of laminin receptors per cell was estimated at 100,000. Using laminin affinity chromatography, a 67,000-dalton laminin receptor was isolated from metastatic murine BL6 melanoma cells (13); in other laboratories the receptor was isolated from muscle cells (14) and fibrosarcoma cells (15).

The method used to isolate the laminin receptor involves isolation of tumor cell plasma membranes which are then extracted with detergent, followed by isolation by laminin affinity chromatography (13,16). The high affinity of the laminin receptor for its ligand has been used in the strategy to isolate purified laminin receptor in chemical amounts (16). This has involved several high salt washes on a laminin affinity column, followed by elution of specific protein with 1 M NaCl (16). When isolated in the presence of the detergent Nonidet P-40 (NP-40), the receptor has a mobility on reduced sodium dodecyl sulfate (SDS)–polyacrylamide gels of a polypeptide with a mass slightly under 70,000 daltons. The receptor contains an intrachain disulfide bond, as evidenced by a faster migration on nonreduced SDS–polyacylamide gels (17).

ROLE OF LAMININ AND LAMININ RECEPTOR IN TUMOR CELL METASTASIS

Several investigators have shown that carcinoma cells may use laminin during the process of attachment and metastasis (18–24). Tumor cells exposed to laminin produce 10 times more experimental metastases than control cells (18). Highly malignant cell lines express laminin or lamininlike moieties on their surface (19,20). When tumor cells were preincubated with either laminin or protease-generated laminin fragments, there were significant effects on metastasis (21). Preincubation with whole laminin increased the number of metastases, whereas a small chymotrypsin-generated fragment of laminin inhibited formation of metastases. This suggests that laminin bound to cell surfaces via the laminin receptor stimulates tumor metastasis, in part due to the interaction between the bound laminin and the other components of the basement membrane of the host (e.g., type IV collagen or proteoglycan), an effect that requires the presence of the globular end regions of the laminin molecule. The chymotrypsin-generated fragment, on the other hand, inhibited metastasis because the laminin receptors were occupied by regions of the laminin molecule not capable of binding to type IV collagen or heparan sulfate proteoglycan in the basement membranes of the host.

The number and/or degree of exposure of laminin receptors on the cell surface may be altered in human carcinomas (5,22). Breast carcinoma tissue contains a higher number of laminin receptors than benign breast tissue (23).

MONOCLONAL ANTIBODIES TO HUMAN LAMININ RECEPTOR

A panel of monoclonal antibodies against the laminin receptor extracted from human breast carcinoma plasma membranes has been developed (24). The antibodies fall into two classes, based on their ability to inhibit laminin binding to the receptor. When added together with labeled laminin, the LR1 group of antibodies, but not the LR2 group, inhibited specific laminin binding to human breast carcinoma cells in a dose-dependent manner (24). Furthermore, the LR1 group of monoclonal antibodies was found to inhibit the attachment of human breast carcinoma cells and human melanoma cells to the surface of authentic human amnion basement membrane (25). Presumably, the LR1 monoclonal antibodies recognize the laminin binding domain of the receptor, or another domain with which the binding domain interacts. Both the LR1 and LR2 groups of antibodies recognize the same size protein on immunoblots of human breast carcinoma cells (24).

ISOLATION OF HUMAN LAMININ RECEPTOR cDNA

The isolation of protein encoding genes from recombinant cDNA libraries can be achieved by using antibodies to detect antigen produced by specific recombinants. In a λgt11 library, the cDNAs are inserted into the *Eco*RI site of the phage vector in such a way that they are fused to the end of truncated *lac*Z (β-galactosidase) gene (26–28). Depending on orientation and reading frame, one in six phages should synthesize a hybrid protein consisting of a shortened β-galactosidase fused to the protein encoded by the cDNA.

We set out to obtain a cDNA clone of the human laminin receptor in such an expressing library (29). Our goals were to elucidate the primary structure of the laminin receptor via DNA sequencing, to predict functional domains of the receptor, and to use the cDNA as a probe to determine control elements regulating laminin receptor gene expression. Figure 1 presents a diagramatic summary of the approach we took. We started with a cDNA library which was constructed at Meloy Laboratories (Springfield, Virginia) by Dr. George Ricca and Dr. William Drohan. This library was constructed using RNA template from human endothelial cells (30). The library was amplified, and 1.5 million plaques were screened at Meloy Laboratories by Dr. Michael Jaye, using monoclonal antibody 2H5. This antibody is one of the LR1 group of monoclonal antibodies which blocks laminin binding to the laminin receptor (24). Six plaques were initially selected. After plaque purification, all six clones showed an intense reaction with the antilaminin receptor monoclonal antibody but showed no reactivity toward a class-matched (IgM) monoclonal antibody directed against α-amylase, nor toward antibodies directed against laminin or against proteins similar in size to laminin receptor such as albumin (Fig. 2).

OVERLAPPING DOMAIN OF LAMININ RECEPTOR cDNA INSERTS

The overlapping domain of the cDNAs that we isolated from the library should encode the epitope of the monoclonal antibody 2H5. Since that antibody inhibited binding of laminin to the laminin receptor, it was likely that the overlapping DNA sequence of the six clones would encode the ligand binding site of the receptor. Restriction endonuclease mapping of the six clones, designated λELR1–6, was performed using single, double, and triple digests of *Sac*I, *Kpn*I, *Hin*dIII, and *Eco*RI. These experiments, summarized in Fig. 3, indicate that the left *Eco*RI site in all six clones was missing. The sizes of the cDNA inserts ranged from 450 to 950 base pairs. The inserts of λELR2 and λELR3 were identical in size, suggesting that they are duplicate clones arising from amplification of the library. All of the clones shared a common 450-base pair DNA sequence, extending from just to the 5′ of the *Sac*I site to the 3′ end of each cDNA insert. This region apparently encodes the antigenic domain recognized by the monoclonal antibody, possibly the laminin binding domain of the laminin receptor (see Fig. 4).

Southern blot experiments were performed in which *Kpn*I–*Eco*RI and *Kpn*I–*Sac*I double digests of the DNA from each of the six recombinant phage were stringently hybridized to a *Pst*I–*Sph*I restriction fragment iso-

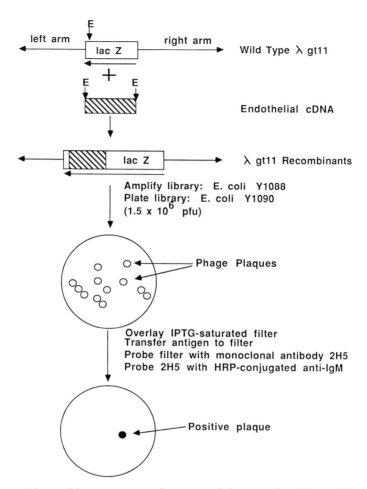

Fig. 1. Scheme of the construction and screening of a human endothelial λgt11 library. The top line shows a map of the wild-type λgt11 phage. The phage has two arms, delineated by an EcoRI (E) restriction site. The *lacZ* gene which encodes β-galactosidase is shown in the box. When the phage is digested with EcoRI, two arms are produced. Endothelial cell cDNA was synthesized, and oligonucleotides containing EcoRI sites were ligated to the ends (shown in the hatched box). When the EcoRI-linkered cDNA was digested with EcoRI, staggered EcoRI sites were available for ligation to the EcoRI-digested wild-type phage arms. The resulting recombinant phage contained a fused region which encoded a fusion protein. The amino domain of the fusion protein encodes most of the bacterial β-galactosidase gene (*lacZ*), while the carboxy domain encodes the endothelial cell protein. Since the fused gene is under the regulatory control of the *lac* operon in the phage, it can be induced to produce large quantities of the fusion protein by the inducer isopropyl-β-D-thiogalactopyranoside (IPTG). This was accomplished by infecting *Escherichia coli* Y1090 cells with approximately 1.5 million plaque-forming units (pfu) of the λgt11 phage library. A nitrocellulose filter which was saturated with the inducing agent IPTG was overlayed on the plaques. As the phage produced fusion protein, it was bound to the nitrocellulose filter. The resulting antigen was probed with the monoclonal antibody 2H5. Binding of the antilaminin receptor antibody was detected using horseradish peroxidase (HRP)-conjugated affinity purified rabbit anti-mouse IgM. Positive plaques were detected by their blue color.

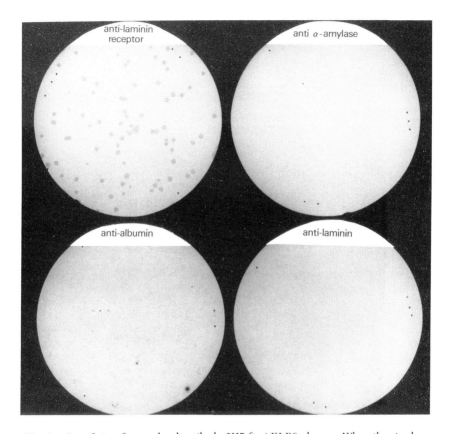

Fig. 2. Specificity of monoclonal antibody 2H5 for λELR6 plaques. When the six clones were purified, filters were prepared containing approximately 50 recombinant plaques, each producing recombinant β-galactosidase–laminin receptor fusion protein. Four filters were prepared from each plate and reacted with antilaminin receptor monoclonal antibody 2H5, anti-α-amylase, antialbumin, or antilaminin antibodies. Binding of antibody was detected using HRP-conjugated second antibody.

lated from λELR4 (29). These results confirmed that the six clones contained a 450-base pair common sequence.

NUCLEOTIDE AND DEDUCED AMINO ACID SEQUENCE OF THE 2H5 EPITOPE OF HUMAN LAMININ RECEPTOR

The nucleotide sequence of the cDNA insert of the λELR4 was determined (29). The sequence included a stop codon (UAA), and the 3' untranslated region extended for 66 base pairs, including a canonical polyadenyla-

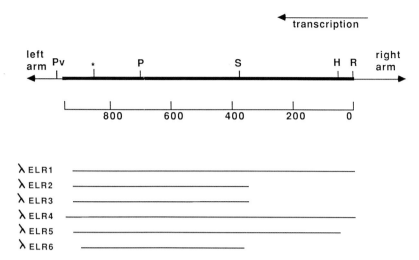

Fig. 3. Overlapping laminin receptor cDNA inserts. The top line is a diagram of the largest clone, λELR4, designating internal restriction sites for *Hin*dIII (H), *Sac*I (S), and *Pst*I (P), as well as the beginning of a poly(A) tail (*). The *Eco*RI (R) site linking the cDNA insert to the right arm of λgtll is shown; however, the left *Eco*RI site is not present in any of the six clones. A *Pvu*II (Pv) site close to the linker to the left arm was used in subcloning experiments (29). The direction of transcription from the *lacZ* gene is also shown. Below the restriction map is a numerical scale in base pairs designating the length of the cDNA insert of λELR4 from the right *Eco*RI site. The size and overlapping regions of the six cDNA inserts are shown below the numerical scale.

tion signal (AAUAAA), 17 bases upstream from a long poly(A) stretch. Thus, we can narrow down the largest possible extent of the 2H5 epitope, as defined by the common cDNA region of λELR1–6 which encodes protein, to 393 base pairs, or 131 amino acids (Fig. 4). The deduced amino acid sequence of the 2H5 epitope is presented in Fig. 5.

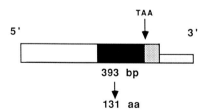

Fig. 4. Common domain of λELR1–6 cDNA inserts (putative laminin binding domain). The predicted size of a full-length mRNA for laminin receptor is shown, with the black box designating the common domain of the six recombinant laminin receptor clones selected on the basis of antigenicity to the 2H5 monoclonal antibody. The common domain also includes a translation stop codon and a 3′ untranslated region, as well as a poly(A) tail. Thus, the common domain of the six clones which encodes protein is 393 base pairs, encoding a 131 amino acid region.

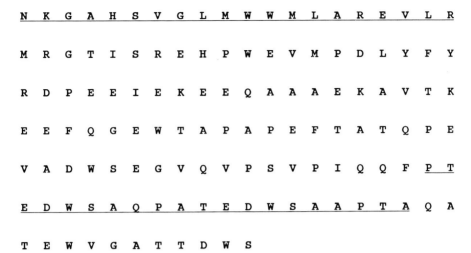

Fig. 5. Amino acid sequence of the 2H5 epitope of the laminin receptor. The nucleotide sequence of the human laminin receptor cDNA clones was determined (29), and the amino acid sequence was derived. The one-letter amino acid code is shown. The underlined regions designate synthetic peptides used in subsequent experiments that are areas of high reverse turn probability as described in the text.

We were particularly interested in predicting the sequence of cyanogen bromide fragments from the cDNA sequence, since the sequence of several such fragments of purified laminin receptor from human placenta had been determined (29). A unique octapeptide, MLAREVLR, matched completely (Fig. 6). It is somewhat unusual for the methionine residue of cyanogen bromide fragments to be clearly eluted from the protein sequencer in large quantity, as occurred with the laminin receptor octapeptide. We noted that the cDNA-predicted sequence of this region (Fig. 5) includes a Met-Trp-Trp tripeptide immediately amino terminal to the octapeptide. Cleavage of tryptophanyl peptide bonds by cyanogen bromide has been reported (31). We therefore believe that cyanogen bromide digestion of placental laminin receptor resulted in the cleavage of the Trp-Trp peptide bond, resulting, in fact, in a nonapeptide, with the tryptophanyl residue retained on the protein sequencer. This argument is strengthened by our finding a very short peptide derived from cyanogen bromide digestion of purified tumor laminin receptor containing only one tryptophanyl residue. The clean pattern of the "nonapeptide" microsequence, with no trail after the final arginine, suggests that an arginine–methionine peptide bond was cleaved by the cyanogen bromide treatment. Thus, complete homology between the cDNA clone and purified laminin receptor extends over a 12 amino acid region, MWWMLAREVLRM.

```
cDNA      ATG TGG TGG ATG CTG GCT CGG GAA GTT CTG CGC ATG
 ↓                 ↓                                  ↓
derived   met trp trp met leu ala arg glu val leu arg met

CB peptide            met leu ala arg glu val leu arg
```

Fig. 6. cDNA-derived and cyanogen bromide peptides. The top line shows the cDNA sequence derived from the laminin receptor clone (29). The derived amino acid sequence is shown below, with arrows designating potential cyanogen bromide cleavage sites, as described in the text. The third line shows the sequence of an authentic cyanogen bromide-derived fragment from purified human placental laminin receptor (29).

SYNTHETIC PEPTIDES OF THE PUTATIVE LIGAND BINDING DOMAIN OF LAMININ RECEPTOR

The amino acid sequence of the 2H5 epitope was analyzed for the probability of reverse turn occurrence using the parameters of Chou and Fasman (32). Such secondary structures are often found on the external surface of proteins which are associated with receptor–ligand binding domains (33). Two different 20-mer peptides with high probability for β-turn formation were synthesized (see underlined areas of Fig. 5). Polyclonal antisera to these peptides was demonstrated by enzyme-linked immunosorbent assay (ELISA) (34). Using such antisera in immunofluorescence staining experiments, it was demonstrated that the carboxy-terminal domain of the laminin receptor is localized on the outside of the cell membrane of tumor cell lines (34). It was also shown that antiserum raised against the synthetic peptide closest to the carboxy terminus (PTEDWSAQPATEDWSAAPTA) had an inhibitory effect on laminin-mediated attachment of A2058 human melanoma cells (34).

Haptotaxis is the directed migration of cells along a gradient of substratum-bound insolubilized factor (35). It can be distinguished from chemotaxis, in which the cell responds to a soluble gradient of attractant. The antisera against the synthetic peptides shown in Fig. 5 inhibited the laminin haptotaxis of the human melanoma cells (34). In contrast, haptotaxis to fibronectin was not affected. These results support the concept that different receptors are involved in laminin versus fibronectin haptotaxis.

EXPRESSION OF LAMININ RECEPTOR mRNA

The amount and surface distribution of laminin receptor are different in various carcinomatous human tissues (23). In general, malignant tissues have more unoccupied laminin receptors on their cell surface and bind to more laminin than do their more normal counterparts. To determine if laminin receptor mRNA levels play a role in determining the amount of cell surface laminin receptors available for ligand binding, we performed a Northern hybridization experiment (Fig. 7). The level of hybridized laminin receptor mRNA from a variety of human breast carcinoma-derived cells differed. In

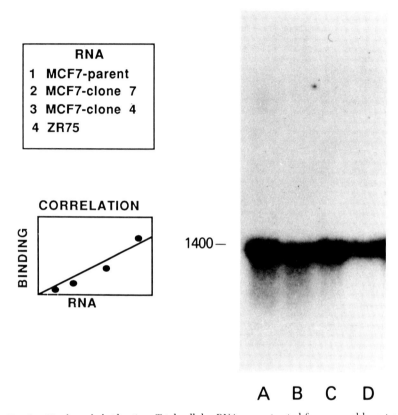

Fig. 7. Northern hybridization. Total cellular RNA was extracted from several breast carcinoma cell lines (29), separated on methylmercury/agarose gels, and transferred to diazobenzyloxymethyl cellulose paper as described (29). The filter was hybridized to a ^{32}P-labeled EcoRI–SacI fragment of λELR4. The blot was counterscreened with other cDNA probes to ensure that equal amounts of RNA were transferred (29). Lane 1, MCF-7 parent; lane 2, MCF-7-clone 3E5; lane 3, MCF-7-clone 5H7; lane 4, ZR-75. Various time exposures of autoradiographs were measured densitometrically to determine a linear response range. The lowest amount of RNA hybridized in the series was assigned the number 1.0, and all other values were calculated relative to that value. The number of laminin receptors per cell for each cell line was calculated from Scatchard plots of specifically bound ^{125}I-labeled laminin to logarithmic growth phase cells as described (29). A linear regression analysis performed on the Northern and Scatchard data revealed a correlation coefficient of 0.97.

particular, there was greater hybridization to RNA from the metastatic breast carcinoma cell line MCF-7 than to RNA from the breast carcinoma cell line ZR-75 (29). The level of hybridized RNA correlated directly with laminin binding assays (Fig. 7), suggesting that the amount of laminin receptor mRNA available for the biosynthesis of receptor may be a rate-limiting control step in the regulation of cellular attachment to laminin in the basement membrane. The correlation between laminin receptor mRNA levels and the number of laminin receptors determined by Scatchard analysis of laminin receptor binding assays was also extended to other carcinomatous cell lines, including renal and pancreatic cell lines (29).

CONCLUSIONS

We have been studying the interaction between tumor cells and laminin to understand the mechanism by which the metastatic tumor cell interacts with the basement membrane. A critical step in the metastatic process involves the attachment of the tumor cell to the basement membrane. The high affinity laminin receptor appears to play an essential role in this process.

We used an antilaminin receptor monoclonal antibody that specifically blocks the binding of cells or plasma membranes to laminin or to amnion basement membrane to isolate an expressing clone of the human laminin receptor from a cDNA library. The laminin receptor epitope recognized by the monoclonal antibody presumably contains the laminin binding site or interacts closely with the ligand binding site. The overlapping domain of the cDNAs we isolated from a human endothelial λgt11 library should, therefore, encode the epitope of the laminin receptor involved in ligand binding. Sequence analysis of the laminin receptor cDNA clones allowed us to define a region of no more than 131 amino acids that may be the laminin binding domain of the receptor. We subsequently synthesized oligopeptides derived from the cDNA sequence and showed that antisera to the peptides could inhibit haptotaxis of tumor cells to laminin.

Studies on the levels of laminin receptor mRNA from a variety of carcinoma-derived cell lines suggest that the increased content of laminin receptors on highly metastatic cells may be due to increased biosynthesis of the laminin receptor in the more metastatic tissue. The amount of laminin receptor mRNA available for the biosynthesis of receptor may be a rate-limiting control step in the regulation of laminin-mediated cellular attachment to basement membranes and, hence, in the metastatic process.

ACKNOWLEDGMENTS

We are grateful to the following former members of the Laboratory of Pathology, Drs. U. Wewer, C. N. Rao, S. Barsky, G. Bryant, and G. Taraboletti, whose data contributed significantly to our current understanding of the laminin receptor. We are also grateful to Drs. G. Ricca, M. Jaye, and W. Drohan from Meloy Laboratories for their assistance in the isolation of the cDNA clone for laminin receptor, as well as to Dr. J. Coligan and Mr. M. Raum for protein sequence data. We also thank Ms. A. Claysmith for hybridization and nucleotide sequence data.

REFERENCES

1. Liotta, L. A., Rao, C. N., and Wewer, U. M. (1986). *Annu. Rev. Biochem.* **55**, 1037–1057.
2. Timpl, R., and Martin, G. R. (1982). In "Immunochemistry of the Extracellular Matrix" (H. Furthmayr, ed.), Vol. 2. pp. 120–150. CRC Press, Boca Raton, Florida.
3. Stanley, J. R., Woodley, D. T., Katz, J. I., and Martin, G. R. (1982). *J. Invest. Dermatol.* **79**, 69s–74s.
4. Vracko, R. (1974). *Am. J. Pathol.* **77**, 313–346.
5. Liotta, L. A., Rao, C. N., and Barsky, S. H. (1983). *Lab. Invest.* **49**, 636–649.
6. Liotta, L. A. (1986). *Cancer Res.* **46**, 1–7.
7. Kefalides, N. A., Alper, R., and Clark, C. C. (1979). *Int. Rev. Cytol.* **61**, 167–171.
8. Timpl, R., Wiedemann, H., Van Delden, V., Furthmayr, H., and Kuhn, K. (1981). *Eur. J. Biochem.* **120**, 203–212.
9. Kanwar, Y. S., and Farquhar, M. G. (1979). *Proc. Natl. Acad. Sci. U.S.A.* **76**, 4493–4496.
10. Timpl, R., Johansson, S., Van Delden, V., Oberbaumer, I., and Hook, M. (1983). *J. Biol. Chem.* **258**, 8922–8928.
11. Liotta, L. A., Wewer, U., Rao, N. C., Schiffmann, E., Stracke, M., Guirguis, R., Thorgeirrsson, U., Muschel, R., and Sobel, M. (1987). *Anti-Cancer Drug Des.* **2**, 195–202.
12. Terranova, V. P., Rao, C. N., Kalebic, T., Margulies, I. M., and Liotta, L. A. (1983). *Proc. Natl. Acad. Sci. U.S.A.* **80**, 444–448.
13. Rao, C. N., Barsky, S. H., Terranova, V. P., and Liotta, L. A. (1983). *Biochem. Biophys. Res. Commun.* **111**, 804–808.
14. Lesot, H., Kuhl, U., and von der Mark, K. (1983). *EMBO J.* **2**, 861–865.
15. Malinoff, H. L., and Wicha, M. S. (1983). *J. Cell Biol.* **96**, 1475–1479.
16. Wewer, U. M., Albrechtsen, R., Rao, C. N., and Liotta, L. A. (1986). *Rheumatology* **10**, 451–478.
17. Liotta, L. A. (1987). *Clin. Physiol. Biochem.* **5**, 190–199.
18. Terranova, V. P., Liotta, L. A., Russo, R. G., and Martin, G. R. (1982). *Cancer Res.* **42**, 2265–2269.
19. Varani, I., Lovett, E. J., McCoy, J. P., Shibata, S., Maddox, D., Goldstein, I. J., and Wicha, M. S. (1983). *Am. J. Pathol.* **111**, 27–34.
20. Malinoff, H. L., McCoy, J. P., Varani, J., and Wicha, M. S. (1984). *Int. J. Cancer* **33**, 651–655.
21. Barsky, S. H., Rao, C. N., Williams, J. E., and Liotta, L. A. (1984). *J. Clin. Invest.* **74**, 843–848.
22. Hand, P. H., Thor, A., Schlom, J., Rao, C. N., and Liotta, L. A. (1985). *Cancer Res.* **45**, 2713–2719.

23. Barsky, S. H., Rao, C. N., Hyams, D., and Liotta, L. A. (1984). *Breast Cancer Res. Treat.* **4**, 181–188.
24. Liotta, L. A., Hand, P. H., Rao, C. N., Bryant, G., Barsky, S. H., and Schlom, J. (1985). *Exp. Cell Res.* **49**, 636–649.
25. Togo, S., Wewer, U., Margulies, I., Rao, C. N., and Liotta, L. A. (1985). *In* "Basement Membranes" (S. Shibata, ed.), pp. 325–333. Elsevier, New York.
26. Young, R. A., and Davis, R. W. (1983). *Proc. Natl. Acad. Sci. U.S.A.* **80**, 1194–1198.
27. Young, R. A., and Davis, R. W. (1983). *Science* **222**, 778–782.
28. Huynh, T. V., Young, R. A., and Davis, R. W. (1985). *In* "DNA Cloning: A Practical Approach" (D. M. Glover, ed.), Vol. 1, pp. 49–78. IRL Press, Oxford, England.
29. Wewer, U. M., Liotta, L. A., Jaye, M., Ricca, G. A., Drohan, W. N., Claysmith, A. P., Rao, C. N., Wirth, P., Coligan, J. E., Albrechtsen, R., Mudryj, M., and Sobel, M. E. (1986). *Proc. Natl. Acad. Sci. U.S.A.* **83**, 7137–7141.
30. Jaye, M., McConathy, E., Drohan, W., Tong, B., Deuel, T., and Maciag, T. (1985). *Science* **228**, 882–885.
31. Ozols, J., and Gerard, C. (1977). *J. Biol. Chem.* **252**, 5986–5989.
32. Chou, P. Y., and Fasman, G. D. (1978). *Annu. Rev. Biochem.* **47**, 251–276.
33. Nestor, J. J., Newman, S. R., DeLustro, B., Todaro, G. J., and Screiber, A. B. (1985). *Biochem. Biophys. Res. Commun.* **129**, 226–232.
34. Wewer, U. M., Taraboletti, G., Sobel, M. E., Albrechtsen, R., and Liotta, L. A. (1987). *Cancer Res.* **47**, 5691–5698.
35. Carter, S. B. (1965). *Nature (London)* **208**, 1183–1187.

9
Control of Cell Proliferation by Transforming Growth Factors

H. L. MOSES, J. BARNARD, C. C. BASCOM,
R. D. BEAUCHAMP, R. J. COFFEY,
R. M. LYONS, AND N. J. SIPES
Departments of Cell Biology, Medicine, and Pediatrics
Vanderbilt University School of Medicine
Nashville, Tennessee 37232

INTRODUCTION

Transforming growth factors (TGF) were originally defined by their biological effects on murine fibroblastic cell lines. These effects included induction of morphologic transformation in monolayer culture and stimulation of colony formation in soft agar (reviewed in Ref. 1). The early studies with TGF indicated that they may play a role in neoplastic transformation and led to the purification and cloning of two very important growth regulatory molecules, TGFα and TGFβ. Studies with the purified molecules have demonstrated a very important potential role in normal growth and development. Interestingly, one of these factors, TGFα, is a potent mitogen for most cell types (2) while the other, TGFβ, is the most potent growth inhibitory polypeptide known for most cell types (3,4).

TRANSFORMING GROWTH FACTOR-α

First described by DeLarco and Todaro (5), TGFα was originally termed sarcoma growth factor. It was later shown that the preparation called sarcoma growth factor contained both TGFα and TGFβ (6) and that much of the biological activity originally ascribed to sarcoma growth factor was due to the TGFβ. TGFα was purified to homogeneity by Marquardt *et al.* (7) and is a 50 amino acid molecule that has some sequence and significant structural homology to epidermal growth factor (EGF). The gene for TGFα has been cloned and indicates a precursor of 160 amino acids that is processed in a complex manner to yield the mature molecule (8,9). TGFα binds to the EGF receptor and appears to mediate all of its biological effects through this EGF receptor binding. No evidence has been presented for a TGFα receptor separate from the EGF receptor. The cell culture effects of TGFα are virtually identical to those of EGF (6). However, in *in vivo* and organ culture assays, differences between the biological effects of TGFα and EGF have been observed (reviewed in Ref. 2). These differences were quantitative, with TGFα tending to be more potent than EGF, and no qualitative differences in the biological effects of EGF and TGFα have been reported.

TGFα was originally shown to be produced by murine sarcoma virus-transformed mouse 3T3 cells but not by the nontransformed parent cells (5). It was later found in medium conditioned by human carcinoma cells in culture (10). Shortly thereafter, TGFα was identified in embryonic tissue (11). These findings, in addition to the absence of reports of finding TGFα in normal adult tissue, led to the widespread view that TGFα is an embryonic molecule inappropriately expressed in some cancer cells. It was in conjunction with the discovery of TGFα that the autocrine hypothesis was first published as an explanation for the excessive growth that occurs in neoplastic cells (12).

We have recently demonstrated that TGFα does occur in normal adult cells which comprise the largest organ, the skin (13). In these studies, normal keratinocytes in secondary culture were grown in low calcium, serum-free medium (14). These cells require EGF/TGFα for proliferation and retain the ability to differentiate under high calcium conditions. It was demonstrated that TGFα mRNA and protein are produced by both normal neonatal and adult keratinocytes in culture (13). Additionally, TGFα expression in these cells was found to be dependent on the presence of EGF; both EGF and TGFα induced significant levels of TGFα mRNA. EGF was demonstrated to stimulate the release of TGFα protein into the culture medium. It was not possible to examine for TGFα induction of protein release because of the exogenously added factor which would interfere in the enzyme-linked immunosorbent assay (ELISA) assay. However, TGFα did induce TGFα mRNA, demonstrating autoinduction, and we have speculated that such autoinduction may be a mechanism of signal amplification for finer control of the proliferative response.

Evidence for TGFα mRNA and protein expression *in vivo* was obtained using *in situ* hybridization and immunocytochemistry, respectively (13). This demonstrated that TGFα production does occur in normal adult epithelial cells that are also capable of responding to the factor, suggesting the possibility of normal autocrine regulation of cell proliferation. These studies along with those from other laboratories demonstrating normal autocrine stimulation by platelet-derived growth factor, insulin-like growth factor-I, and interleukin 2 suggest that the autocrine hypothesis should now be altered (reviewed in Ref. 1). It can no longer be considered as a phenomenon occurring only in neoplastic cells producing excessive proliferation, but rather is a normal regulated phenomenon important in normal growth control.

TRANSFORMING GROWTH FACTOR-β

TGFβ was first described as a growth stimulatory molecule by its ability to induce soft agar colony formation of mouse embryo-derived fibroblastic AKR-2B cells (15) and shortly thereafter by its biological effects in combination with EGF on rat fibroblastic NRK cells (16). Although originally described as being produced by neoplastically transformed cells (15,16), it is now known that TGFβ is a highly ubiquitous molecule, and it has been purified from several normal tissues (reviewed in Ref. 17). Platelets, which give rise to the TGFβ found in serum (18), are the most abundant source for purification of TGFβ. The intact, active TGFβ molecule has a molecular weight of 25,000 and is composed of two identical disulfide-linked subunits of 12,000 (19). The gene for TGFβ has been cloned and the amino acid sequence deduced from the cDNA sequence (20). This indicates a subunit of 112 amino acids and suggests a precursor encoded in a 390-residue open reading frame. The precursor is processed by proteolytic cleavage to yield the active molecule (21). The murine TGFβ has also been cloned, and comparison with the human sequence shows an exceptionally high degree of evolutionary conservation (22).

A second TGFβ (called TGFβ2 to distinguish it from the originally described TGFβ, now called TGFβ1) has been identified in porcine platelets and bovine bone (23). TGFβ2 apparently binds to the same receptor as TGFβ1 and has similar biological activities (4,23), although some differences in the biological activities of TGFβ1 and TGFβ2 have been found in hematopoietic stem cells (24). A molecule apparently identical to TGFβ2 has been identified as an immunosuppressive agent produced by glioma cells (25). The growth inhibitor from African green monkey (BSC-1) cells originally described by Holley *et al.* (26) has now been cloned, and the cDNA sequence indicates that it is identical to TGFβ2 (27). The term polyergin has been suggested by these authors to designate this molecule and to distinguish it from the TGFβ1 isolated from human platelets. Other molecules with structural and some sequence homology to TGFβ have been purified or identified by gene cloning and DNA sequencing. These include Müllerian inhibiting substance (28), inhibins (and their β chain dimers, activins) (29), the *Drosophila* decapentaplegic gene (30), and the *Xenopus Vg-1* gene (31). Müllerian inhibiting substance and inhibins/activins have receptors that are distinct from the TGFβ receptor (32,33).

TGFβ has its own specific cell surface receptors which, like the TGFβ molecule itself, are highly ubiquitous (34). Specific binding of ^{125}I-TGFβ to various mesenchymal and epithelial cells in primary and secondary culture and continuous cell lines, both normal and neoplastic, has been reported. The dissociation constants reported have ranged from 25 to 140 pM and receptor number per cell from 10,000 to 40,000 (34–36). The TGFβ receptor is apparently quite different from other growth factor receptors. Recently, at least three types of receptors for TGFβ have been proposed on the basis of chemical cross-linking studies (23). No kinase or other enzymatic activities have been reported for the TGFβ receptor thus far.

TGFβ has been reported to have numerous and diverse biological activities. It is only mitogenic for fibroblastic and selected other mesenchymal cells (3). Data have been presented indicating that TGFβ mitogenesis in monolayer culture is indirect through induction of c-*sis* and autocrine stimulation by endogenous platelet-derived growth factor (37). TGFβ stimulation of soft agar growth has been postulated to be secondary to induction of fibronectin synthesis and release (38). TGFβ is a potent stimulator of extracellular matrix production by increasing synthesis of matrix components including procollagen type I and fibronectin (38–40) and by diminishing matrix degradation through stimulating protease inhibitor (41) and decreasing protease production (42,43). TGFβ also is a very potent chemotactic factor for dermal fibroblast (44). All of these actions probably contribute to the ability of TGFβ to stimulate connective tissue formation *in vivo* (45). TGFβ also inhibits differentiation of adipocytes and myoblasts (46).

We demonstrated that the growth inhibitor originally described by Holley *et al.* (26) from BSC-1 cells (now called polyergin and TGFβ2) is similar to human platelet-derived TGFβ (now called TGFβ1) (4). In addition to demonstrating a similarity between TGFβ1 and TGFβ2, it was further shown that human platelet-derived TGFβ1 was a highly potent inhibitor of the BSC-1 and CCL-64 (mink lung) epithelial cells. This led to studies of the inhibitory effects of TGFβ on a variety of normal cells, demonstrating that TGFβ1 and TGFβ2 are the most potent growth inhibitory polypeptides known for a wide variety of cell types including epithelial, lymphoid, and myeloid cells (for review, see Ref. 47).

TGFβ is a potent growth inhibitor for secondary cultures of human foreskin keratinocytes (3,48). The keratinocytes were found to be reversibly inhibited in their growth by TGFβ, with the majority of cells blocked in the G_1 phase of the cell cycle. Half-maximal inhibition was obtained at 12 pM TGFβ. There was no induction of any of several differentiation markers examined, indicating that the mechanism of growth inhibition is not through induction of terminal differentiation. The keratinocytes were also demonstrated to synthesize and release TGFβ into the medium, with confluent cultures producing as much as 80 pM per 24 hr (48). However, all the detectable TGFβ released was in a latent form detectable only after acid treatment of the medium. Whether the latent TGFβ activates spontaneously or can be activated by the cells with subsequent binding to cell surface receptors is not known. However, since the keratinocytes have receptors for TGFβ (48), are capable of responding to the factor, and secrete relatively large quantities into conditioned medium, the possibility of negative autocrine regulation by TGFβ in normal keratinocytes seems highly probable.

TGFβ is released by cells in culture (including the human keratinocytes) and by platelets in a latent form that is irreversibly activated by acid treatment (48–50). Various treatments known to dissociate hydrogen bonding such as high salt, urea, and detergents also irreversibly activate TGFβ (50). These and other data suggest that the activation of the latent form is separate from the proteolytic processing steps known to be necessary to generate the 25-kDa molecule (21). Processing appears to have already occurred by the time TGFβ is released from the cells. Recent studies on possible physiological mechanisms of activation have demonstrated that plasmin, a ubiquitous serine protease, can activate at least a portion of the TGFβ released by cells in culture (51). Plasmin activation of TGFβ as a physiological mechanism is attractive because of the ubiquity of plasmin and because it has been shown that plasmin induces the endothelial cell type plasminogen activator inhibitor, PAI-1 (41). This would provide a negative feedback control of TGFβ activation.

AUTOCRINE REGULATION OF EPITHELIAL CELLS BY TRANSFORMING GROWTH FACTORS-α AND -β

The data summarized above demonstrate that control of keratinocyte proliferation may involve both positive and negative polypeptide regulators that bind to cell surface membrane receptors. The keratinocytes produce the same peptides that stimulate or inhibit their proliferation. TGFα stimulates proliferation, and the keratinocytes produce TGFα; this production is autoregulated (13). It is postulated that such autoregulation may provide needed amplification of the growth stimulatory signal. TGFβ reversibly inhibits proliferation of keratinocytes which also secrete this factor in a latent form (3,48). The latent material could be activated in the medium or at the cell membrane by proteolytic action (51), resulting in a negative effect on cell proliferation. It is hypothesized that autocrine regulation by both stimulators and inhibitors of proliferation is an important physiological mechanism in the regulation of keratinocyte proliferation. A similar mechanism may be involved in the regulation of proliferation of other cell types as well and may involve peptide growth factors and growth inhibitors other than TGFα and TGFβ. The presence of opposing regulatory pathways should allow for a more precise control of the very important process of cell proliferation than an on/off stimulatory pathway provided by the growth factors. This suggests that the growth inhibitory polypeptides may play as important a role in the control of cell proliferation as do the growth stimulatory factors.

REFERENCES

1. Goustin, A. S., Leof, E. B., Shipley, G. D., and Moses, H. L. (1986). *Cancer Res.* **46**, 1015–1029.
2. Derynck, R. (1986). *J. Cell. Biochem.* **32**, 293–304.
3. Moses, H. L., Tucker, R. F., Leof, E. B., Coffey, R. J., Halper, J., and Shipley, G. D. (1985). *In* "Growth Factors and Transformation" (J. Feramisco, B. Ozanne, and C. Stiles, eds.), Vol. 3: Cancer Cells, pp. 65–71. Cold Spring Harbor Laboratory, Cold Spring Harbor, New York.
4. Tucker, R. F., Shipley, G. D., Moses, H. L., and Holley, R. W. (1984). *Science* **226**, 705–707.
5. DeLarco, J. E., and Todaro, G. J. (1978). *Proc. Natl. Acad. Sci. U.S.A.* **75**, 4001–4005.
6. Anzano, M. A., Roberts, A. B., Smith, J. M., Sporn, M. B., and DeLarco, J. E. (1983). *Proc. Natl. Acad. Sci. U.S.A.* **80**, 6264–6268.
7. Marquardt, H., Hunkapiller, M. W., Hood, L. E., and Todaro, G. J. (1984). *Science* **223**, 1079–1082.
8. Derynck, R., Roberts, A. B., Winkler, M. E., Chen, E. Y., and Goeddel, D. V. (1984). *Cell (Cambridge, Mass.)* **38**, 287–297.
9. Bringman, T. S., Lindquist, P. B., and Derynck, R. (1987). *Cell (Cambridge, Mass.)* **48**, 429–440.
10. Todaro, G. J., Fryling, C., and DeLarco, J. E. (1980). *Proc. Natl. Acad. Sci. U.S.A.* **77**, 5258–5262.
11. Twardzik, D. R., Ranchalis, J. E., and Todaro, G. J. (1982). *Cancer Res.* **42**, 590–593.
12. Sporn, M. B., and Todaro, G. J. (1980). *N. Engl. J. Med.* **303**, 878–880.
13. Coffey, R. J., Derynck, R., Wilcox, J. N., Bringman, T. S., Goustin, A. S., Moses, H. L., and Pittelkow, M. R. (1987). *Nature (London)* **328**, 817–820.
14. Wille, J. J., Pittlekow, M. R., Shipley, G. D., and Scott, R. E. (1984). *J. Cell Physiol.* **121**, 31–44.
15. Moses, H. L., Branum, E. B., Proper, J. A., and Robinson, R. A. (1981). *Cancer Res.* **41**, 2842–2848.

16. Roberts, A. B., Anzano, M. A., Lamb, L. C., Smith, J. M., and Sporn, M. B. (1981). *Proc. Natl. Acad. Sci. U.S.A.* **78,** 5339–5343.
17. Roberts, A. B., Frolik, C. A., Anzano, M. A., and Sporn, M. B. (1983). *Fed. Proc., Fed. Am. Soc. Exp. Biol.* **42,** 2621–2626.
18. Childs, C. B., Proper, J. A., Tucker, R. F., and Moses, H. L. (1982). *Proc. Natl. Acad. Sci. U.S.A.* **79,** 5312–5316.
19. Assoian, R. K., Komoriya, A., Meyers, C. A., Miller, D. M., and Sporn, M. B. (1983). *J. Biol. Chem.* **258,** 7155–7160.
20. Derynck, R., Jarrett, J. A., Chen, E. Y., Eaton, D. H., Bell, J. R., Assoian, R. K., Roberts, A. B., Sporn, M. B., and Goeddel, D. V. (1985). *Nature (London)* **316,** 701–705.
21. Gentry, L. E., Webb, N. R., Lim, G. J., Brunner, A. M., Ranchalis, J. E., Twardzik, D. R., Lioubin, M. N., Marquardt, H., and Purchio, A. F. (1987). *Mol. Cell. Biol.* **7,** 3418–3427.
22. Derynck, R., Jarrett, J. A., Chen, E. Y., and Goeddel, D. V. (1986). *J. Biol. Chem.* **261,** 4377–4379.
23. Cheifetz, S., Weatherbee, J. A., Tsang, M. L., Anderson, J. K., Mole, J. E., Lucas, R., and Massague, J. (1987). *Cell (Cambridge, Mass.)* **48,** 409–415.
24. Ohta, M., Greenberger, J. S., Anklesaria, P., Bassola, A., and Massague, J. (1987). *Nature (London)* **329,** 539–541.
25. Wrann, M., Bodmer, S., de Martin, R., Siepl, C., Hofer-Warbinek, R., Frei, K., Hofer, E., and Fontana, A. (1987). *EMBO J.* **6,** 1633–1636.
26. Holley, R. W., Armour, R., and Baldwin, J. H. (1978). *Proc. Natl. Acad. Sci. U.S.A.* **75,** 1864–1866.
27. Hanks, S. K., Armour, R., Baldwin, J. H., Maldonado, F., Spiess, J., and Holley, R. W. (1988). *Proc. Natl. Acad. Sci. U.S.A.* **85,** 79–82.
28. Cate, R. L., Mattaliano, R. J., Hession, C., Tizard, R., Farber, N. M., Cheung, A., Ninfa, E. G., Frey, A. Z., Gash, D. J., Chow, E. P., Fisher, A., Bertonis, J. M., Torres, G., Wallner, B. P., Ramachandran, K. L., Ragin, R. C., Managanaro, T. F., MacLaughlin, D. T., and Donahoe, P. K. (1986). *Cell (Cambridge, Mass.)* **45,** 685–698.
29. Mason, A. J., Hayflick, J. S., Ling, N., Esch, F., Ueno, N., Ying, S. Y., Guillemin, R., Niall, H., and Seeburg, P. H. (1985). *Nature (London)* **318,** 659–663.
30. Padgett, R. W., St. Johnston, R. D., and Gelbart, W. M. (1987). *Nature (London)* **325,** 81–84.
31. Weeks, D. L., and Melton, D. A. (1987). *Cell (Cambridge, Mass.)* **51,** 861–867.
32. Ying, S., Becker, A., Ling, N., Ueno, N., and Guillemin, R. (1986). *Biochem. Biophys. Res. Commun.* **136,** 969–975.
33. Coughlin, J. P., Donahoe, P. K., Budzik, G. P., and MacLaughlin, D. T. (1987). *Mol. Cell. Endocrinol.* **49,** 75–86.
34. Tucker, R. F., Branum, E. L., Shipley, G. D., Ryan, R. J., and Moses, H. L. (1984). *Proc. Natl. Acad. Sci. U.S.A.* **81,** 6757–6761.
35. Frolik, C. A., Wakefield, L. M., Smith, D. M., and Sporn, M. B. (1984). *J. Biol. Chem.* **259,** 10995–11000.
36. Massague, J., and Like, B. (1985). *J. Biol. Chem.* **260,** 2636–2645.
37. Leof, E. B., Proper, J. A., Goustin, A. S., Shipley, G. D., DiCorleto, P. E., and Moses, H. L. (1986). *Proc. Natl. Acad. Sci. U.S.A.* **83,** 2453–2457.
38. Ignotz, R. A., and Massague, J. (1986). *J. Biol. Chem.* **261,** 4337–4345.
39. Ignotz, R. A., Endo, R., and Massague, J. (1987). *J. Biol. Chem.* **262,** 6443–6446.
40. Raghow, R., Postlethwaite, A. E., Keski-Oja, J., Moses, H. L., and Kang, A. H. (1987). *J. Clin. Invest.* **79,** 1285–1288.

41. Laiho, M., Saksela, O., Andreasen, P. A., and Keski-Oja, J. (1986). *J. Cell Biol.* **103**, 2403–2410.
42. Matrisian, L. M., Leroy, P., Ruhlmann, C., Gesnel, M., and Breathnach, M. (1986). *Mol. Cell. Biol.* **6**, 1679–1686.
43. Edwards, D. R., Murphy, G., Reynolds, J. J., Whitham, S. E., Docherty, J. P., Angel, P., and Heath, J. K. (1987). *EMBO J.* **6**, 1899–1904.
44. Postlethwaite, A. E., Keski-Oja, J., Moses, H. L., and Kang, A. H. (1987). *J. Exp. Med.* **165**, 251–256.
45. Roberts, A. B., Sporn, M. B., Assoian, R. K., Smith, J. M., Roche, N. S., Wakefield, L. M., Heine, U. I., Liotta, L. A., Falanga, V., Kehrl, J. H., and Fauci, A. S. (1986). *Proc. Natl. Acad. Sci. U.S.A.* **83**, 4167–4171.
46. Massague, J. (1987). *Cell (Cambridge, Mass.)* **49**, 437–438.
47. Moses, H. L., and Leof, E. B. (1986). *In* "Oncogenes and Growth Control" (P. Kahn and T. Graf. eds.), pp. 51–57. Springer-Verlag, Heidelberg.
48. Shipley, G. D., Pittelkow, M. R., Wille, J. J., Scott, R. E., and Moses, H. L. (1986). *Cancer Res.* **46**, 2068–2071.
49. Lawrence, D. A., Pircher, R., Kryceve-Martinerie, C., and Jullien, P. (1984). *J. Cell. Physiol.* **121**, 184–188.
50. Pircher, R., Jullien, P., and Lawrence, D. A. (1986). *Biochem. Biophys. Res. Commun.* **136**, 30–37.
51. Lyons, R. M., Keski-Oja, J., and Moses, H. L. (1988). *J. Cell Biol.* **106**, 1659–1665.

PART II

POLYPEPTIDE GROWTH FACTORS AND THEIR CELL MEMBRANE RECEPTORS: RELATION TO ONCOGENES

10
The EGF Receptor Protooncogene: Structure, Evolution of Properties of Receptor Mutants

JOSEPH SCHLESSINGER
Department of Pharmacology
New York University School of Medicine
New York, New York 10016

INTRODUCTION

Epidermal growth factor (EGF) is a small protein of 53 amino acids which acts as a mitogen for various cell types *in vitro* and *in vivo* (1). Several lines of evidence suggest that the receptor for this factor can also play a role in the uncontrolled proliferation characteristic of neoplastic cells. First, the v-*erbB* oncogene of avian erythroblastosis virus encodes a truncated EGF receptor (2), and we proposed that the v-*erbB* protein transforms by functioning as an activated growth factor receptor. Second, various animal and human tumor cells produce a growth factor called transforming growth factor-α (TGF-α) (3). This growth factor is related to EGF; both factors bind to the EGF receptor with similar affinities and induce the proliferation of cells bearing the EGF receptor. It has been suggested that TGF-α plays a role in oncogenesis by inducing autocrine growth (3). Finally, the EGF receptor gene is amplified and rearranged in a significant proportion of human brain tumors of glial origin (4). The resultant overexpression of the EGF receptor may play a role in the development or progression of these tumors. Hence, it is clear that the investigation of the structure of the EGF receptor and the mechanism of its activation may provide important clues to fundamental questions underlying the mechanisms of normal growth control and neoplasia.

STRUCTURE OF THE EGF RECEPTOR

Following the purification of the human EGF receptor by immunoaffinity chromatography (5) and its partial sequencing (2), the complete amino acid sequence of the EGF receptor was deduced from the nucleotide sequence of cDNA clones (6). The mature receptor is composed of 1186 amino acid residues which are preceded at the N-terminal end by a signal peptide of 24 hydrophobic amino acids.

The mature receptor is composed of three major structural elements. The extracellular EGF-binding domain is composed of 621 amino acid residues and is anchored in the plasma membrane by a single transmembrane region of 23 hydrophobic amino acids. The transmembrane region is followed by a sequence of mostly basic residues, a feature common to many membrane proteins. The cytoplasmic domain of the EGF receptor contains 542 amino acids. We have used immunological probes to demonstrate that the N-terminal end of the receptor is extracellular while its C-terminal end faces the cytoplasm (7). The cytoplasmic domain contains a region of approximately 300 amino acid residues which show a high degree of homology to the catalytic domain of the protein tyrosine kinases encoded by the *src*-related oncogenes (6). Like the other protein-tyrosine kinases, the catalytic domain of the EGF receptor kinase contains a lysine residue which is located 15 residues to the C-terminal side of the consensus sequence Gly-X-Gly-X-Phe-Gly-X-Val. The lysine residue, together with the consensus sequence, probably functions as part of the ATP binding site (8).

The binding of EGF to its receptor leads to the activation of the receptor tyrosine kinase, which phosphorylates various cellular proteins, as well as the EGF receptor itself. In intact cells, autophosphorylation occurs mainly on Tyr-1173, a residue which is deleted in the v-*erbB* protein (9). However, at least two additional tyrosine residues are autophosphorylated when EGF is added to solubilized membranes or to the pure receptor.

A remarkable feature of the EGF receptor extracellular domain is the high proportion of cysteine residues. Most cysteines are clustered in two regions each 160 residues long. The two cysteines can be aligned, forming internal repeats with similar spacing between the cysteine residues. Cysteine-rich domains have also been found in the extracellular domains of the low density lipoprotein (LDL) receptor, insulin receptor, and a human protein highly homologous to the EGF receptor, termed HER2 (10). It is possible that HER2 functions as a membrane receptor for an as yet unknown growth factor or hormone. The HER2 protein is probably the human counterpart of the rat *neu* oncogene product, an oncogene associated with neuroblastomas (11). Interestingly the cysteine-rich domains of the EGF receptor and HER2 can be aligned with the cysteine-rich domain of the insulin receptor, suggesting that these domains probably evolved from a common ancestor.

Insight into the evolution of the various EGF receptor domains was obtained from the deduced amino acid sequence of the *Drosophila* homolog of the EGF receptor gene (12). The *Drosophila* homolog shows extensive sequence homology to the human EGF receptor and to HER2. Like the human receptor, the *Drosophila* protein shows three distinct domains: an extracellular EGF-binding domain, a hydrophobic transmembrane region, and a cytoplasmic kinase domain. The overall amino acid homology is 55% in the kinase domain and 40% in the extracellular domain. Sequence analysis of the *Drosophila* receptor extracellular domain reveals three cysteine-rich clusters. The additional cysteine-rich domain in the *Drosophila* protein was probably generated by a duplication of one of the two cysteine-rich domains. The striking conservation of the cysteine-rich domains and the multiple duplication events they have undergone suggest that these regions play an important role in the function of the receptor. However, they probably do not form the ligand-combining site, since receptors with different specificities share the same sequences, and since these clusters are more rigid than would be expected if they are part of a ligand-combining site.

PROPERTIES OF EGF RECEPTOR MUTANTS

Questions concerning the mechanism of action and regulation of the EGF receptor were addressed by exploring the properties and cellular effects of various EGF receptor mutants introduced into animal cells. Cultured CHO or NIH 3T3 cells devoid of endogenous EGF receptor were transfected with cDNA constructs encoding human EGF receptor and various EGF receptor deletion mutants (13). The conclusion from this study was that the cytoplasmic domain of the EGF receptor is required for the expression of high affinity receptors, mitogenic responsiveness, and receptor-mediated endocytosis (13).

The role of the protein-tyrosine kinase domain of the EGF receptor was explored using two mutants: (a) an insertional mutant which contains four additional amino acids in the kinase domain after residue 708 (14,15) and (b) a point mutant in which Lys-721, that is, a part of the ATP binding site, was substituted by an alanine residue (15). Unlike the "wild-type" receptor expressed in these cells, which exhibits EGF-stimulatable protein-tyrosine kinase activity, the mutated receptors lack protein-tyrosine kinase activity both *in vitro* and *in vivo*. Despite this deficiency the mutant receptors are properly processed, bind EGF, and exhibit both high and low affinity binding sites. Moreover, the mutated receptors undergo efficient EGF-mediated endocytosis. However, EGF fails to stimulate DNA synthesis and is unable to stimulate the phosphorylation of S6 ribosomal protein in cells expressing these receptor mutants. Hence, it is proposed that the protein-tyrosine kinase activity of the EGF receptor is essential for the initiation of S6 phosphorylation and for DNA synthesis induced by EGF. However, EGF receptor processing and the expression of high and low affinity surface receptors appear to require neither kinase activity nor receptor autophosphorylation. Interestingly, phorbol ester (TPA) fails to abolish the high affinity state of the insertional mutant and is also unable to stimulate the phosphorylation of this receptor mutant (16). This result is consistent with the notion that kinase C phosphorylation of the EGF receptor is essential for the loss of the high affinity EGF receptors caused by TPA.

The tumor promoter phorbol ester (TPA) modulates the binding affinity of the epidermal growth factor receptor and the mitogenic capacity of EGF. Moreover, TPA-induced kinase C phosphorylation occurs mainly on Thr-654 of the EGF receptor, suggesting that the phosphorylation state of this residue may regulate the ligand binding affinity and kinase activity of the EGF receptor. To examine the role of this residue we have prepared a Tyr-654 EGF receptor cDNA construct by *in vitro* site-directed mutagenesis (17). The addition of TPA to NIH 3T3 cells expressing a "wild-type" human EGF receptor blocked the mitogenic capacity of EGF. However, this inhibition did not occur in NIH 3T3 cells expressing the Tyr-654 EGF receptor mutant. In the latter cells, EGF is able to stimulate DNA synthesis even in the presence of inhibitory concentrations of TPA. Hence, we propose that the phosphorylation of Thr-654 by kinase C may provide a negative control mechanism for EGF-induced mitogenesis in mouse NIH 3T3 fibroblasts (17).

REFERENCES

1. Carpenter, G., and Cohen, S. (1979). *Annu. Rev. Biochem.* **48**, 193–216.
2. Downward, J., Yarden, Y., Mayes, E., Scrace, G., Totty, N., Stockwell, P., Ullrich, A., Schlessinger, J., and Waterfield, M. D. (1984). *Nature (London)* **311**, 483–485.
3. Todaro, G. J., Fryling, C., and DeLarco, J. E. (1980). *Proc. Natl. Acad. Sci. U.S.A.* **77**, 5258–5262.
4. Libermann, T. A., Nussbaum, H. R., Razon, N., Kris, R., Lax, I., Soreq, M., Whittle, N., Waterfield, M. D., Ullrich, A., and Schlessinger, J. (1985). *Nature (London)* **313**, 144–147.
5. Yarden, Y., Harari, I., and Schlessinger, J. (1985). *J. Biol. Chem.* **260**, 315–319.
6. Ullrich, A., Coussens, L., Hayflick, J. S., Dull, T. J., Gray, A., Tam, A. W., Lee, J., Yarden, Y., Libermann, T. A., Schlessinger, J., Downward, J., Mayes, E. L. V., Whittle, N., Waterfield, M. D., and Seeburg, P. H. (1984). *Nature (London)* **309**, 418–425.
7. Kris, R., Lax, I., Gullick, M., Waterfield, M., Ullrich, A., Fridkin, M., and Schlessinger, J. (1985). *Cell (Cambridge, Mass.)* **40**, 619–625.
8. Hunter, T., and Cooper, J. A. (1985). *Annu. Rev. Biochem.* **54**, 897–930.
9. Yamamoto, T., Nishida, T., Miyajima, N., Kawai, S., Ooi, T., and Toyoshima, K. (1983). *Cell (Cambridge, Mass.)* **35**, 71–78.
10. Coussens, L., Yan-Feng, T. L., Liao, Y.-C., Chen, E., Gray, A., McGrath, J., Seeburg, P.

H., Libermann, T. A., Schlessinger, J., Franke, U., Levinson, A., and Ullrich, A. (1985). *Science* **230,** 1132–1139.
11. Schechter, A. L., Stern, D. F., Vaidyanathan, L., Decker, S. J., Drebin, J. A., Greene, M. I., and Weinberg, R. A. (1984). *Nature (London)* **312,** 513–516.
12. Livneh, E., Glaser, L., Segal, D., Schlessinger, J., and Shilo, B. Z. (1985). *Cell (Cambridge, Mass.)* **40,** 599–607.
13. Livneh, E., Prywes, R., Kasheles, O., Reiss, N., Sasson, I., Mory, Y., Ullrich, A., and Schlessinger, J. (1986). *J. Biol. Chem.* **261,** 12490–12497.
14. Prywes, R., Livneh, E., Ullrich, A., and Schlessinger, J. (1986). *EMBO J.* **5,** 2179–2190.
15. Honegger, A. M., Dull, T., Felder, S., Van Obberghen, E., Bellot, F., Szapary, D., Schmidt, A., Ullrich, A., and Schlessinger, J. (1987). *Cell* **51,** 199–209.
16. Livneh, E., Berent, E., Reiss, N., Ullrich, A., and Schlessinger, J. (1987). *EMBO J.* **6,** 2669–2676.
17. Livneh, E., Dull, T., Prywes, R., Ullrich, A., and Schlessinger, J. (1987). *Mol. Cell Biol.* **8,** 2302–2308.

11
Biological Effects of the v-*erbA* Oncogene in Transformation of Avian Erythroid Cells

BJÖRN VENNSTRÖM,[1,5] HARTMUT BEUG,[1,6]
KLAUS DAMM,[1,4] DOUGLAS ENGEL,[2]
ULLRICH GEHRING,[3] THOMAS GRAF,[1]
ALBERTO MUÑOZ,[1,7] JAN SAP,[1,8] AND MARTIN ZENKE[1,6]

1. Differentiation Programme
European Molecular Biology Laboratory
D-6900 Heidelberg, Germany
and
2. Northwestern University
Department of Biochemistry
Molecular and Cellular Biology
Evanston, Illinois 60201
and
3. Institute of Biochemistry
University of Heidelberg
D-6900 Heidelberg, Germany

4. Present address: The Salk Institute, P.O. Box 85800, San Diego, California 92138-9216.
5. Present address: Department of Molecular Biology, CMB, Karolinska Institute, Box 60400, S-104 01 Stockholm, Sweden.
6. Present address: I.M.P., Dr Bohrgasse 7, A-1030 Vienna, Austria.
7. Present address: Instituto de Investigaciones Biomedicas, c/Arturo Duperier 4, 28029 Madrid, Spain.
8. Present address: Weizmann Institute of Science, Dept. of Chem. Immun., Rehovot 76100, Israel.

BIOLOGICAL EFFECTS OF THE v-erbA AND v-erbB ONCOGENES

The avian erythroblastosis virus (AEV) strain ES4 contains the v-erbA and v-erbB oncogenes and transforms fibroblasts and erythroblasts *in vitro*, a property paralleled *in vivo* by its induction of erythroblastosis and sarcomas (for review, see Graf and Beug, 1978). v-erbB, which represents a truncated version of the epidermal growth factor (EGF) receptor (Ullrich et al., 1984), is sufficient for transformation of both types of cells. The v-erbA gene, which is fused to the viral structural gene gag, is nononcogenic on its own but enhances leukemic cell transformation *in vivo* and *in vitro* induced by v-erbB and other oncogenes such as v-src, v-sea, or v-H-ras (Frykberg et al., 1983; Kahn et al., 1986a). More specifically, two distinct and measurable effects of v-erbA in erythroblasts have been observed: it completely blocks the spontaneous differentiation exhibited by erythroblasts transformed by v-erbB, and it allows cell proliferation at pH and ionic conditions otherwise toxic to the cells (Beug et al., 1985; Kahn et al., 1986b; Damm et al., 1987). The effects of v-erbA in fibroblasts are less pronounced, but an enhancement of the transformed phenotype including focus formation has been observed (Frykberg et al., 1983; Jansson et al., 1987).

The v-erbA protein was recently shown to bear homology with receptors for steroid hormones. The highest amino acid homology, 45%, is located in the putative DNA region (Fig. 1) which contains many cysteines and positively charged residues. The domain corresponding to the ligand-binding region is much less conserved (20% homology). Subsequent experiments demonstrated that the c-erbA protein is a high affinity nuclear receptor for the thyroid hormones triiodothyronine (T_3) and thyroxine (T_4) (Sap et al., 1986; Weinberger et al., 1986) and that the v-erbA protein binds no ligand (Sap et al., 1986). The results suggest that the v-erbA and v-erbB proteins act by distinct mechanisms, v-erbA presumably by affecting transcription at the promoter level and v-erbB by mediating mitogenic signals from its location in the plasma membrane.

COMPARISON OF v-erbA AND c-erbA

Nucleotide sequence comparison between the viral and cellular *erbA* genes revealed several amino acid differences between the encoded proteins. v-erbA exhibits 13 amino acid substitutions, 2 of which are located in the putative DNA-binding domain and 11 in other regions (see Fig. 1). Furthermore, the v-erbA protein has suffered a 9 amino acid internal deletion located 3 residues before the C terminus. Finally, the N terminus of the c-erbA polypeptide is 12 amino acids longer than the *erbA*-specific part of the v-erbA protein, since the latter is fused to gag at this position.

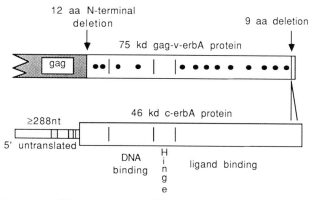

Fig. 1. Comparison of the *erbA* domains of the viral and cellular proteins. The black dots in v-*erbA* indicate the positions of point mutations.

The identification of the cellular homolog for v-*erbA* has made it possible to investigate the oncogenic effects of v-*erbA*: How do the many mutations in the viral protein alter its binding of ligand, its specificity in DNA sequence recognition, and its positively and/or negatively regulating effects on transcription? To identify which of the mutation(s) in v-*erbA* conferred the inability to bind hormone, a series of recombinants between the viral and cellular genes was constructed utilizing restriction sites common to both genes (Fig. 2). The sites were chosen so that the influence on hormone binding of the C-terminal deletion, the *gag* domain, and the internal amino acid substitutions in v-*erbA* could be tested.

The chimeric genes, cloned into plasmids containing a prokaryotic promoter, were transcribed with T7 RNA polymerase and translated in rabbit reticulocyte lysates *in vitro*. The resulting proteins were then tested for binding of $[^{125}I]T_3$ at a ligand concentration of 1 nM. The results (Fig. 2) show that all the proteins having the entire ligand-binding domain from v-*erbA* failed to bind hormone. Insertion of the 9 amino acid deletion from the C terminus of v-*erbA* into c-*erbA* considerably diminished but did not completely abolish ligand binding; likewise, reconstitution of the C terminus of v-*erbA* with that of c-*erbA* increased binding but did not fully restore it to wild-type c-*erbA* levels. Only proteins with the complete ligand-binding domain from c-*erbA* bound hormone at high levels.

To determine accurately the effect of the mutations on hormone binding we performed Scatchard analyses with those chimeric proteins that bound hormone. Figure 2 shows that proteins with either the 9 amino acid deletion or the internal point mutations had dissociation constants that were 10-fold higher than those containing the entire ligand-binding domain from c-*erbA*. Furthermore, the results demonstrate that the *gag* region or the mutations in the DNA-binding region of v-*erbA* do not affect the affinity for ligand.

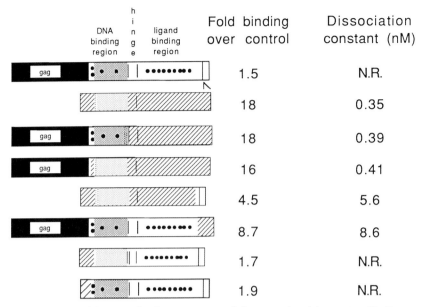

Fig. 2. Thyroid hormone-binding properties of chimeric v/c-erbA proteins. Binding was measured by coretention on nitrocellulose filters of protein–hormone complexes formed at a ligand concentration of 1 nM[^{125}I]T_3 or by Scatchard analysis.

Taken together, the results suggest that both the amino acid substitutions and the C-terminal deletion in v-erbA contribute to its loss of T_3-binding activity.

EFFECTS OF MUTANT v-erbA GENES ON ERYTHROBLASTS

Important clues as to the respective roles of v-erbA and v-erbB in erythroblast transformation came from the analysis of td359 AEV, a transformation-defective mutant of AEV unable to transform erythroblasts, and its revertant (r12), obtained after in vivo passage of the td359 mutant. Nucleotide sequence analysis showed that the v-erbA protein of td359 contained among other substitutions one amino acid change (proline to arginine) located in the short "hinge" region (Fig. 3) at position 144 (Damm et al., 1987). The v-erbA protein of r12 had reverted to a proline in this position and had in addition acquired a 31 amino acid insertion of unknown origin in its gag domain. In contrast, the v-erbB genes of both the mutant and the revertant carried identical deletions in their 3' ends: a 306 nucleotide deletion had removed most of the region located after the domain of v-erbB homologous to tyrosine kinases (Damm et al., 1987, and data not shown).

To analyze the importance of these mutations in v-erbA and v-erbB for erythroblast transformation, we constructed recombinant viruses containing all possible combinations of wild-type, mutant, and revertant v-erbA and

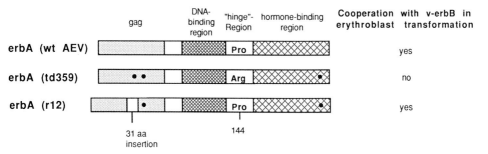

Fig. 3. Comparison of v-*erbA* proteins from the wild-type, *td*359, and *r*12 AEV strains. The black dots indicate positions of point mutations.

v-*erbB* genes. The results of subsequent bone marrow transformation experiments with the resulting viruses (Fig. 3) showed first that both the *td*359 and the *r*12 v-*erbB* genes were unable to transform erythroblasts in the absence of an active v-*erbA* gene: v-*erbA* of *r*12 or wild type but not *td*359 were able to restore the erythroblast-transforming activity to the mutant v-*erbB* genes, confirming that the reversion in the *r*12 genome had occurred in the v-*erbA* oncogene. To unambiguously identify the inactivating mutation in the *td*359 v-*erbA* gene, chimeras between the *td*359 and *r*12 v-*erbA* genes were combined in retrovirus constructs with the *td*359 v-*erbB* oncogene; consequently, only viruses with active v-*erbA* genes would transform erythroblasts. The results (not shown) confirmed that the respective changes at position 144 were responsible for the inactivation of the *td*359 gene and the reactivation in *r*12, and they also suggest that the hinge region serves an important role in the structure and function of the v-*erbA* protein.

HOW DOES v-*erbA* CONTRIBUTE TO TRANSFORMATION OF ERYTHROBLASTS?

The thyroid hormone receptor is known to either positively or negatively regulate the expression of a wide variety of genes, presumably by binding to their respective promoter regions in a manner similar to those of steroid receptors (for review, see Oppenheimer and Samuels, 1983). To test if the v-*erbA* protein also binds DNA, labeled nuclear extracts from AEV-transformed erythroblasts were passed over DNA-cellulose columns and then eluted with increasing concentrations of KCl. The results (not shown) of subsequent immunoprecipitation and sodium dodecyl sulfate–polyacrylamide gel electrophoresis (SDS–PAGE) analyses demonstrated that the v-*erbA* protein eluted at salt concentrations between 0.2 and 0.4 M, that is, it bound DNA with an affinity similar to that of the thyroid hormone receptor (MacLeod and Baxter, 1977).

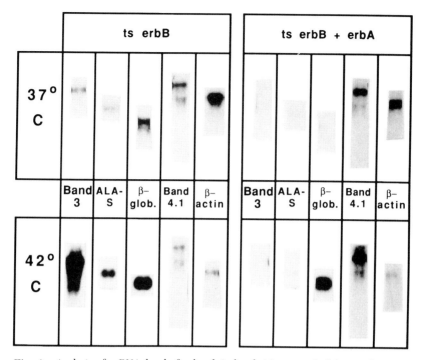

Fig. 4. Analysis of mRNA levels for band 3, band 4.1, actin, β-globin, and ALA-S in undifferentiated and partially differentiated erythroblasts. The cells were transformed by a *ts* oncogene, expanded, and half of the culture was superinfected with a virus expressing v-*erbA* and resistance to G418. The cells were then incubated for 2 days at the temperature permissive (37°C) or nonpermissive (42°C) for the *ts* oncogene. Poly(A)$^+$ RNA was isolated and subjected to Northern analysis.

We therefore tested if v-*erbA* affects the expression of genes that play important roles in the maturation of erythroid cells. Accordingly, poly(A)$^+$ RNA was isolated from erythroblasts transformed by AEV strains expressing a temperature-sensitive (*ts*) v-*erbB* gene plus an active ("+ v-*erbA*") or inactive ("− v-*erbA*") *erbA* gene, grown at the permissive temperature (37°C) or differentiated for 2–3 days at the nonpermissive temperature (42°C). A Northern analysis (Fig. 4) shows that irrespective of their differentiation state the v-*erbA*-containing cells contained no mRNA for band 3, the major anion transporter in erythroid cells which also serves as an anchor for the red cell cytoskeleton. A decrease of mRNA for δ-aminolevulinic acid synthetase (ALA-S, a key enzyme in hemin synthesis) was also observed, whereas other RNAs like those for β-globin, actin, and band 4.1 (a red cell cytoskeleton protein) were unaffected by the oncogene. The expression of the Na$^+$/K$^+$-ATPase mRNA was also unaffected (data not shown). To determine if this effect of v-*erbA* was at the transcriptional or posttranscriptional level we performed run-on analyses with nuclei from cells treated in the same way. The results showed that the decrease in mRNA for band 3 and ALA-S in the v-*erbA*-containing cells was due to an arrest of transcriptional initiation by the oncogene protein (data not shown).

We also wanted to know whether the v-*erbA* oncogene arrested anion transporter expression in maturing cells were actively transcribing this gene. This became possible by exploiting our earlier observation that temperature-induced *wt-erbA/ts-erbB* erythroblasts were blocked in differentiation in normal medium (pH 7.2) but readily differentiated into erythrocytes in a specific, alkaline medium (pH 8.1) (Beug et al., 1982; Beug and Hayman, 1984). Figure 5 shows that these cells expressed high levels of anion transporter protein after temperature induction in alkaline medium, but not at all in normal medium. This suggested that some aspect of v-*erbA* function was reversibly blocked by treatment of cells with alkaline medium and allowed us to monitor how quickly the oncogene turns off band 3 expression in maturing cells. Consequently, v-*erbA/ts* v-*erbB* erythroblasts were induced to differentiate at 42°C in pH 8.1 medium and then shifted to pH 7.2 medium for increasing time periods. The cells were then labeled, extracted, and immunoprecipitated with antiband 3 antibodies. The data in Fig. 6 show that band 3 synthesis was already shut off 3 hr after the shift to lower pH, suggesting that v-*erbA* can down regulate transcription of the anion transporter in partially differentiated cells.

Taken together, our results indicate that erythroblasts transformed by only v-*erbB* synthesize significant amounts of anion transporter, a process completely blocked by v-*erbA*. We therefore hypothesized that the production of anion transporter could be the cause of the stringent pH requirement for growth of the v-*erbB* transformed cells, namely, that the probably abnor-

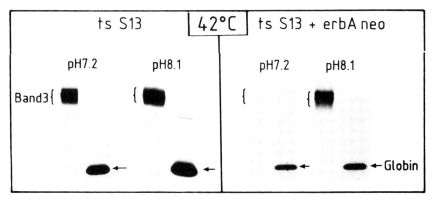

Fig. 5. Analysis of band 3 expression *ts* v-*sea*-transformed erythroblasts, containing or not containing v-*erbA*, and grown at pH 7.2 or 8.1. See text for explanations.

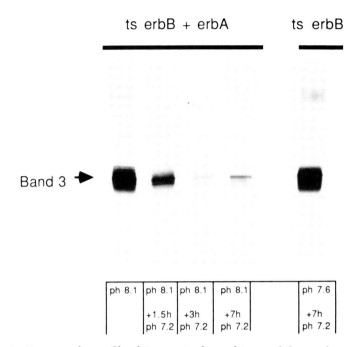

Fig. 6. Down regulation of band 3 expression by v-*erbA* upon shift to medium of pH 8.1. Erythroblasts, transformed by *ts* v-*erbB* and v-*erbA*, were grown at pH 7.2 for 2 days, which permits synthesis of band 3, and were then shifted to pH 8.1 for the times indicated. Extracts from [^{35}S]methionine-labeled cells were immunoprecipitated with antiband 3 antibodies as indicated and analyzed by SDS–PAGE.

mally high activity of band 3 in the immature cells would be toxic to the cells unless this activity was either "balanced" by corresponding pH and ion concentrations in the culture medium or suppressed by v-*erbA*. To test this, erythroblasts transformed by v-*erbB* alone were incubated with an inhibitor of band 3 activity (DIDS, 4,4′-diisothiocyanostilbene-2,2′-disulfonate; Falke and Chan, 1986), using a medium that normally does not allow propagation of these cells. Figure 7B shows that in the presence of the drug the cells incorporated [^{3}H]thymidine at levels comparable to those of the untreated cells containing v-*erbA* and v-*erbB*. In contrast, the cells with only v-*erbB* exhibited a declining [^{3}H]thymidine incorporation in the absence of the drug. Figure 7A also shows that these cells become badly vacuolated in the absence but not the presence of the drug. The results therefore suggest that the narrow pH requirements for growth of v-*erbB*-transformed erythroblasts is alleviated by v-*erbA* through suppression of synthesis of band 3, a highly active ion transporter apparently toxic to immature erythroid cells when expressed at high levels.

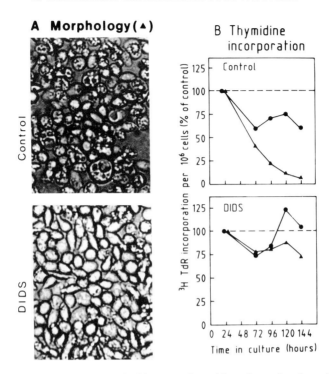

Fig. 7. Growth of pH 7.2 of erythroblasts transformed by only v-*erbB* after inhibition of band 3 activity with DIDS. The proliferation was monitored by [^{3}H]thymidine incorporation per 10^6 cells as indicated.

CONCLUSIONS

Our results show that the v-*erbA* protein represents a ligand-independent form of the thyroid hormone receptor. The mutations conferring the lack of binding consequently identify the domain required for hormone binding. This region has a structure and location similar to those of the ligand-binding domains of receptors for steroid hormones, suggesting that these receptors share a common ancestral gene. The fact that both the deletion in v-*erbA* and the point mutations are required for abolishing hormone binding suggests that they have accumulated gradually, and that the acquisition of hormone independence has given the virus a selective advantage by increasing the oncogenic activity of v-*erbA*.

The v-*erbA* protein binds DNA, suggesting that it acts like the thyroid hormone receptor by binding to regulatory elements of specific target genes. Consequently, the v-*erbA* protein might constitutively either up or down regulate the expression of its target genes. The nucleotide sequences recog-

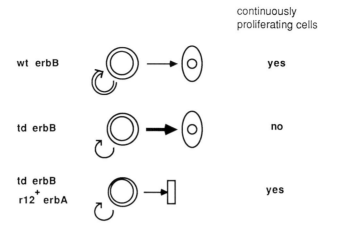

Fig. 8. Model for the cooperativity between wild-type and mutant v-*erbA* and v-*erbB* genes.

nized by v-*erbA* might, however, be distinct from those binding to the c-*erbA* protein, since the viral protein contains two amino acid substitutions in its putative DNA-binding domain. At any rate, if the expression of such genes (or, alternatively, absence thereof) is important for differentiation or growth of erythroblasts, the effects of v-*erbA* on erythroid differentiation would be explained. We therefore investigated the effects of v-*erbA* on the expression of several genes potentially important for red cell differentiation. The down regulation of band 3 synthesis not only suggests a mechanism by which the oncogene broadens the pH range permissible for erythroblast growth, but could also have other effects on the erythroblasts. The ion transporter serves as an anchor for the rigid red cell cytoskeleton as well as several enzymes. A lack of proper anchor would naturally disturb the organization of the proteins in the cytoskeletal network and might subsequently contribute to arresting the cell in an immature stage of differentiation, for example, by leading to rapid turnover of red cell skeleton proteins. It is likely, however, that v-*erbA* exerts its effects in erythroblasts pleiotropically, for instance, by affecting the expression of other genes. Hemin has been proposed to be a coregulator of erythroid cell differentiation, and the clear but low suppression of ALA-S expression by v-*erbA* might decrease the synthesis of hemin, thereby contributing to the block of differentiation (Schmidt *et al.*, 1986). Clearly, a more extensive survey of the expression of genes in erythroblasts is required to further clarify this issue.

How do two oncogenes acting in concert, such as the wild-type v-*erbA* and the $td359$ or $r12$ v-*erbB* genes, transform erythroblasts when neither of them by itself is capable of transformation? The mechanism for the cooperativity between v-*erbA* and v-*erbB* was elucidated by the analysis of the differentiation states and the proliferation capacities of erythroblasts transformed by the $td359$ and $r12$ AEV strains and their recombinants (Fig. 8). The mutant v-*erbB* genes only induced a weak self-renewal of the erythroblasts, and if no active v-*erbA* gene was present the cells differentiated terminally, leading to eventual loss of proliferating cells. v-*erbA* blocked this cell loss and thus enabled the mutant v-*erbB* genes to slowly generate a mass population of transformed erythroblasts. It is unclear why the $td359/r12$ v-*erbB* genes fail to induce an erythroblast self-renewal strong enough to maintain the undifferentiated state while they are capable of fully transforming fibroblasts, but it is possible that the C-terminal domain of v-*erbB* is required for proper mitogenic signaling in erythroblasts but not in fibroblasts.

REFERENCES

Beug, H., and Hayman, M. J. (1984). *Cell (Cambridge, Mass.)* **36**, 963–972.
Beug, H., Palmier, S., Freudenstein, C., Zentgraf, H., and Graf, T. (1982). *Cell (Cambridge, Mass.)* **28**, 907–919.
Beug, H., Kahn, P., Döderlein, G., Hayman, M. J., and Graf, T. (1985). In "Modern Trends in Human Leukaemia VI" (R. Neth, R. Gallo, M. Greaves, and S. Janka, eds.), Vol. 29, pp. 290–297. Springer-Verlag, Berlin and Heidelberg.
Damm, K., Beug, H., Graf, T., and Vennström, B. (1987). *EMBO J.* **6**, 375–382.
Falke, J., and Chan, S. (1986). *Biochemistry* **25**, 7888–7894.
Frykberg, L., Palmieri, S., Beug, H., Graf, T., Hayman, M. J., and Vennström, B. (1983). *Cell (Cambridge, Mass.)* **32**, 227–238.
Graf, T., and Beug, H. (1978). *Biochim. Biophys. Acta* **516**, 269–299.
Jansson, M., Beug, H., Gray, C., Graf, T., and Vennström, B. (1987). *Oncogene* **1**, 167–173.
Kahn, P., Frykberg, L., Brady, C., Stanley, I. J., Beug, H., Vennström, B., and Graft, T. (1986a). *Cell (Cambridge, Mass.)* **45**, 349–356.
Kahn, P., Frykberg, L., Graf, T., Vennström, B., and Beug, H. (1986b). In "Modern Trends in Leukaemia VII" (R. Neth, ed.), in press. Springer-Verlag, Berlin and Heidelberg.
MacLeod, K., and Baxter, J. (1977). *J. Biol. Chem.* **251**, 7380–7387.
Oppenheimer, J. H., and Samuels, M. H., eds. "Molecular Basis of Thyroid Hormone Action." Academic Press, New York, 1983.
Sap, J., Munoz, A., Damm, K., Goldberg, Y., Ghysdael, J., Leutz, A., Beug, H., and Vennström, B. (1986). *Nature (London)* **324**, 635–640.
Schmidt, J. A., Marshall, J., Hayman, M. J., Pouka, P., and Beug, H. (1986). *Cell (Cambridge, Mass.)* **46**, 41–51.
Ullrich, A., Coussens, L., Hayflick, J. S., Dull, T. J., Gray, A., Tam, A. W., Lee, J., Yarden, Y., Liberman, T. A., Schlessinger, J., Downward, J., Mayes, E. L. V., Whittle, N., Waterfield, M. D., and Seeburg, P. H. (1984). *Nature (London)* **309**, 418–425.
Weinberger, C., Thompson, C., Ong, E., Lebo, R., Gruol, D., and Evans, R. (1986). *Nature (London)* **324**, 641–646.

12

ras Transformation of 3T3 Cells Induces Basic Fibroblast Growth Factor Synthesis *in Vitro* and *in Vivo*

MICHAEL KLAGSBRUN, PAUL FANNING,
AND SNEZNA ROGELJ
*Departments of Surgery and Biological Chemistry
Children's Hospital and Harvard Medical School
Boston, Massachusetts 02115*

ras-Transformed NIH 3T3 cells and the tumors they produce express basic fibroblast growth factor (bFGF) but not acidic fibroblast growth factor (aFGF). *ras*-Transformed NIH 3T3 cells produce about 25-fold more bFGF than do their parental NIH 3T3 cell counterparts. It is suggested that transformation of cells by the *ras* oncogene induces bFGF synthesis and that the bFGF may play a role in tumor growth.

INTRODUCTION

Oncogenes stimulate cells to proliferate continuously by permanently activating a step along the mitogenic pathway. Transformation might be a result of activation of an intracellular second message pathway [e.g., *ras* (1)] or activation of a cell surface receptor [e.g., *erb*B (2), *neu* (3), and v-*fms* (4)]. Alternatively, an oncogene can encode a growth factor, as is the case with platelet-derived growth factor (PDGF) (5,6). Growth factors that are a product of an oncogene or whose expression is induced by oncogenes have the potential of transforming those cells which are responsive to the growth factor via an autocrine loop (7). To date PDGF, transforming growth factor-α (TGF-α), and transforming growth factor-β (TGF-β) have been shown to be expressed in oncogene-transformed cell lines (5,8–10).

bFGF has been found in tumors such as chondrosarcoma (11). However, it is not clear whether tumor-derived bFGF is a product of transformed cells or of the endothelial cells which populate highly vascularized tumors. Aortic, umbilical vein, capillary, and corneal endothelial cells all synthesize significant amounts of bFGF (12–14). bFGF has been shown previously to be oncogenic when fused to a signal peptide (15,16). Given the oncogenic potential of bFGF we wanted to know whether oncogene-mediated transformation *in vivo* or *in vitro* was accompanied by increased expression of bFGF in transformed cells, such as has been shown for other growth factors. In this chapter we demonstrate that tumors produced by *ras*-transformed cells express bFGF and that bFGF expression is markedly elevated in *ras*-transformed cells *in vitro* when compared to parental cell lines. Neither the tumors nor the transformed cells express aFGF.

RESULTS

NIH 3T3 cells transformed by the EJ *ras* gene were injected into syngeneic animals. Tumors appeared after 2 weeks and were analyzed for bFGF expression. Heparin affinity chromatography of tumor extracts, using 3T3 DNA synthesis and endothelial cell proliferation as bioassays, showed one peak of growth factor activity eluting at about 1.5 M NaCl, characteristic of bFGF (Fig. 1). The growth factor was shown to be bFGF by Western blot analysis (Fig. 2). Antibodies directed against bFGF reacted with control bFGF (lane 2) but not control aFGF markers (lane 1). These antibodies also reacted strongly with an 18-kDa protein in the tumor extract (lane 3). In the presence of excess bFGF immunizing peptide no immunoblotting of the 18-kDa protein was observed, indicating that the 18-kDa protein was bFGF (not shown). In contrast, antibodies directed against aFGF, which could detect aFGF (lane 4) but not bFGF (lane 5), did not detect any aFGF in the *ras* tumor extract (lane 6).

We wished to determine whether the source of bFGF in the *ras* tumors were *ras*-transformed fibroblasts. Thus, equal amounts of *ras*-transformed NIH 3T3 cells (EJ-62Bam6a) (19) and parental NIH 3T3 cells were prepared, lysed, and analyzed by heparin affinity chromatography (Fig. 3). A large peak of heparin-binding growth factor activity was found in the *ras*-transformed 3T3 cells (Fig. 3B) compared to the parental NIH 3T3 cells (Fig. 3A). The heparin-binding growth factor activity in *ras*-transformed 3T3 cells was

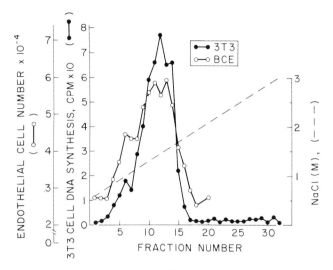

Fig. 1. Heparin affinity chromatography of extracts of *ras*-transformed NIH 3T3 cells. Syngeneic NIH/NFS mice were injected subcutaneously with 2×10^6 *ras*-transformed NIH 3T3 cells (EJ-62Bam6a) (19). After 2 weeks, tumors of about 1 cm^3 were harvested, placed in a solution of 1 M NaCl, 10 mM Tris-HCl, pH 7.0, diced into small pieces, and sonicated. The sonicated extracts were centrifuged to remove debris, and the supernatant fraction was applied to a 3-ml column of heparin-Sepharose (11). Fractions were eluted with a gradient of NaCl (0.1–3 M) and tested for the ability to stimulate DNA synthesis in BALB/c 3T3 cells (solid circles) and proliferation of capillary endothelial cells (open circles) as previously described (11).

shown to be authentic bFGF by Western blotting (not shown). *ras*-Transformed 3T3 cells produce about 0.5 units/10^4 cells compared to about 0.02 units/10^4 for NIH 3T3 cells. Thus, the induction of bFGF synthesis after *ras* transformation of cells *in vitro* is about 25-fold. There was no apparent production of aFGF in either cell line.

DISCUSSION

We have found that tumors produced by *ras*-transformed cells synthesize substantial amounts of bFGF. Although these tumors contain heterogeneous cell populations, it is highly probable that the bFGF is synthesized by the *ras*-transformed cells themselves. This conclusion is based on the observation that an approximate 25-fold induction in bFGF is observed in cultured *ras*-transformed NIH 3T3 cells as compared to the parental NIH 3T3 cells. In contrast to the elevated levels of bFGF, there appears to be no induction of aFGF synthesis either in the *ras*-transformed 3T3 cells or in the tumors

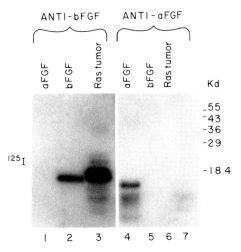

Fig. 2. Western blot analysis of tumor extracts purified by heparin-Sepharose affinity chromatography. aFGF, bFGF, and fractions containing heparin-binding growth factor as shown in Fig. 1 were analyzed by sodium dodecyl sulfate (SDS)–polyacrylamide gel electrophoresis (20,21). The proteins were transferred electrophoretically to nitrocellulose paper (20,21) and probed with either anti-aFGF or anti-bFGF antibody as previously described (20,21). To detect aFGF and bFGF the nitrocellulose paper was incubated first with either anti-aFGF or anti-bFGF antibody which had been prepared in rabbits and then with ^{125}I-labeled donkey anti-rabbit antibody. To visualize aFGF and bFGF, autoradiography was performed by applying X-ray film to the nitrocellulose paper at $-70°C$ overnight. Lanes 1–3, probed with anti-bFGF antibodies; lanes 4–6, probed with anti-aFGF antibodies; lanes 1 and 4, aFGF; lanes 2 and 5, bFGF; lanes 3 and 6, pooled fractions 7–15 from heparin affinity chromatography of *ras*-transformed 3T3 tumors as shown in Fig. 1.

they produce. At this point, the significance of bFGF induction in *ras*-transformed cells is not clear. bFGF is not a secreted protein, but remains cell-associated (17). Thus, a paracrine role for bFGF such as the stimulation of angiogenesis remains unlikely. However, if cell lysis occurred, as is possible in necrotic tumors, bFGF could be released to stimulate neighboring cells. Alternatively, bFGF might act to stimulate cell proliferation by an autocrine mechanism. NIH 3T3 cells have bFGF receptors, and if the bFGF in *ras*-transformed NIH 3T3 cells had access to these receptors, an autocrine loop would lead to constitutive 3T3 cell proliferation. Autocrine loops involving growth factors and their receptors have been previously shown for *sis*-transformed cells (18) and for cells transformed by bFGF to which a signal peptide has been fused (15). To ascertain whether bFGF plays a role in *ras*-mediated tumorgenicity, it will be necessary to specifically inhibit bFGF synthesis by expression of antisense bFGF RNA or by injection of antibodies that neutralize this growth factor.

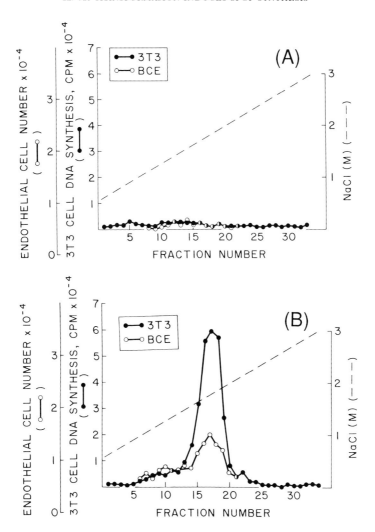

Fig. 3. Heparin-Sepharose affinity chromatography of extracts of NIH 3T3 cells and *ras*-transformed NIH 3T3 cells. Parental NIH 3T3 cells and *ras*-transformed NIH 3T3 cells (EJ-62Bam6a) were grown to confluence (2×10^7 cells/T150 culture flask). Extracts were sonicated in 1 M NaCl, 10 mM Tris-HCl, pH 7.0, clarified by centrifugation, and applied to columns of heparin-Sepharose as in Fig. 1. The fractions were analyzed for the ability to stimulate DNA synthesis in 3T3 cells (closed circles) or endothelial cell proliferation (open circles). (A) NIH 3T3 cells; (B) EJ-62Bam6a cells.

It has been previously shown that transformation by various oncogenes induces expression of growth factors such as PDGF, TGF-α, and TGF-β (5,8–10), and it has been suggested that these growth factors could play a significant supportive role in tumor growth. Given our observation that bFGF synthesis is induced *in vitro* and *in vivo* as a result of *ras* transformation, we suggest that bFGF is an important growth factor to be considered as a possible mediator of oncogenic transformation.

ACKNOWLEDGMENTS

This work was supported by a grant from National Cancer Institute (CA 37392). We thank Carlene Pavlos for typing the manuscript.

REFERENCES

1. Barbacid, M. (1987). *Annu. Rev. Biochem.* **56**, 779–827.
2. Downward, J., Yarden, Y., Mayes, E., Scrace, G., Totty, N., Stockwell, P., Ullrich, A., Schlessinger, J., and Waterfield, M. D. (1984). *Nature (London)* **307**, 521–527.
3. Schechter, A. L., Hung, M. C., Vaidyanathan, L., Weinberg, R. A., Yang-Feg, T. L., Francke, U., Ullrich, A., and Coussens, L. (1985). *Science* **229**, 976–978.
4. Wheeler, E. F., Rettenmier, C. W., Look, A. T., and Scherr, C. J. (1986). *Nature (London)* **324**, 377–379.
5. Doolittle, R. F., Hunkapiller, M. W., Hood, L. E., Devare, S. G., Robbins, K. C., Aaronson, S. A., and Antoniades, H. N. (1983). *Science* **221**, 275–277.
6. Waterfield, M. D., Scrace, G. T., Whittle, N., Stroobant, P., Johnson, A., Wasteson, A., Westermark, B., Heldin, C. H., Huang, J. S., and Deuel, T. F. (1983). *Nature (London)* **304**, 35–39.
7. Sporn, M. B., and Todaro, G. J. (1980). *N. Engl. J. Med.* **303**, 878–808.
8. DeLarco, J. E., and Todaro, G. J. (1978). *Proc. Natl. Acad. Sci. U.S.A.* **75**, 4001–4005.
9. Ozanne, B., Fulton, R. J., and Kaplan, P. L. (1980). *J. Cell. Physiol.* **105**, 163–180.
10. Roberts, A. B., Anzano, M. A., Lab, L. C., Smith, J. M., Frolick, C. A., Marquardt, H., Todaro, G. J., and Sporn, M. B. (1982). *Nature (London)* **295**, 1417–1419.
11. Shing, Y., Folkman, J., Sullivan, R., Butterfield, C., Murray, J., and Klagsbrun, M. (1984). *Science* **223**, 1296–1299.
12. Schweigerer, L., Neufeld, G., Friedman, J., Abraham, J. A., Fiddes, J. C., and Gospodarowicz, D. (1987). *Nature (London)* **325**, 257–259.
13. Hannan, R. L., Kourembanas, S., Flanders, K. C., Rogelj, S., Roberts, A., Faller, D. V., and Klagsbrun, M. (1988). *Growth Factors* **1**, 1–17.
14. Vlodavsky, I., Folkman, J., Sullivan, R., Fridman, R., Ishai-Michaeli, R., Sasse, J., and Klagsbrun, M. (1987). *Proc. Natl. Acad. Sci. U.S.A.* **84**, 2292–2296.
15. Rogelj, S., Weinberg, R. A., Fanning, P., and Klagsbrun, M. (1988). *Nature (London)* **331**, 173–175.
16. Blam, S. B., Mitchell, R., Ticher, E., Rubin, J. S., Silva, M., Silver, S., Fiddes, J. C., Abraham, J. A., and Aaronson, S. A. (1988). *Oncogene* **3**, 129–136.

17. Vlodavsky, I., Fridman, R., Sullivan, R., Sasse, J., and Klagsbrun, M. (1987). *J. Cell. Physiol.* **131,** 402–408.
18. Keating, M. T., and Williams, L. T. (1988). *Science* **239,** 914–916.
19. Bernstein, S. C., and Weinberg, R. A. (1985). *Proc. Natl. Acad. Sci. U.S.A.* **82,** 1726–1730.
20. Klagsbrun, M., Sasse, J., Sullivan, R., Smith, S., and Smith, J. A. (1986). *Proc. Natl. Acad. Sci. U.S.A.* **83,** 2448–2452.
21. Wadzinski, M., Folkman, J., Sasse, J., Ingber, D., Devey, K., and Klagsbrun, M. (1987). *Clin. Physiol. Biochem.* **5,** 200–209.

13
Cellular Oncogenes Conferring Growth Factor Independence on NIH 3T3 Cells

MITCHELL GOLDFARB
Department of Biochemistry and Molecular Biophysics
Columbia University College of Physicians and Surgeons
New York, New York 10032

We have devised conditions for the efficient growth of NIH 3T3 cells cultured in defined medium containing platelet-derived growth factor (PDGF) or fibroblast growth factors (FGFs). We observed that these cells die in the absence of PDGF or FGFs, whereas 3T3 cells transformed by various oncogenes grow effectively without these growth factors. This finding has been exploited in the form of a new assay for the identification of new oncogenes. NIH 3T3 cells are transfected with tumor DNAs and, after a procedure to enrich for cells that have taken up DNA, the cells are maintained in PDGF/FGF-free defined medium. The appearance of growing colonies indicates the transfer of an oncogene from the tumor DNA. This procedure has detected activated *ras* oncogenes in several tumor cell lines. More significantly, a new oncogene has been detected by this assay. This gene, called FGF-5, encodes a growth factor structurally and functionally related to other fibroblast growth factors.

INTRODUCTION

DNA-mediated gene transfer (DNA transfection) into NIH 3T3 murine fibroblast cells has been a fruitful means for detecting oncogenes in mammalian cellular DNA. The most extensively used transformation assay is the NIH 3T3 focus assay, in which transformed 3T3 cells are detected by their growth into dense foci (Shih et al., 1979). The focus assay has detected activated oncogenes in the genomes of many tumors and tumor-derived cell lines (Krontiris and Cooper, 1981; Murray et al., 1981; Perucho et al., 1981; Pulciani et al., 1982; Yuasa et al., 1983). The detected oncogenes are usually members of the ras gene family (Der et al., 1982; Parada et al., 1982; Pulciani et al., 1982; Santos et al., 1982; Hall et al., 1983; Shimizu et al., 1983; Yuasa et al., 1983), although other genes have been characterized as well (Pulciani et al., 1982; Eva and Aaronson, 1985; Fukui et al., 1985; Martin-Zanca et al., 1986). The alternative oncogene assay monitors tumor formation in immunodeficient mice following injection with transfected NIH 3T3 cells. The oncogenes met (Dean et al., 1985) and mas (Young et al., 1986) were discovered by this method.

We sought to devise a new transformation assay which might detect NIH 3T3 cells transformed with different oncogenes which do not induce a scoreable phenotype with other assays. Such an assay would be based on the growth factor requirements of normal and transformed cells. The growth of fibroblasts in culture is most easily achieved in basal medium supplemented with serum as a source of substratum attachment factors and polypeptide growth factors. The principal serum mitogen is platelet-derived growth factor (Pledger et al., 1977). Other identified serum mitogens are the insulin-like growth factors or somatomedins (Stiles et al., 1979). Various oncogenes which induce morphological transformation of fibroblasts also reduce the serum requirement for growth; specifically, it is the PDGF requirement which is reduced (Scher et al., 1978; McClure, 1983; Powers et al., 1984). We report here the defined medium components required for the growth of normal and oncogene-transformed NIH 3T3 cell lines. These findings have established culture conditions in medium lacking PDGF which can detect transformation of 3T3 cells by an assortment of oncogenes.

Transfection of DNAs from human tumor cell lines into NIH 3T3 cells frequently gives transformants detectable in this assay. ras oncogenes account for transforming activities in three of these tumor cell DNAs. We have molecularly cloned another detected oncogene which lacks homology to 18 known oncogenes. This oncogene encodes a protein related to fibroblast growth factors. This protein, which we term FGF-5, is distinct from four previously described members of the FGF gene family. Preliminary data regarding FGF-5 gene expression and mitogenic activities have been obtained.

RESULTS

Our cell culture procedure is a modification of previously described methods (McClure, 1983; Powers et al., 1984). Culture dishes (35 mm diameter) were precoated with 1mg/ml polylysine for 2 hr, then rinsed with buffered saline and incubated overnight in Dulbecco's modified Eagle's medium containing human fibronectin (15 μg/ml). These coated dishes were used for plating freshly trypsinized fibroblasts in medium containing calf serum. After a 3-hr cell attachment period, the cells were refed with a 3:1 mixture of Dulbecco's medium and Ham's F12 medium supplemented with sodium bicarbonate (8 mM), HEPES, pH 7.4 (15 mM), manganese chloride (4 μM), histidine (3 mM), ethanolamine (10 μM), sodium selenite (0.1 μM), transferrin (5 μg/ml), serum albumin–linoleic acid complex (500 μg/ml), and hydrocortisone (2 μM). The supplements and attachment factors were purchased from Sigma and Collaborative Research. The next day (Day 0), cells were refed with the above medium supplemented with mitogens and were refed accordingly on a 3-day schedule. The medium change on Day 0 also served to remove residual serum components which might otherwise gratuitously stimulate cell growth.

We first explored the growth of sparsely plated NIH 3T3 cells in media variably supplemented with PDGF, FGF, and insulin. Cell counts at Days 0 and 5 were used to compute the number of cell doublings in the assay

TABLE I
Growth of NIH 3T3 Cells in Defined Media[a]

FGF (ng/ml), PDGF (ng/ml), or serum	No. of cells doublings in 5-day assay supplemented with insulin (μ/ml)		
	0	10	30
None	<0	<0	<0
FGF (10)	2.4	3.6	3.5
FGF (30)	3.4	4.6	4.6
FGF (100)	4.4	5.6	5.5
PDGF (5)	3.2	3.9	3.8
PDGF (15)	4.3	5.1	5.0
PDGF (50)	4.3	5.0	5.1
10% calf serum	6.3		

[a] NIH 3T3 cells were sparsely plated on fibronectin-coated dishes. On Day 0, cells from two plates were trypsinized and counted. Duplicate cultures were refed with basal defined media supplemented with different concentrations of insulin, FGF, or PDGF. Two other cultures were refed with serum-containing medium. All cultures were appropriately refed on Day 3, and cells were trypsinized and counted on Day 5. Cell counts from duplicate cultures always varied less than 6%. Cell doublings were calculated as $\log_2$ (Day 5 cell number ÷ Day 0 cell number). In cultures maintained in defined media without FGF nor PDGF, extensive cell death had occurred by Day 5.

period. These cells grew in appropriate defined media virtually as well as in serum-containing medium (Table I). PDGF or FGF was essential for growth, and in their absence the cells slowly lost viability and detached from the plate. Insulin was not essential, but it potentiated the activity of PDGF or FGF 3- to 4-fold.

We next examined the effects of fibroblast transformation on mitogen requirements for cell growth. We transformed NIH 3T3 cells with four retroviral oncogenes (v-*src*, v-*mos*, v-*sis*, v-*fos*) and one cellular oncogene [c-H-*ras*(Val-12)] chosen for the functional diversity of their encoded proteins. We transfected NIH 3T3 cells by the standard protocol (Wigler *et al.*, 1979) with plasmids containing these oncogenes. In some cases, the oncogene-bearing plasmids also contained a gene (*neo*) conferring resistance to the neomycin analog G418; in other cases, a second plasmid containing *neo* was cotransferred with the oncogene-bearing plasmid. A clone of each resultant G418-resistant, morphologically transformed cell type was expanded for analysis of growth in defined medium. We found that clones of 3T3 cells transformed with *mos*, *ras*, *src*, or *sis* could grow in the absence of FGF or PDGF (Table II). Insulin still enhanced the growth of these transformed cells. By contrast, the v-*fos* transformant failed to grow appreciably in FGF-free medium. We examined the levels of v-*fos* RNA in these cells in the presence and absence of FGF and found no appreciable difference. Furthermore, when pools of v-*fos*-transfected cells were cultured in defined medium lacking FGF, no colonies developed (data not shown). It is therefore likely that the *fos* gene product, unlike those of the other oncogenes, does not abrogate the PDGF/FGF growth requirement.

The absolute requirement of NIH 3T3 cells for PDGF or FGF, and the abrogation of this requirement by oncogene transformation, forms the basis of our new transformation assay. As shown in Fig. 1, NIH 3T3 cells are transfected with cellular DNA and pLTRneo and are first selected with G418 to enrich for cells with stably acquired foreign DNA. Pools of resultant colonies are then maintained in PDGF-free defined medium to select for transformed cells. We have performed transfections with DNAs from 17 human tumor-derived cell lines and have used human placental DNA and NIH 3T3 DNAs as negative controls. The results are tabulated in Table III. DNAs from 8 of these tumor cell lines gave transformants upon transfection with variable frequencies greater than that observed with control DNAs.

DNA was prepared from independent transformants derived by transfection with human tumor cell line DNAs. These transformant DNAs were used in a second cycle of 3T3 cell transfection, G418 selection, and defined medium selection. As shown in Table IV, most transformant DNAs generated secondary transformants upon transfection.

We wished to determine the oncogenes in the tumor DNAs responsible

TABLE II
Growth of Oncogene-Transformed Cells in Defined Media[a]

Exp. no. (days of growth)	Cell line	No. of cell doublings			
		Without FGF, without insulin	Without FGF, with insulin	With FGF, with insulin	Without FGF/with FGF ratio
1(5)	NIH v-sis	2.9	5.1	6.1	0.84
	NIH c-H-ras	3.9	5.2	5.4	0.95
2(8)	NIH v-src	2.4	4.1	5.8	0.71
	NIH v-mos	1.5	6.1	6.0	1.02
3(5)	NIH v-fos	<0	0.4	4.3	0.10

[a] Transformed cells were cultured in defined medium containing 30 µg/ml insulin and 25 ng/ml FGF, containing insulin alone, or containing no supplements.

1) Transfect each 100mm plate with 30μg tumor DNA + 1μg pLTRneo. 8hr.

2) Refeed with DME + serum.O/N

3) Trypsinize.Split 1:2.

4) DME + serum + G418. 12 Days.

5) Trypsinize.

6) Plate 10⁵ cells on fibronectin coated 35mm dish. Refeed and maintain in defined medium w/o PDGF/FGF.

7) 8 days.

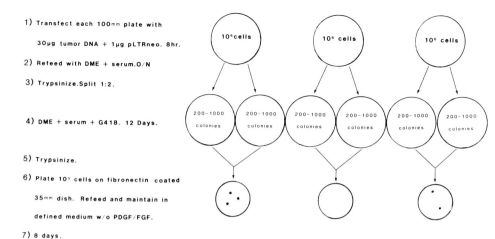

Fig. 1. Protocol for detecting oncogenes by defined medium culture.

TABLE III
Selection of Transformed Cells Following Transfection of NIH 3T3 Cells with DNAs from Human Tumor Cell Lines[a]

Tumor cell line	Tissue of origin	No. G418r colonies screened	No. of independent tranformants	Tranformants per 10,000 G418r colonies	ras homology
Transformed cells					
Calu-4	Lung	8000	>8	>10	K-ras
KNS 62	Lung	9500	4	4.1	H-ras
SAOS-2	Bone	26,000	6	2.3	None
639 V	Bladder	15,000	2	1.3	K-ras
SK-N-MC	Neural	29,000	4	1.3	None
VM-CUB-2	Bladder	19,000	2	1.1	None
SK-MES-1	Lung	53,000	4	0.8	None
SW 1088	Astrocyte	22,000	1	0.5	None
A 172	Neural	9000	0		
BT-20	Breast	9000	0		
HT 1376	Bladder	8000	0		
5637	Bladder	6500	0		
MCF-7	Breast	8000	0		
SK-HEP-1	Liver	7000	0		
VM-CUB-1	Bladder	6500	0		
IMR-32	Neural	9000	0		
MDA-MB-453	Breast	8000	0		
Nontransformed cells					
NIH 3T3		95,000	0	<0.1	
Human placenta		84,000	1	0.1	

[a] NIH 3T3 cells transfected with tumor cell DNA and pLTRneo were first selected for G418 resistance in serum-containing medium. The number of G418r colonies were approximated prior to pooling and culturing in PDGF/FGF-free defined medium. Transformed colonies were considered to derive from independent transfection events if they arose on separate defined media dishes. Transformants containing human ras oncogenes are indicated.

TABLE IV
Selection of Transformed Cells Following Transfection of NIH 3T3 Cells with DNAs from Primary Tranformant Cell Lines[a]

Primary tranformant cell line	No. of G418r colonies screened	No. of secondary transformants	Transformants per 10,000 G418r colonies
SAOS2-1	10,000	2	2.0
SAOS2-2	16,000	0	<1
SAOS2-3	5000	2	4.0
VMCUB2-1	6500	3	4.7
VMCUB2-2	13,000	4	3.1
SKNMC-1	3000	3	10.0
SKNMC-2	6500	3	4.7
SKNMC-3	9500	0	<1
SKMES-1	6000	1	1.7
SKMES-2	7000	4	5.7
SKMES-3	8500	3	3.6

[a] Primary transformant cell lines were derived from transfection and defined selection as listed in Table III; for example, VMCUB2-2 derived from transfection with DNA from human tumor cell line VMCUB2.

for the observed transformants. Since activated *ras* oncogenes have been observed in certain tumors, and since *ras*-transformed fibroblasts can grow in PDGF/FGF-free medium, we first screened transformants for the presence of human *ras* genes acquired from the tumor DNAs. Using radiolabeled probes for the H-*ras*, K-*ras*, and N-*ras* genes and the technique of Southern blotting, we could show that transformants elicited by DNAs from three of the tumor cell lines contained *ras* genes (Table III). We turned our further attention to transformants presumably bearing human oncogenes other than *ras*. For the purpose of this chapter, only transformants obtained with DNA from the VMCUB2 bladder carcinoma cell line are discussed.

DNAs from the two primary transformants of the VMCUB2 tumor cell line were assayed for the presence of human *Alu* repeat sequences. As shown in Fig. 2, four secondary transformants derived from the primary transformant VMCUB2-2 lack *Alu* repeats (lanes b–e) while three transformants derived from the primary transformant VMCUB2-1 each have two *Alu* sequences (lanes g–i), one of which is within a conserved 2.3-kilobase pair (kbp) *Eco*RI DNA fragment. Hence, different molecular events gave rise to the two primary VMCUB2 transformants. The human oncogene in transformant VMCUB2-1 has been cloned from a phage genomic library of secondary transformant VMCUB2-1-1. Initial clones were obtained by *Alu* sequence homology, and further clones were obtained by genomic walking.

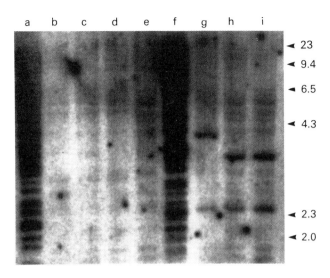

Fig. 2. Human *Alu* repeat sequences in transformants derived from VMCUB2 tumor DNA transfections. *Eco*RI-digested DNAs from primary and secondary transformants of VMCUB2 were analyzed by filter blot hybridization with a SP6 polymerase-derived ^{32}P-labeled RNA probe containing the human *Alu* repeat sequence. DNAs from primary transformant VMCUB2-2 (lane a) and its secondary transformants (lanes b–e), primary transformant VMCUB2-1 (f) and its secondary transformants (g–i) are shown. Arrows denote size markers (in kilobase pairs).

A physical map of the VMCUB2-1 oncogene is shown in Fig. 3. The genomic inserts in the RA3 and R4 phage DNA isolates lack transforming activity, but a ligated mixture of RA3 and R4 partial *Eco*RI digests gave many transformed colonies following transfection (data not shown), demonstrating that these two overlapping clones span the entire oncogene. We have iso-

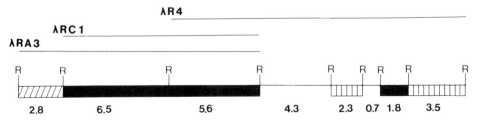

Fig. 3. Physical map of the VMCUB2-1 oncogene. The inserts of three overlapping genomic clones are shown. Lengths (in kilobase pairs) of *Eco*RI (R) restriction fragments are indicated. Fragments containing *Alu* repeats are indicated with vertical hash marks, fragments homologous to the oncogene cDNA clone are indicated as solid boxes, and the 2.8-kbp fragment derived from pLTRneo is marked with diagonal stripes.

lated a 1.1-kbp cDNA clone (termed 1-2-2) which is homologous to two transcripts of the VMCUB2-1 oncogene. This cDNA lacks native *Eco*RI cleavage sites, and it hybridizes to the 6.5-, 5.6-, and 1.8-kbp *Eco*RI fragments of the VMCUB2-1 oncogene. Hence, this oncogene contains at least three exons.

We have inserted the 1-2-2 cDNA clone into the pvcos-7 retroviral expression vector. The resultant construct, pLTR122, has potent transforming activity when transfected cells are assayed by defined medium selection (data not shown). Interestingly, the transforming activity of pLTR122 is much weaker when assayed by the morphological focus assay.

In order to identify this cloned oncogene, we first tried hybridizing the 1-2-2 cDNA clone to a panel of 18 known oncogenes. Such analysis failed to detect homology between our new isolate and any of these genes. Hence, we characterized this cDNA clone by determining its DNA sequence. The complete nucleotide sequence of the cDNA clone was determined by the dideoxynucleotide chain termination method (Sanger *et al.*, 1977), and the deduced sequence is shown in Fig. 4. The 1120-base pair sequence lacks the 3' poly(A) tract because the cDNA cloning procedure involved *Eco*RI digestion, which cut the native cDNA at least once. The single strand shown corresponded to that of the RNA, as demonstrated by the ability of an oligonucleotide of complementary sequence to prime reverse transcription of the oncogene-encoded RNA. The cDNA sequence contained two ATG-initiated open reading frames (ORFs); ORF-1 and ORF-2 can specify polypeptides of 38 and 267 amino acid residues, respectively. The two reading frames slightly overlapped, with the ORF-1 termination codon, TGA, situated one nucleotide downstream from the ORF-2 initiator ATG.

The protein specified by ORF-2 bore a leucine-rich hydrophobic amino terminus, which may serve as a signal sequence for contranslational transport into the endoplasmic reticulum (Blobel *et al.*, 1979). The lack of other extensive tracts of hydrophobic residues in the protein suggests that the ORF-2 product can be secreted. The predicted protein also bears a consensus sequence for N-linked glycosylation, Asn-Gly-Ser (residues (109–111). When the ORF-2 protein sequence was compared with sequences in the PIR-NBRF data base, substantial homology was detected between the ORF-2 protein and both acidic and basic fibroblast growth factors (Jaye *et al.*, 1986; Abraham *et al.*, 1986). The recently described *int*-2 and *hst*/KS3 predicted protein sequences (Moore *et al.*, 1986; Taira *et al.*, 1987) are homologous to, but distinct from, the ORF-2 protein, which we have termed FGF-5.

A comparison of the FGF-5 amino acid sequence with those for other FGF family proteins is shown in Fig. 5. Two blocks of FGF-5 amino acid residues (90 to 180 and 187 to 207) showed substantial homology to the other proteins,

```
ORF-1..................METSerThrArgCysGlyGluAlaGlyGlnSerArgGlyThrGlnProHisArgGlyTyrArg      20
CCTCTCCCCTTCTCTTCCCCGAGGCTATGTCCACCCGGTGCGGCGAGGCGGGCCAGAGCAGAGGCACGCAGCCGCACAGGGGCTACAGA    89

AlaGlnAsnGlnProTyrLysMetHisLeuGlyProProArgLeuGluGluEND                                       38
GCCCAGAATCAGCCCTACAAGATGCACTTAGGACCCCCGCGGCTGGAAGAATGAGCTTGTCCTTCCTCCTCCTCCTCTTCTTCAGCCAC   178
ORF-2.........................................METSerLeuSerPheLeuLeuLeuLeuPhePheSerHis        13

CTGATCCTCAGCGCCTGGGCTCACGGGGAGAAGCGTCTCGCCCCCAAAGGGCAACCCGGACCCGCTGCCACTGATAGGAACCCTAGAGGC  268
LeuIleLeuSerAlaTrpAlaHisGlyGluLysArgLeuAlaProLysGlyGlnProGlyProAlaAlaThrAspArgAsnProArgGly   43

TCCAGCAGCAGACAGAGCAGCAGTAGCGCTATGTCTTCCTCTTCTGCCTCCTCCTCCCCCGCAGCTTCTCTGGGCAGCCAAGGAAGTGGC  358
SerSerSerArgGlnSerSerSerSerAlaMetSerSerSerSerAlaSerSerSerProAlaAlaSerLeuGlySerGlnGlySerGly   73

TTGGAGCAGAGCAGTTTCCAGTGGAGCCTCGGGGCGCGGACCGGCAGCCTCTACTGCAGAGTGGGCATCGGTTTCCATCTGCAGATCTAC  448
LeuGluGlnSerSerPheGlnTrpSerLeuGlyAlaArgThrGlySerLeuTyrCysArgValGlyIleGlyPheHisLeuGlnIleTyr  103

CCGGATGGCAAAGTCAATGGATCCCACGAAGCCAATATGTTAAGTGTTTTGGAAATATTTGCTGTGTCTCAGGGGATTGTAGGAATACGA  538
ProAspGlyLysValAsnGlySerHisGluAlaAsnMetLeuSerValLeuGluIlePheAlaValSerGlnGlyIleValGlyIleArg  133

GGAGTTTTCAGCAACAAATTTTTAGCGATGTCAAAAAAAGGAAAACTCCATGCAAGTGCCAAGTTCACAGATGACTGCAAGTTCAGGGAG  628
GlyValPheSerAsnLysPheLeuAlaMetSerLysLysGlyLysLeuHisAlaSerAlaLysPheThrAspAspCysLysPheArgGlu  163

CGTTTTCAAGAAAATAGCTATAATACCTATGCCTCAGCAATACATAGAACTGAAAAAACAGGGCGGGAGTGGTATGTTGCCCTGAATAAA  718
ArgPheGlnGluAsnSerTyrAsnThrTyrAlaSerAlaIleHisArgThrGluLysThrGlyArgGluTrpTyrValAlaLeuAsnLys  193

AGAGGAAAAGCCAAACGAGGGTGCAGCCCCCGGGTTAAACCCCAGCATATCTCTACCCATTTTCTTCCAAGATTCAAGCAGTCGGAGCAG  808
ArgGlyLysAlaLysArgGlyCysSerProArgValLysProGlnHisIleSerThrHisPheLeuProArgPheLysGlnSerGluGln  223

CCAGAACTTTCTTTCACGGTTACTGTTCCTGAAAAGAAAAATCCACCTAGCCCTATCAAGTCAAAGATTCCCCTTTCTGCACCTCGGAAA  898
ProGluLeuSerPheThrValThrValProGluLysLysAsnProProSerProIleLysSerLysIleProLeuSerAlaProArgLys  253

AATACCAACTCAGTGAAATACAGACTCAAGTTTCGCTTTGGATAATATTAATCTTGGCCTTGTGAGAAACCATTCTTTCCCCTCAGGAGT  988
AsnThrAsnSerValLysTyrArgLeuLysPheArgPheGlyEND                                               267

TTCTATAGGTGTCTTCAGAGTTCTGAAGAAAAATTACTGGACACAGCTTCAGCTATACTTACACTGTATTGAAGTCACGTCATTTGTTTC 1078

AGTGTGACTGAAACAAAATGTTTTTTGATAGGAAGGAAACTG                                                  1120
```

Fig. 4. Sequence of the 1-2-2 FGF-5 cDNA clone. The nucleotide sequence is shown (1120 bases) along with the predicted amino acid sequences specified by the two open reading frames, ORF-1 and ORF-2.

ranging from 40% versus acidic FGF to 50% versus *hst*/KS3. Within these homology blocks, the five proteins were identical at 20% of the residues, and, allowing for conservative amino acid substitutions, the five proteins shared 29% homology. Nucleotide sequence homology between the coding sequences of FGF-5 and those of related genes was minimal.

The sequences of the FGF-related proteins differed in several respects. First, the five sequences differed in the length and sequence of residues between the two homology blocks and distal to them. Second, the FGF-5 sequence was unique in bearing an insertion within the second homology block (Cys-Ser, residues 201 and 202). Last, the amino-terminal sequences of the FGF-5, *hst*/KS3, and *int*-2 proteins were extensively hydrophobic, while those of acidic and basic FGF precursor proteins were not, suggesting differences in posttranslational trafficking among the FGF-like proteins.

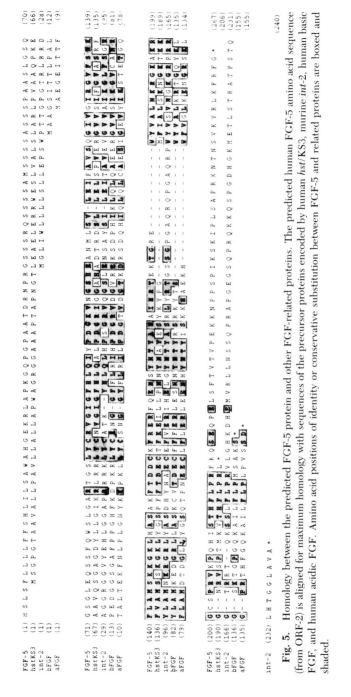

Fig. 5. Homology between the predicted FGF-5 protein and other FGF-related proteins. The predicted human FGF-5 amino acid sequence (from ORF-2) is aligned for maximum homology with sequences of the precursor proteins encoded by human *hst*/KS3, murine *int*-2, human basic FGF, and human acidic FGF. Amino acid positions of identity or conservative substitution between FGF-5 and related proteins are boxed and shaded.

Our sequence data suggested that the FGF-5 gene encodes a secreted growth factor. We have assayed for mitogenic activity secreted from 3T3 cells transformed by the FGF-5 gene or by FGF-5 cDNA linked to retroviral promoter sequences. Conditioned media from such transformed cell cultures (termed VMCUB2-1 and 3T3-LTR122, respectively) were serially diluted and assayed for the ability to stimulate DNA synthesis in quiescent BALB/c 3T3 fibroblast cultures. These transformed cells secreted a mitogenic activity that was detectable at 1:8 dilutions (Table V). Secretion of mitogenic activity is not a property of transformed cells per se, as NIH 3T3 cells transformed by activated human H-*ras* or v-*src* oncogenes released little or no mitogenic activity (Table V).

As a means of assessing whether the mitogen secreted by FGF-transformed cells is, indeed, FGF-5, we tested whether this mitogen has proper-

TABLE V
Stimulation of quiescent BALB/c 3T3 Cells with Conditioned Media[a]

Conditioned medium or extract	Dilution	[^{3}H]Thymidine incorporation (cpm/10,000 cells)
Conditioned medium from		
3T3-*ras*	1:2	1.1
	1:4	1.5
	1:8	1.0
3T3-*src*	1:2	1.5
	1:4	1.1
	1:8	1.1
VMCUB2-1	1:2	23.0
	1:4	9.5
	1:8	3.2
3T3-LTR122	1:2	130.0
	1:4	99.1
	1:8	40.8
Without supplements		0.9
With 10% calf serum		146.0

[a] BALB/c 3T3 cells were plated in culture wells in serum-containing medium and maintained for 5 days without refeeding, allowing the cells to form quiescent, serum-exhausted monolayers. Cultures were refed with serum-free medium containing dilutions of conditioned medium from transformed cells. Tritiated thymidine was added 15 hr later, and after 3 hr of incubation, the cultures were fixed in 15% trichloroacetic acid, the DNA dissolved in 0.5 N NaOH, and incorporated label assayed by liquid scintillation.

ties diagnostic for FGFs. One property of acidic and basic FGFs is their ability to strongly bind to the glycosaminoglycan heparin (Conn and Hatcher, 1984; Maciag *et al.*, 1984; Shing *et al.*, 1984). Elution of FGFs from heparin affinity resins requires NaCl concentrations of 1.0 M or greater. By contrast, PDGF, a basic protein which binds heparin by weaker ionic interactions, elutes at approximately 0.5 M (Vlodavsky *et al.*, 1987). Mitogenic conditioned medium from FGF-5 cDNA-transformed 3T3 cells was passed directly over a heparin-Sepharose column, which was washed extensively with buffered 0.45 M NaCl, and then eluted with a stepwise increase in salt concentrations. Dilutions of column fractions were assayed for mitogenic activity on BALB/c 3T3 cells. The peak of mitogenic activity eluted in the 1.0 and 1.5 M NaCl fractions (Fig. 6).

A second property of FGFs is their broad spectrum of mitogenicity, including their activity toward vascular endothelial cells. Conditioned medium from FGF-5-transformed cells was tested for the ability to stimulate the proliferation of fetal bovine heart endothelial cells. Fetal bovine heart endothelial cell cultures were fed with conditioned medium diluted into medium

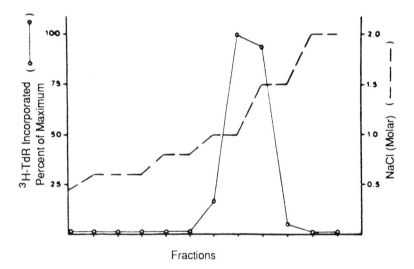

Fig. 6. Heparin-Sepharose chromatography of mitogenic activities secreted from transformed cells expressing FGF-5. Conditioned medium (300 ml) from cells transformed by the FGF-5 cDNA was applied to a 1-ml heparin-Sepharose column, and the column was washed with 20 ml of 0.45 M NaCl in 20 mM Tris, pH 7.5. Bound material was eluted stepwise with buffered NaCl solutions (0.6, 0.8, 1.0, 1.5, 2.0 M). Ten microliters of the 1-ml fractions was assayed for the ability to stimulate DNA synthesis in quiescent BALB/c 3T3 cell cultures. Tritiated thymidine incorporation is expressed as a percentage of the maximum incorporation attainable in the assay by using 10% calf serum.

containing 1.5% calf serum and 5 μg/ml heparin. Cell growth over 6 days was determined by counting trypsinized cells. Whereas cell cultures without conditioned medium underwent only 1.5 population doublings during the assay, 1:2 and 1:8 dilutions of conditioned medium stimulated growth to 3.3 and 3.4 population doublings, respectively. This stimulation is comparable with that induced by partially purified basic FGF (3.1 doublings). These data strongly suggest a functional similarity between FGF-5 and the well-characterized FGFs.

We have looked for expression of FGF-5 within a panel of human tumor cell lines of solid tumor origin. Cytoplasmic RNAs were prepared from 13 such cell lines, and the samples were assayed for FGF-5 transcripts by gel electrophoresis and filter blot hybridization. None of the cell lines had shown any evidence for FGF-5 gene rearrangement (data not shown). Two of the cell lines, hepatoma SKHEP1 and bladder carcinoma 639V, expressed two RNA species homologous to the FGF-5 cDNA probe (Fig. 7, lanes h and k). A third cell line, endometrial carcinoma HEC-1A, expressed FGF-5 RNA at lower levels (lane j), while the other tumor cell lines did not express

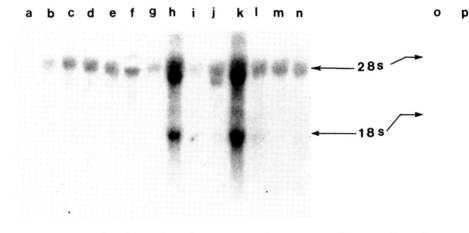

Fig. 7. Northern blot analysis of FGF-5 RNA in human tumor cell lines. Total cytoplasmic RNAs (10 μg; lanes a–n) or poly(A)-selected RNAs (1 μg, lanes o and p) were subjected to electrophoresis through 1.5% agarose gels containing 2.2 M formaldehyde. Gel-embedded RNA was transferred to nitrocellulose, hybridized with nick-translated FGF-5 cDNA, and autoradiographed. Total RNAs were from NIH 3T3 cells (lane a), human tumor cell lines VMCUB1 (lane b), VMCUB2 (c), Calu4 (d), KNS62 (e), BT20 (f), MDAMB469 (g), SKHEP1 (h), MCF7 (i), HEC-1A (j), 639V (k), 253J (l), HT29 (m), and SH1 (n). Poly(A) RNAs were from 639 V (o) and from NIH 3T3 transformant VMCUB2-1 bearing the FGF-5 gene (p). The positions of 18 and 28 S rRNAs are indicated.

FGF-5 detectably. (The cDNA probe also hybridized weakly to human 28 S rRNA.)

Quite remarkably, the VMCUB2 cell line was negative for FGF-5 expression (Fig. 7, lane c), despite the fact that a 3T3 transformed cell line derived with VMCUB2 DNA was used to isolate the FGF-5 oncogene. We have found that the FGF-5 gene had become fortuitously activated during the original VMCUB2 DNA transfection by a DNA rearrangement which juxtaposed a retroviral enhancer (present in the plasmid pLTRneo) just 5' to the FGF-5 gene (data not shown). Despite this transfection artifact, our transformation assay has allowed us to identify a new growth factor gene with oncogenic potential. The expression of FGF-5 in certain tumors (Fig. 7) suggests a role for this gene in neoplasia.

DISCUSSION

Our motivation for developing a new defined medium transformation assay was a desire to identify new oncogenes and to characterize their roles in malignant as well as normal cell growth. The human FGF-5 oncogene represents the first gene to be discovered by this approach. This transformation assay is likely to unearth other oncogenes as well. For example, many of the other transformants derived in the transfections summarized in Table III may contain other novel oncogenes. Furthermore, we have found that normal human placental DNA yields many transformants if the transfected DNA is supplemented with a retroviral DNA promoter/enhancer element; these transfected retroviral sequences induce random transcriptional activation of the cotransfected placental DNA (M. Goldfarb, unpublished data). Hence, we feel that future exploration with this methodology will yield many new oncogenes.

The FGF-5 gene represents the fifth member of a gene family which likely plays many roles in normal growth, cellular differentiation, tissue repair, and neoplasia. Acidic and basic FGFs stimulate growth or differentiation of a broad spectrum of mesoderm- and neuroectoderm-derived cell types (Thomas and Gimenez-Gallego, 1986). Mitogenicity of FGFs toward vascular endothelial cells also accounts for the potent angiogenic activity of these factors (Esch et al., 1985; Thomas et al., 1985). Furthermore, basic FGF has been shown to promote survival and differentiation of cortical neurons cultured in vitro (Morrison et al., 1986) and to induce the conversion of cultured Xenopus embryonic ectoderm to mesoderm (Slack et al., 1987). To determine the roles which FGF-5 plays in normal and neoplastic processes, we shall have to define where and when FGF-5 is expressed and to characterize the range of biological responses which FGF-5 can elicit. The data described in this presentation have appeared in three recent publications from our laboratory (Zhan and Goldfarb, 1986; Zhan et al., 1987, 1988).

REFERENCES

Abraham, J. A., Whang, J. L., Tumulo, A., Mergia, A., Friedman, J., Gospadaorwicz, D., and Fiddes, J. C. (1986). *Embo J.* **5**, 2523–2528.
Blobel, G., Walter, P., Change, G. N., Goldman, B. M., Erickson, A. H., and Lingappa, V. R. (1979). *Symp. Soc. Exp. Biol.* **33**, 9–36.
Conn, G., and Hatcher, V. B. (1984). *Biophys. Res. Commun.* **124**, 262–268.
Dean, M., Park, M., Lebeau, M. M., Robins, T. S., Diaz, M. O., Rowley, J. D., Blair, D. G., and Vande Woude, G. F. (1985). *Nature (London)* **318**, 385–388.
Der, C. J., Krontiris, T. G., and Cooper, G. M. (1982). *Proc. Natl. Acad. Sci. U.S.A.* **79**, 3637–3640.
Esch, F., Baird, A., Ling, N., Ueno, N., Hill, F., Denoroy, L., Kleper, R., Gospadarowicz, D., Bohlen, P., and Guillemin, R. (1985). *Proc. Natl. Acad. Sci. U.S.A.* **82**, 6507–6511.
Eva, A., and Aaronson, S. A. (1985). *Nature London* **316**, 273–275.
Fukui, M., Yamamoto, T., Kawai, S., Maruo, K., and Toyoshima, K. (1985). *Proc. Natl. Acad. Sci. U.S.A.* **82**, 5954–5958.
Hall, A., Marshall, C. J., Spurr, N. K., and Weiss, R. A. (1983). *Nature (London)* **303**, 396–400.
Jaye, M., Howk, R., Burgess, W., Ricca, G. A., Chiu, I.-M., Ravera, M. W., O'Brien, S. J., Modi, W. S., Maciag, T., and Drohan, W. N. (1986). *Science* **233**, 541–545.
Krontiris, T. G., and Cooper, G. M. (1981). *Proc. Natl. Acad. Sci. U.S.A.* **78**, 1181–1184.
McClure, D. B. (1983). *Cell (Cambridge, Mass.)* **32**, 999–1006.
Maciag, T., Mehlman, T., Freisel, R., and Schreiber, A. (1984). *Science* **225**, 932–935.
Martin-Zanca, D., Hughes, S. H., and Barbacid, M. (1986). *Nature (London)* **319**, 743–748.
Morrison, R. S., Sharma, A., de Vellis, J., and Bradshaw, R. A. (1986). *Proc. Natl. Acad. Sci. U.S.A.* **83**, 7537–7541.
Murray, M., Shilo, B., Shih, C., Cowing, D., Hsu, H. W., and Weinberg, R. A. (1981). *Cell (Cambridge, Mass.)* **25**, 355–361.
Parada, L. F., Tabin, C. J., Shih, C., and Weinberg, R. A. (1982). *Nature (London)* **297**, 474–478.
Pledger, W. J., Stiles, C. D., Antoniades, H. N., and Scher, C. D. (1977). *Proc. Natl. Acad. Sci. U.S.A.* **74**, 4481–4485.
Powers, S., Fisher, P. B., and Pollack, R. (1984). *Mol. Cell. Biol.* **4**, 1572–1576.
Pulciani, S., Santos, E., Lauver, A. V., Long, L. K., Aaronson, S. A., and Barbacid, M. (1982). *Nature (London)* **300**, 539–542.
Sanger, F., Nicklen, S., and Coulson, A. R. (1977). *Proc. Natl. Acad. Sci. U.S.A.* **74**, 5463–5467.
Santos, E., Tronick, S. R., Aaronson, S. A., Pulciani, S., and Barbacid, M. (1982). *Nature (London)* **298**, 343–347.
Scher, C. D., Pledger, W. J., Martin, P., Antoniades, H., and Stiles, C. D. (1978). *J. Cell Physiol.* **97**, 371–380.
Shih, C., Shilo, B.-Z., Goldfarb, M. P., Dannenberg, A., and Weinberg, R. A. (1979). *Proc. Natl. Acad. Sci. U.S.A.* **76**, 5714–5718.
Shimizu, K., Goldfarb, M., Suard, Y., Perucho, M., Li, Y., Kamata, T., Feramisco, J., Stavnezer, E., Fogh, J., and Wigler, M. (1983). *Proc. Natl. Acad. Sci. U.S.A.* **80**, 2112–2116.

Shing, Y., Folkman, J., Sullivan, R., Butterfiled, C., Murray, J., and Klagsbrun, M. (1984). *Science* **223**, 1296–1299.

Slack, J. M. W., Darlington, B. G., Heath, J. K., and Godsave, S. F. (1987). *Nature (London)* **326**, 197–200.

Stiles, C. D., Capone, G. T., Scher, C. D., Antoniades, H. N., Van Wyk, J. J., and Pledger, W. J. (1979). *Proc. Natl. Acad. Sci. U.S.A.* **76**, 1279–1283.

Taira, M., Yoshida, T., Miyagawa, K., Sakamoto, H., Terada, M., and Sugimura, T. (1987). *Proc. Natl. Acad. Sci. U.S.A.* **84**, 2980–2984.

Thomas, K. A., and Gimenez-Gallego, G. (1986). *Trends Biochem. Sci.* **11**, 81–84.

Thomas, K. A., Rios-Candelore, M., Gimienez-Gallego, G., DiSalvo, J., Bennet, C., Rodkey, J., and Fitzpatrick, S. (1985). *Proc. Natl. Acad. Sci. U.S.A.* **82**, 6409–6413.

Vlodavsky, I., Folkman, J., Sullivan, R., Friedman, R., Ishai-Michaeli, R., Sasse, J., and Klagsbrun, M. (1987). *Proc. Natl. Acad. Sci. U.S.A.* **84**, 2292–2296.

Wigler, M., Sweet, R., Sim, G. K., Wold, B., Pellicer, A., Lacy, E., Maniatis, T., Silverstein, S., and Axel, R. (1979). *Cell (Cambridge, Mass.)* **16**, 777–785.

Young, D., Waitches, G., Birchmeier, C., Fasano, O., and Wigler, M. (1986). *Cell (Cambridge, Mass.)* **45**, 711–719.

Yuasa, Y., Srivastava, S., Dunn, D. Y., Rhim, J. S., Reddy, E. P., and Aaronson, S. A. (1983). *Nature (London)* **303**, 775–779.

Zhan, X., Bates, B., Hu, X., and Goldfarb, M. (1988). *Mol. Cell. Biol.* **8**, 3487–3495.

Zhan, X., Culpepper, A., Reddy, M., Loveless, J., and Goldfarb, M. (1987). *Oncogene* **1**, 369–376.

Zhan, X., and Goldfarb, M. (1986). *Mol. Cell. Biol.* **6**, 3541–3544.

14
Potential Function of the *mos* Protooncogene in Germ Cell Differentiation and Early Development

GEOFFREY M. COOPER
*Dana-Farber Cancer Institute, and
Department of Pathology
Harvard Medical School
Boston, Massachusetts 02115*

INTRODUCTION

Oncogenes were initially defined as retroviral genes which induced neoplastic transformation of virus-infected cells. Subsequently, cellular oncogenes were identified as (a) homologs of retroviral oncogenes, (b) cellular genes which induce transformation upon transfection of appropriate recipient cells, and (c) genes which are frequently altered in neoplasms by virus integration, chromosome translocations, or amplification. To date, more than 50 cellular oncogenes have been isolated by one or more of these approaches. Those oncogenes which have been characterized so far encode proteins which can be divided into five different functional groups: extracellular growth factors, plasma membrane-associated tyrosine kinases, plasma membrane-associated guanine nucleotide-binding proteins, cytoplasmic serine/threonine kinases, and nuclear proteins. These genes can thus be viewed as elements in signal transduction pathways which can act to induce abnormal cell growth.

Activated oncogenes, which induce cell transformation, are altered versions of the normal cell homologs (protooncogenes). At least three different mechanisms of oncogene activation have been described: (a) changes in the regulation of gene expression, (b) point mutations which result in single amino acid substitutions, and (c) formation of recombinant fusion proteins consisting of a protein kinase catalytic domain which has been dissociated by recombination from its normal amino terminus. Although activated oncogenes induce abnormal cell proliferation, it does not necessarily follow that the normal function of protooncogenes is also concerned with cell proliferation, particularly since many activated oncogenes encode proteins which are significantly different than their normal cell progenitors.

The normal cellular role of four protooncogenes has been indicated by nucleotide sequence analysis. Thus, the *sis* oncogene is derived from platelet-derived growth factor, the *erbB* oncogene from epidermal growth factor receptor, the *fms* oncogene from macrophage colony-stimulating factor receptor, and the *erbA* oncogene from thyroid hormone receptor (1–6). The normal functions of other protooncogenes, however, remain unknown. Some protooncogenes, such as *myc*, are expressed in proliferating but not in quiescent cells, suggesting that these protooncogenes may function in control of normal cell proliferation. Other protooncogenes, however, may normally function in developmental processes which are distinct from cell growth.

The *src* protooncogene, which is the progenitor of the tyrosine kinase oncogene of Rous sarcoma virus, is normally expressed at high levels during neuronal differentiation (7–9). In addition, the *src* protein expressed in postmitotic neurons appears to differ both structurally and functionally from that expressed at lower levels in other cell types (7). Further, introduction of activated *src* into the PC12 rat phaeochromocytoma cell line results in terminal neuronal differentiation of these cells (10). These findings suggest that the *src* tyrosine kinase protooncogene may normally play a role in differentiation of neurons, which are postmitotic cells, rather than in cell proliferation.

The *ras* protooncogenes are also expressed at high levels in neurons, and introduction of activated *ras* genes into PC12 cells results in neuronal differentiation rather than cell proliferation (11–13). Moreover, microinjection of PC12 cells with an anti-*ras* monoclonal antibody will block the differentiation of these cells induced by nerve growth factor (14). In this cell system, therefore, the *ras* oncogene also acts to induce differentiation rather than transformation. The results of the antibody microinjection experiments further suggest the possibility that *ras* may normally function in PC12 cells to transduce a differentiation signal initiated by nerve growth factor.

An example of a protooncogene which may normally function in reproductive processes is *abl*, which encodes a tyrosine kinase. High levels of a unique *abl* transcript of 4.7 kilobases (kb) are expressed during spermatogenesis, whereas *abl* is expressed in somatic cells as transcripts of 6.2 and 8.0 kb (15). The 4.7-kb transcript is specifically expressed in haploid postmeiotic germ cells, the round and condensing spermatids (15). The specific transcription of *abl* in these cells suggests a function for this protooncogene in the differentiation of haploid spermatids to mature spermatozoa.

A role for another protooncogene, *mos*, in reproductive processes has also been suggested by the finding of specific *mos* transcripts of 1.7 and 1.4 kb, respectively, in the testes and ovaries of adult mice (16). To begin to define the possible function(s) of this protooncogene we have investigated the cell types in testis and ovary in which *mos* is transcribed.

EXPRESSION OF *mos* IN MALE GERM CELLS

To identify the cell type expressing *mos* in the testis, initial experiments utilized mutant mice with defects in germ cell development (17). Expression of *mos* was analyzed in testes of four strains of mice, each of which lacked mature germ cells but retained their complement of somatic cells, including Leydig and Sertoli cells. The testes of male W/W^v and S1/S1^d mice are devoid of germ cells as a result of failure of the primordial germ cells to migrate to and/or proliferate in the gonadal ridge during embryogenesis. X/X sex-reversed (Sxr) mice are phenotypic males in which germ cells do not survive in the testes past 10 days of age. Male mice carrying the X-linked testicular-feminization (Tfm) gene develop small testes but no male accessory organs as a consequence of a lack of androgen receptors. The testes of Tfm mice contain spermatogonia in the mitotic stages of stem cell proliferation, but spermatogenesis does not progress past meiotic prophase.

mos RNA was not detected in the testes of W/W^v, S1/S1^d, Sxr, and Tfm mice, whereas a 1.7-kb *mos* transcript was detected in testes of matched controls (17). Likewise, the 4.7-kb *abl* transcript was detected in the testes of normal control mice but not in the testes of W/W^v, S1/S1^d, Sxr, or Tfm mice, as was the 1.5-kb actin transcript specific for postmeiotic germ cells (18). Expression of *mos* in the testis is thus correlated with normal germ cell differentiation.

We next took advantage of the temporal development of spermatogenic cell types in the testes of normal mice (19,20). *mos* expression was analyzed in RNA from testes of 6-, 20-, and 35-day-old mice. At 6 days of age, over 80% of the cells within the seminiferous epithelium are the somatic Sertoli cells, and the germ cells are represented solely by the mitotically proliferating spermatogonia. By 20 days of age the germ cells are more prominently represented, comprising about 70% of the cells within the seminiferous epithelium. At this age, the germ cell population consists of spermatogonia and spermatocytes in a continuum of developmental stages of meitotic prophase. A large proportion of the spermatocytes (~50%) are in the pachytene stage of late meiotic prophase. By 35 days of age, postmeiotic round and condensing spermatids comprise about 70% of the cells within the seminiferous epithelium and represent the major germ cell types within the testis. The *mos* transcript was detected in 35-day-old but not in 20- or 6-day-old mouse testes (17). Similarly, the 4.7-kb *abl* transcript and the 1.5-kb actin transcript were expressed in 35-day-old but not in 6- or 20-day-old testes. Thus, *mos* expression is temporally correlated with the development of postmeiotic germ cells.

The specific spermatogenic cell types expressing *mos* were then determined by purifying RNA from populations of germ cells enriched for specific stages of development (17). Neither *mos* RNA, the 4.7-kb *abl* transcript, nor the 1.5-kb actin transcript were detected in pachytene spermatocytes. In contrast, RNA of postmeiotic round spermatids contained the 1.7-kb *mos* transcript as well as the 4.7-kb *abl* and 1.5-kb actin transcripts. Thus *mos* transcription, like testis-specific *abl* and actin transcription, is first detected in postmeiotic germ cells.

mos transcription, however, appeared to differ from that of *abl* and actin in later stages of spermatogenesis. *mos* RNA was not detected in residual bodies, whereas the 4.7-kb *abl* and 1.5-kb actin transcripts were present in residual body RNAs at levels comparable to those in whole testis and round spermatid RNAs. Thus, while *abl* and actin RNAs appear to persist throughout the differentiation of postmeiotic spermatids until formation of mature spermatozoa, *mos* RNA is present only transiently during postmeiotic male germ cell development.

EXPRESSION OF *mos* IN OOCYTES

In contrast to the testis, in which the full array of differentiating spermatogenic cell types are represented, germ cells within the ovary are uniformly arrested in the diplotene stage of meiotic prophase. Our first approach to identify the ovarian cell type expressing *mos* involved hormonal stimulation of prepubertal mice. The ovary of prepubertal mice contains oocytes at all stages of oocyte growth, but, in the absence of hormonal stimulation, full follicle development has not occurred. Upon administration of gonadotropins, follicular growth is stimulated, involving proliferation and differentiation of the granulosa cells. Since the number of oocytes remains constant, any change in *mos* transcription as a consequence of hormonal stimulation can most likely be attributed to the proliferation and/or differentiation of granulosa cells in stimulated ovaries. Prepubertal mice were injected with pregnant mare's serum gonadotropin, and ovaries were harvested 36 hr later. RNA was extracted from unstimulated and stimulated ovaries and analyzed by Northern blot hybridization with *mos* probe. The relative concentration of *mos* RNA was 2- to 4-fold higher in unstimulated than in stimulated ovaries (17). Since the effect of hormonal stimulation is to increase substantially the number of granulosa cells while the number of oocytes remains constant, the relative contribution of RNA from the oocyte is greater in unstimulated as compared to stimulated ovaries. Therefore the decrease in relative concentration of *mos* RNA in stimulated ovaries implicates the oocyte as the likely source of *mos* transcription.

We next investigated *mos* expression in oocytes purified from the ovaries of 20- to 23-day-old gonadotropin-stimulated mice. At this age, the ovary contains approximately 6000 oocytes, of which 10% are growing oocytes and 1–2% are large oocytes (21). The large oocytes contain about 200 times more RNA (~0.45 ng RNA/oocyte) than typical somatic cells (22,23). Large germinal vesicle stage oocytes were collected and mechanically denuded of granulosa cells in the presence of dibutyryl cyclic AMP to prevent spontaneous maturation. RNA was extracted in the presence of NIH 3T3 cell carrier RNA, and RNA from 900 oocytes (~0.4 μg) was analyzed by Northern blot hybridization with *mos* probe (17). The *mos* 1.4-kb transcript was detected at high levels in the oocyte RNA, demonstrating directly that *mos* is transcribed in growing oocytes. The concentration of *mos* RNA in oocytes was approximately 100-fold higher than that in round spermatids and is estimated by hybridization intensity to be of the order of 100,000 copies per cell.

POSTTRANSCRIPTIONAL PROCESSING OF *mos* IN EGGS

The accumulation of high levels of *mos* RNA in oocytes suggested the possibility that this protooncogene normally functions as a maternal message. Oocytes synthesize and store maternal proteins and mRNAs which function in oocyte maturation, fertilization, and the early stages of embryonic development. In the mouse, the maternal RNA pool supports new protein synthesis and development to the two-cell stage (24–29). Specific maternal mRNAs have not been identified in mammals, but it is known that approximately 50% of the bulk mRNA of the mouse oocyte is retained in mature ovulated eggs, which have undergone the first meiotic division, and in fertilized eggs (zygotes), which have completed meiosis (30–32). However 90% of the maternal mRNA is degraded by the mid to late two-cell embryo stage, when active transcription of the embryonic genome becomes predominant (30–32). In addition, cytoplasmic polyadenylation of preexisting maternal mRNA has been found to occur in ovulated eggs and/or zygotes of both the mouse and lower organisms, potentially correlated with the recruitment of these mRNAs for translation (31,33–36). We therefore investigated the stability and polyadenylation of *mos* RNA during oocyte maturation, fertilization, and early development in order to begin to elucidate the possible function of *mos* in these processes (37).

We initially compared the levels of *mos* RNA in germinal vesicle stage oocytes, mature ovulated eggs, zygotes, and mid to late two-cell embryos by Northern blot hybridization. *mos* RNA was retained in eggs and zygotes, although the amount of *mos* transcript was reduced 2- to 4-fold from that present in germinal vesicle stage oocytes. In contrast, no *mos* RNA was detected in the two-cell embryos. The *mos* RNA accumulated in oocytes is thus retained in mature ovulated eggs and zygotes but is degraded at the two-cell embryo stage, consistent with the fate of bulk maternal mRNA in the mouse.

The *mos* transcript from mature eggs and zygotes migrated more diffusely and somewhat more slowly, equivalent to the addition of approximately 100 nucleotides, than the transcript from premeiotic germinal vesicle stage oocytes. Since transcription ceases at the time of chromosome condensation, dissolution of the nucleus, and the first meiotic division, this finding suggested posttranscriptional processing of *mos* RNA in concert with oocyte maturation. We therefore investigated whether the increased size of *mos* RNA in eggs and zygotes resulted from polyadenylation of the oocyte transcript.

RNAs from ovaries, eggs, and zygotes were hybridized with oligo(dT) followed by digestion with RNase H to destroy RNA–DNA hybrids, thereby resulting in hydrolysis of poly(A) tracts (37). The sizes of control and digested RNAs were subsequently analyzed by electrophoresis in agarose gels and by Northern blot hybridization with *mos* probe. RNase H digestion did not affect the electrophoretic mobility of the *mos* transcript in ovary RNA, indicating that *mos* RNA is not significantly polyadenylated in oocytes. However, the *mos* transcript in both eggs and zygotes migrated more quickly and less diffusely after RNase H treatment, now appearing similar in size to the oocyte-derived transcript. These results indicate that *mos* RNA is posttranscriptionally modified by polyadenylation in mature ovulated eggs (37).

Polyadenylation of stored maternal mRNAs following maturation and/or fertilization has been found in the mouse as well as in the lower organisms *Spisula*, *Asterias*, *Urechis*, and *Xenopus* and is correlated with mRNA translation (31,33–36). Although specific maternal mRNAs have not been previously identified in the mouse, proteins which are synthesized in mature ovulated eggs and zygotes but not in germinal vesicle stage oocytes have been identified by two-dimensional gel electrophoresis. These proteins are translated from maternal mRNAs which were transcribed and stored prior to meiosis, since they are synthesized in the absence of new transcription and can be identified as *in vitro* translation products of RNA isolated from germinal vesicle stage oocytes (24–29). The posttranscriptional polyadenylation of *mos* RNA in ovulated eggs suggests that *mos* may be a member of the class of proteins which are actively translated in ovulated eggs or zygotes from stored maternal message and may function in meiosis or early embryonic events.

A number of proteins are specifically modified by phosphorylation and glycosylation in mature eggs and zygotes (24,28,29,38,39). Protein phosphorylation is thought to play a regulatory role, allowing rapid activation and deactivation of proteins essential for meiosis and early cleavages (38–40). Since *mos* is a member of the serine/threonine protein kinase family, it is possible that *mos* functions as a regulatory kinase in these early developmental events.

The protooncogenes *mos*, *abl*, and *int*-1 display unique patterns of transcription during spermatogenesis in the mouse (2,3,15,17,41–43). Of these genes, *mos* is distinct, in that it is also transcribed in mouse oocytes (17,41). This suggests the possibility that *mos* functions in a process common to both male and female germ cell development, presumably meiosis. This notion is compatible with transcription of *mos* in premeiotic oocytes followed by polyadenylation of *mos* RNA in ovulated eggs, which have undergone the first meiotic division and will complete meiosis after fertilization. *mos* RNA in male germ cells is detected in early postmeiotic cells but not in meiotic prophase (17,41). However, spermatocytes entering and undergoing meiosis are transiently appearing minor cell populations which cannot be isolated for biochemical analysis. Therefore, it is possible that *mos* RNA detected in postmeiotic male germ cells was transcribed late in meiotic prophase and functioned during meiosis. Whether *mos* acts as a regulatory kinase during meiosis of male and female germ cells and/or during early embryonic cleavages awaits detection and functional analysis of the *mos* protein.

REFERENCES

1. Doolittle, R. F., Hunkapiller, M. W., Hood, L. E., Devare, S. G., Robbins, K. C., Aaronson, S. A., and Antoniades, H. N. (1983). *Science* **221**, 275–277.
2. Downward, J., Yaden, Y., Mayes, E., Scrace, G., Trotty, N., Stockwell, P., Ullrich, A., Schlessinger, J., and Waterfield, M. D. (1984). *Nature (London)* **307**, 521–527.
3. Sap, J., Munoz, A., Damm, K., Goldberg, Y., Ghysdael, J., Leutz, A., Beug, H., and Vennström, B. (1986). *Nature (London)* **324**, 635–640.
4. Sherr, C. J., Rettenmier, C. W., Sacca, R., Roussel, M. F., Look, A. T., and Stanley, E. R. (1985). *Cell (Cambridge, Mass.)* **41**, 665–676.
5. Waterfield, M. D., Scrace, G. T., Whittle, N., Stroobant, P., Johnsson, A., Wasteson, A.,

Westermark, B., Heldin, C. H., Huang, J. S., and Deuel, T. F. (1983). *Nature (London)* **304**, 35–39.
6. Weinberger, C., Thompson, C. C., Ong, E. S., Lebo, R., Gruol, D. J., and Evans, R. M. (1986). *Nature (London)* **324**, 641–646.
7. Brugge, J. S., Cotton, P. C., Queral, A. E., Barrett, J. N., Nonner, D., and Keane, R. W. (1985). *Nature (London)* **316**, 554–557.
8. Cotton, P. C., and Brugge, J. S. (1983). *Mol. Cell. Biol.* **3**, 1157–1162.
9. Sorge, L. K., Levy, B. T., and Maness, P. F. (1984). *Cell (Cambridge, Mass.)* **36**, 249–256.
10. Alema, S., Casalbore, P., Agostini, E., and Tato, F. (1985). *Nature (London)* **316**, 558–559.
11. Bar-Sagi, D., and Feramisco, J. R. (1985). *Cell (Cambridge, Mass.)* **42**, 841–848.
12. Guerrero, I., Wong, H., Pellicer, A., and Burstein, D. E. (1986). *J. Cell. Physiol.* **129**, 71–76.
13. Noda, M., Ko, M., Ogura, A., Liu, D. G., Amano, T., Takano, T., and Ikawa, Y. (1985). *Nature (London)* **318**, 73–75.
14. Hagag, N., Halegoua, S., and Viola, M. (1986). *Nature (London)* **319**, 680–682.
15. Ponzetto, C., and Wolgemuth, D. J. (1985). *Mol. Cell. Biol.* **5**, 1791–1794.
16. Propst, F., and Vande Woude, G. F. (1985). *Nature (London)* **315**, 516–518.
17. Goldman, D. S., Kiessling, A. A., Millette, C. F., and Cooper, G. M. (1987). *Proc. Natl. Acad. Sci. U.S.A.* **84**, 4509–4513.
18. Waters, S. H., Distel, R. J., and Hecht, N. B. (1985). *Mol. Cell. Biol.* **5**, 1649–1654.
19. Bellve, A. R., Cavicchia, J. C., Millette, C. F., O'Brien, D. A., Bhatnagar, Y. M., and Dym, M. (1977a). *J. Cell Biol.* **73**, 68–85.
20. Bellve, A. R., Millette, C. F., Bhatnager, Y. M., and O'Brien, D. A. (1977b). *J. Histochem. Cytochem.* **25**, 480–494.
21. Peters, H. (1969). *Acta Endocrinol.* **62**, 98–116.
22. Bachvarova, R. (1985). In "Developmental Biology: A Comprehensive Synthesis" (L. W. Browder, ed.), Vol. 1, pp. 453–524. Plenum, New York and London.
23. Wassarman, P. M. (1983). In "Mechanism and Control of Fertilization" (J. F. Hartman, ed.), pp. 1–54. Academic Press, New York.
24. Pratt, H. P. M., Bolton, V. N., and Gudgeon, K. A. (1983). *Ciba Found. Symp.* **98**, 127–133.
25. Braude, P., Pelham, H., Flach, G., and Lobatto, R. (1979). *Nature (London)* **282**, 102–105.
26. Van Blerkom, J. (1981). In "Cellular and Molecular Aspects of Implantation" (S. Glasser and D. Bullock, eds.), pp. 155–176. Plenum, New York.
27. Cascio, S. M., and Wassarman, P. M. (1982). *Dev. Biol.* **89**, 397–408.
28. Van Blerkom, J. (1981). *Proc. Natl. Acad. Sci. U.S.A.* **78**, 7629–7633.
29. Howlett, S. K., and Bolton, V. N. (1985). *J. Embryol. Exp. Morphol.* **87**, 175–206.
30. Piko, L., and Clegg, K. B. (1982). *Dev. Biol.* **89**, 362–378.
31. Clegg, K. B., and Piko, L. (1983). *Dev. Biol.* **95**, 331–341.
32. Bachvarova, R., De Leon, A. J., Kaplan, G., and Paynton, B. V. (1985). *Dev. Biol.* **108**, 325–331.
33. Rosenthal, E. T., and Ruderman, J. V. (1987). *Dev. Biol.* **121**, 237–246.
34. Rosenthal, E. T., and Wilt, F. H. (1986). *Dev. Biol.* **117**, 55–63.
35. Colot, H. V., and Rosbash, M. (1982). *Dev. Biol.* **94**, 79–86.
36. Dworkin, M. B., and Dworkin-Rastl, E. (1985). *Dev. Biol.* **112**, 451–457.
37. Goldman, D. S., Kiessling, A. A., and Cooper, G. M. (1988). *Oncogene*, **3**, 159–162.
38. Bornslaeger, E. A., Mattei, P., and Schultz, R. M. (1986). *Dev. Biol.* **114**, 453–462.

39. Poueymirou, W. T., and Schultz, R. M. (1987). *Dev. Biol.* **121,** 489–498.
40. Howlett, S. K. (1986). *Cell (Cambridge, Mass.)* **45,** 387–396.
41. Mutter, G. L., and Wolgemuth, D. J. (1987). *Proc. Natl. Acad. Sci. U.S.A.* **84,** 5301–5305.
42. Propst, F., Rosenberg, M. T., Iyer, A., Kaul, K., and Vande Woude, G. F. (1987). *Mol. Cell. Biol.* **7,** 1629–1639.
43. Shackleford, G. M., and Varmus, H. E. (1987). *Cell (Cambridge, Mass.)* **50,** 89–95.

PART III

POLYPEPTIDE GROWTH FACTORS AND THEIR CELL MEMBRANE RECEPTORS: ROLE IN DIFFERENTIATION AND DEVELOPMENT

15
Pleiotypic Actions of the Seminiferous Growth Factor

ANTHONY R. BELLVÉ[1] AND WENXIN ZHENG[2]
*Departments of Anatomy and Cell Biology,[1]
Obstetrics and Gynecology,[2] and Urology[1], and
Center for Reproductive Sciences[1,2]
College of Physicians and Surgeons
Columbia University
New York, New York 10032*

INTRODUCTION

Development of the mammalian testes requires the proliferation and differentiation of multiple cell types. The principal of these are Leydig, Sertoli, and germ cells. Presumptive Sertoli cells are believed to migrate from the major glomerulus of the mesonephros (Zamboni and Upadhyay, 1982), whereas germ cells are derived from the yolk sac endoderm of the primitive embryo (Clark and Eddy, 1975). After migrating from their respective origins, the two cell types continue to proliferate to form the principal components of the bilaterally disposed gonadal anlagen. Progenitor epithelial cells assume a spatial organization and polarity to form the seminiferous cords, later to become the seminiferous tubules (Magre and Jost, 1980). During the postnatal period Sertoli cells cease dividing (Nagy, 1972; Orth, 1982). Shortly after formation of the gonadal anlagen, presumptive Leydig cells arise in the interstitium from undifferentiated mesenchyme to form steroid-secreting tissue. Primordial germ cells become localized within the seminiferous cords and continue to proliferate for a few days before undergoing considerable growth to yield quiescent gonocytes. These germ cells, on initiating mitosis and migrating to the periphery of the cords, form the progenitors of spermatogenesis of puberty and adulthood. Ongoing spermatogenesis is continuous and highly synchronized to ensure the coordinate production of large numbers of differentiated spermatozoa (Bellvé, 1979).

The proliferation of these three predominant cell types, and the synchronous renewal of spermatogenesis, requires the singular and/or concerted actions of gonadotropic hormones and growth factors, each acting at discrete phases of testicular development. The identity of the growth factors and the regulatory processes that must be involved is known cursorily for an increasing number of peptides (Bellvé and Zheng, 1989a). First among these to be identified was the seminiferous growth factor (SGF), the focus of this chapter.

EXPRESSION OF MITOGENIC ACTIVITY

Mitogenic activity in mouse testes was first detected by quantifying the stimulation of [^{3}H]thymidine incorporation into DNA of confluent, quiescent, BALB/c 3T3 cells (Feig et al., 1980). This assay then and since has proved to be valuable for detecting growth factors in other tissues, particularly fibroblast growth factor (Klagsbrun et al., 1977). Based on this assay, a homogenate of adult testes contains a protease-sensitive, mitogenic activity that is detected in a dose-dependent pattern, with a half-maximal reponse occurring at approximately 400 μg protein/ml (Fig. 1) (Feig et al., 1980). Protein from testes or seminiferous tubules exhibits equivalent specific activities, indicating that most of the mitogenicity is located in the seminiferous epithelium, rather than the interstitium. Commensurate levels of mitogen have been detected in the testes of several mammalian species, suggesting that the activity is ubiqituous (Feig et al., 1983; Bellvé and Feig, 1984).

During mouse development testes express the highest levels of mitogenic activity for confluent, BALB/c 3T3 cells during the immediate postnatal period, on Days 6–8 and 12–14 (Feig et al., 1980). These levels of expression gradually decline during development to adulthood. Based on studies with purified cell types, most of the activity present in the Day-6 testes is derived from Sertoli cells (Fig. 2) (Feig et al., 1980). Minor levels of activity are associated with primitive type A spermatogonia and preleptotene spermatocytes, but not with germ cells at other stages of differentiation. Activity of the two germ cell populations cannot be accounted for by their low contamination by Sertoli cells (Feig et al., 1980). Thus, it is possible that one or both forms of "germ cell" mitogenic activity is derived from adjacent Sertoli cells or due to the existence of distinct growth factors.

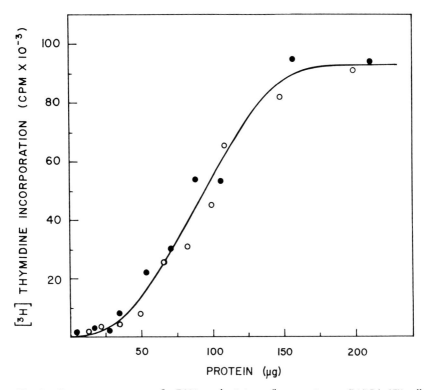

Fig. 1. Dose–response curves for DNA synthesis in confluent, quiescent BALB/c 3T3 cells after being stimulated by the mitogenic activity from mouse testes. Alliquots of perfused, homogenized testes or seminiferous tubules were added, together with [^{3}H]thymidine (4 μCi/ml), to the 3T3 cells. Following a 48-hr culture period, the induction of DNA synthesis was assayed by quantifying the amount of [*methyl*-^{3}H]thymidine incorporated into trichloracetic acid-precipitable material. Data points represent the average of duplicate determinations on homogenates of testes (○) and seminiferous tubules (•). [Reproduced from Feig *et al.* (1980), with permission of the Editor, *Proc. Natl. Acad. Sci. U.S.A.*]

BIOCHEMICAL CHARACTERIZATION

SGF derived from Sertoli cells is a peptide of M_r 15,500 and pI 5.2. Thus, mitogenic activity from mouse and calf seminiferous epithelia elutes on isocratic chromatography, in the presence of 6 M guanidine hydrochloride, as a monomer with an M_r of 16,000–15,000 (Feig *et al.*, 1980). Even after a 350-fold purification, the peak of activity contains peptides having M_r values of 17,000–14,500 on sodium dodecyl sulfate–polyacrylamide gel electrophoresis (SDS–PAGE) (Feig *et al.*, 1983). Further, mitogenic activity from prepubertal calf seminiferous cords elutes in two peaks on applying ion-ex-

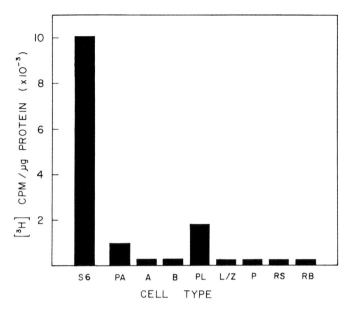

Fig. 2. Growth factor activity in cells isolated from the mouse seminiferous epithelium. The respective cell types were purified from the seminiferous epithelia of prepuberal and adult mice by using sedimentation velocity at unit gravity (Bellvé et al., 1977a,b). The cells were homogenized and aliquots were tested for activity on confluent BALB/c 3T3 cells. The cell types tested include Sertoli cells from 6-day-old mice (S6), primitive type A (PA), type A spermatogonia (A), type B spermatogonia (B), preleptotene spermatocytes (PL), leptotene/zygotene spermatocytes (L/Z), pachytene spermatocytes (P), round spermatids (RS), and residual bodies (RB). [Reproduced from Feig et al. (1980), with permission of the Editor, Proc. Natl. Acad. Sci. U.S.A.]

change chromatography (DE-52, DEAE, pH 6.2); a minor peak is recovered in the flow through, and the other, after binding, elutes with approximately 0.6 M NaCl (Feig et al., 1983). Comparable results are obtained when total activity is subjected to preparative isoelectric focusing under denaturing conditions (6 M urea). In this case, most of the activity is recovered in fractions commensurate with a pI of about 5.2, and the minor peak has a pI of at least 8.2. By contrast, adult mouse testes yield a single peak of activity at a pI of approximately 5.2. The major peak of activity of mouse and calf seminiferous epithelia is denoted as SGF.

Purification of SGF over 80,000-fold has been accomplished by applying a sequence involving affinity chromatography on Sepharose-conjugated Green A and Orange A, both Cibracon dyes, and heparin (Bellvé and Zheng, 1989b). In the last step, SGF activity elutes at about 1.30 M NaCl, intermediate between acidic fibroblast growth factor (aFGF) ($\sim$1.1 M) and basic FGF ($\sim$1.6 M) (Bellvé and Zheng, 1989a), the best characterized of the heparin-binding growth factors (HBGFs) (Folkman and Klagsbrun, 1987). After elution from heparin, the activity is subjected to high-performance liquid chromatography (HPLC) on an octaphenyl (C_8) column (4.6 mm × 25 cm), developed with a linear gradient of 20–45% acetonitrile in 1% trichloroacetic acid. The double peak of activity from HPLC is evident as two polypeptide species of M_r ~16,500 and 15,250 when subjected to SDS–PAGE (Bellvé and Zheng, 1989a). These data are consistent with SGF being a novel growth factor. However, convincing evidence needs to be derived from sequencing of complementary cDNAs and/or induction of distinct biological responses by established cell lines.

TWO TESTICULAR CELL LINES: TM_3 LEYDIG AND TM_4 SERTOLI CELLS

SGF is able to stimulate the incorporation of [^{3}H]thymidine into DNA of confluent BALB/c 3T3 cells. However, these cells are known to respond to a diverse array of HBGFs, and as such do not provide a specific bioassay for SGF activity. Of the testicular cell lines available (Mather, 1980), two are known to express distinct pleiotypic effects in response to SGF (Braunhut et al., 1990; Zheng et al., 1990). Since both cell lines are expected to be valuable for monitoring SGF activity, it is appropriate to document the known characteristics of the TM_3 and TM_4 cells prior to considering evidence justifying their use in specific assays.

Properties of the two epithelial-like, testicular cell lines are defined in a number of diverse studies related to reproductive function (Mather and Phillips, 1984; Vanha-Perttula et al., 1985; Reyes et al., 1986). Both cell lines were derived from the testes of Day-11 to -13 postnatal BALB/c mice, cloned, and propagated in the presence of allohydroxyproline (Mather, 1980). Continued proliferation in the presence of this amino acid analog is consistent with the two lines originating from cells other than fibroblasts (Gilbert and Migeon, 1975; Kao and Prockop, 1977). Further, TM_3 and TM_4 cells have been in continuous culture since 1976 (Mather et al., 1982), and, therefore, by definition (Freshney, 1987) they can be considered transformed cells, although neither are tumorogenic when transplanted to BALB/c nu/nu mice (Mather et al., 1982). Most importantly, both cell lines can be grown in Dulbecco's modified Eagle's and Ham's F12 media (1:1) (DME/F12°), supplemented with insulin, epidermal growth factor, and transferrin (DME/F12*) (Mather, 1980), thereby circumventing the effects of growth-promoting peptides in serum. For present purposes, the two cell lines are denoted, based on their respective characteristics, as TM_3 Leydig and TM_4 Sertoli cells. This will avoid undue confusion in the ensuing discussion.

Characteristically, TM_3 cells have a doubling time of 16 hr, grow to a saturation density of 2.1×10^6 cells/cm^2, and to a volume of approximately 820 μm^3 cell (Mather, 1980; Mather et al., 1982). TM$_3$ cells produce cAMP in response to luteinizing hormone (LH) but not to follicle-stimulating hormone (FSH), convert [^{14}C]cholesterol to progesterone metabolites, and secrete prostaglandin F$_{2\alpha}$ (Mather, 1980). In addition, TM$_3$ cells contain detectable levels of mRNA encoding for a proopiomelanocortin-like gene (Chen et al., 1984) and activin βa and βb subunits but not the α subunit of inhibin (Lee et al., 1989). These cells also express receptors for arginine vasopressin (Maggi et al., 1989).

Comparable studies show TM$_4$ Sertoli cells have a doubling time of 18 hr, grow to a saturation density of 1.2×10^5 cells/cm^2, and to a volume of 1,550 μm^3 cell (Mather, 1980; Mather et al., 1982). These cells have been reported to show increased growth and elevated cAMP levels in response to FSH but not to LH. TM$_4$ cells also secrete transferrin and plasminogen activator, exhibit phagocytic activity (Mather et al., 1982), and form heterotypic aggregates when plated with myoid cells (Mather and Phillips, 1984). Like Sertoli cells in primary culture, TM$_4$ cells are preferentially sensitive to gossypol, a known male infertility drug (Reyes et al., 1984; Robinson et al., 1986).

These studies, although not providing a thorough characterization of the two cell lines, are consistent with the cells having retained characteristics of their presumed cell of origin. Particularly interesting is the notion that the two cell lines may express properties relevant to local autocrine and/or paracrine regulation within the testes. Therefore, it may also be feasible to discern between SGF and related peptides by using the biologically responsive TM$_3$ Leydig and/or TM$_4$ Sertoli cells. Also, purification of SGF requires use of assays capable of discriminating among closely related HBGFs.

PROLIFERATION OF TM$_3$ LEYDIG AND TM$_4$ SERTOLI CELLS

TM$_3$ Leydig and TM$_4$ Sertoli cells express different proliferative responses to SGF, aFGF, and bFGF (Braunhut et al., 1990; Zheng et al., 1990). The two cell lines were stimulated with growth factor up to the level (2 U/ml) that is known to have maximal effects on the proliferation of BALB/c 3T3 cells. TM$_4$ Sertoli cells grown in serum-free DME/F12 (DME/F12°) proliferate in response to SGF but not to aFGF or bFGF (Fig. 3) (Braunhut et al., 1990; Zheng et al., 1990). SGF also stimulates proliferation of TM$_3$ cells at a rate faster than that observed for TM$_4$ cells (Fig. 3) (Zheng et al., 1990). A similar difference in growth rates also occurs in response to 5% fetal calf serum, and therefore the responses probably reflect metabolic properties inherent to

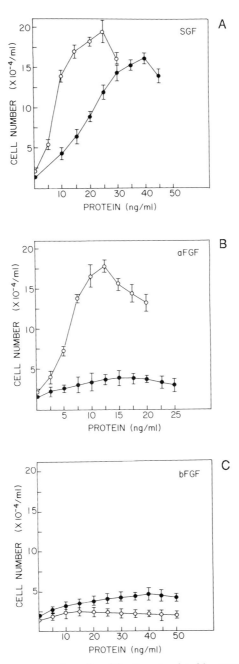

Fig. 3. Dose–response curves for cell proliferation stimulated by SGF, aFGF, and bFGF. TM_3 Leydig and TM_4 Sertoli cells were seeded at 25,000 cells/ml in 24-well plates and cultured for 48 hr in DME/F12* and for 24 hr in DME/F12°. At the end of this time, SGF, aFGF, or bFGF was added, and the cells were cultured for another 48 hr. Cells were detached with trypsin, and their numbers were quantified. The data are derived from triplicate cultures of TM_3 Leydig cells (○) and TM_4 Sertoli cells (•) and are expressed as means ± S.E.M. (A) SGF, (B) aFGF, (C) bFGF. (Reproduced from Zheng et al., 1990, with permission from the editor, *Growth Factors*.)

the two cell lines. Like SGF, bovine aFGF promotes proliferation of TM_3 cells. By contrast, both native and recombinant forms of bFGF are not active on either TM_3 or TM_4 cells (Fig. 3) (Zheng et al., 1990). Also, two oncogene-related HGBFs, hst/KS3 (Delli Bovi et al., 1987) and FGF-5 (Zhan et al., 1988), partially stimulated TM_3 cell divisions but were only weak mitogens for TM_4 cells (W. Zheng, M. Goldfarb, and A. R. Bellvé, unpublished data). These results cannot be explained by a contamination of SGF by another of the known HBGFs. Thus, the proliferative responses of the two testis-derived cell lines differ, and therefore they provide a biological assay to distinguish between the three HBGFs.

EXTRACELLULAR ACCUMULATION OF SULFATED GLYCOPROTEIN

The nonmitogenicity of aFGF and bFGF for TM_4 Sertoli cells may reflect a lack of specific receptors, or the two growth factors could bind but induce different pleiotypic responses. The latter possibility was tested by quantifying the extracellular levels of sulfated glycoprotein-1 (SGP-1) and sulfated glycoprotein-2 (SGP-2) that are accumulated by TM_4 cells when cultured in DME/F12° with the three HBGFs. SGP-1 is known to be a lysosomal constituent secreted by Sertoli cells *in vivo* and then to interact with the differentiating germ cells (Griswold et al., 1989). SGP-1 (also referred to as S70) is processed and secreted by Sertoli cells at Stages VII–VIII of the spermatogenic cycle, coincident with the translocation of leptotene spermatocytes into the central compartment of the seminiferous tubules (Shabanowitz et al., 1986). An internal region of rat SGP-1 shares 78% homology with the human sphingolipid activator protein (SAP-1) of fibroblasts (Collard et al., 1988). The latter is responsible for activating the degradation of glycolipids by water-soluble exohydrolases (Dewji et al., 1987). It is plausible, therefore, that SGP-1 may be involved in activating, modifying, or transporting membrane glycolipids of Sertoli and/or germ cells.

Extracellular SGP-1 is accumulated by TM_4 Sertoli cells in DME/F12° at a rate of approximately 70 ng/hr/10^6 cells between 24 and 48 hr (Fig. 4) (Zheng et al., 1990). This level of SGP-1 production is not enhanced by optimal concentrations of aFGF or a combination of epidermal growth factor, insulin, and transferrin (EIT). However, SGF and bFGF cause approximately 3- and 2-fold increases, respectively, in the serum-free spent media. The total response is more substantial with SGF, particularly since this peptide also causes about a 3-fold increase in cell number during the same 48-hr period. Thus, control TM_4 cells produce approximately 2.1 µg, EIT 5.1 µg, aFGF 2.8 µg, bFGF 5.0 µg, and SGF 20.7 µg SGP-1/flask. The effects of growth

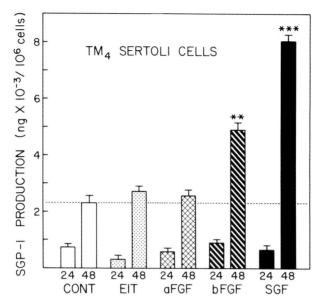

Fig. 4. Stimulation of extracellular, sulfated glycoprotein-1 production by TM_4 Sertoli cells in response to EIT, aFGF, bFGF, or SGF. Cells were prepared as described in the legend to Fig. 3 and cultured in DME/F12° with growth factor for either 24 or 48 hr. Amounts of sulfated glycoprotein-1 (SGP-1) accumulating in the media were quantified by applying a specific micro-ELISA. The data are derived from three independent experiments and are expressed as the means ± S.E. The sensitivity of sulfated glycoprotein-1 detection was approximately 5 ng/well with linearity achieved between 5 and 140 ng/well. No cross-reactivity was detected with sulfated glycoprotein-2 or other proteins. Differences were assessed by applying Student's t tests. **, $p \leq .01$. ***, $p \leq .001$. (Reproduced from Zheng *et al.*, 1990, with permission from the editor, *Growth Factors*.)

factors on SGP-1 production appear to be independent of the mitogenic responses, since EIT and bFGF cause TM_4 cells to produce comparable amounts of extracellular SGP-1. The effects also appear to be specific, as none of the trophic agents increased SGP-1 accumulation by TM_3 Leydig cells, nor did any have an effect on the extracellular production of SGP-2 by either TM_3 or TM_4 cells (Zheng *et al.*, 1990).

The actions of SGF and bFGF on the production of extracellular SGP-1 could occur at the level of transcription, translation, processing, and/or secretion. This question has been resolved, in part, by undertaking Northern analysis with total RNA recovered from about 85% confluent TM_4 cells at 12, 24, 36, and 48 hr after stimulation by SGF (Zheng *et al.*, 1990). The amounts of SGP-1 mRNA were quantitated relative to those of actin mRNA, in both control and stimulated cells. On comparing the relative amount of

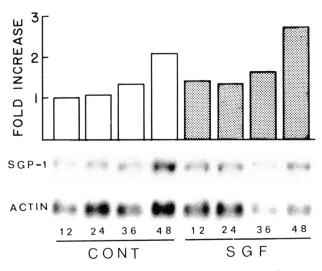

Fig. 5. Northern blots of mRNA encoding sulfated glycoprotein-1 and actin mRNA accumulated by the TM_4 Sertoli cells in response to SGF. Total RNA was recovered from the cells after the specified periods, loaded at 4 μg/lane, fractionated, transferred to nitrocellulose, and hybridized with the ^{32}P-labeled cDNA probes for 12, 24, 36, and 48 hr. Hybridization of the two probes was quantified by densitometry. Note: Steady-state levels of sulfated glycoprotein-1 mRNA, relative to actin mRNA, increased 1.34-fold during the 48-hr period. SGP-1, Sulfated glycoprotein-1 [^{32}P]cDNA; actin, actin [^{32}P]cDNA; cont, control; SGF, seminiferous growth factor. (Reproduced from Zheng et al., 1990, with permission from the editor, Growth Factors.)

hybridization of the two cDNA probes, it is noted that SGF increases the steady-state levels of SGP-1 mRNA approximately 1.34-fold during the 48-hr period (Fig. 5) (Zheng et al., 1990). Since the extracellular levels of the glycoprotein increase about 4-fold during the same period, it is likely that SGF also acts at a posttranscriptional level, perhaps by modifying the targeting of SGP-1 to a secretory pathway rather than to lysosomes. In this context, TM_4 Sertoli cells provide a model for defining the mechanism by which SGF regulates transcription, translation, processing, and secretion of SGP-1 by Sertoli cells *in vivo*.

SYNTHESIS OF SPECIFIC PEPTIDES

The selective effects of HBGFs on the extracellular accumulation of SGP-1 suggest that SGF may alter the synthesis and secretion of other extracellular proteins. This possibility has been examined by comparing the effects of

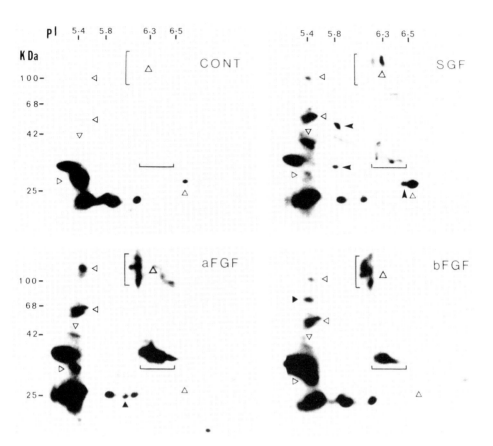

Fig. 6. Immunoelectrofocusing (IEF)/SDS–PAGE fluorograms of [^{35}S]methionine-labeled proteins secreted by TM$_3$ Leydig cells in response to SGF, aFGF, or bFGF. Cells in DME/F12° containing [^{35}S]methionine were stimulated with growth factor, and secreted proteins were collected 12 hr later. Major responses included the following: (brackets) qualitative and quantitative changes in labeling of proteins with M_r values of 210,000 to 115,000 (pI ~6.15) and M_r 33,000 (pI 6.25–6.42); (triangles) up and down regulation of specific proteins; (solid arrowheads) SGF-specific proteins of M_r ~49,000 (pI 5.83), 32,000 (pI 5.80), 28,500 (pI 6.50); aFGF-specific protein of M_r 24,500 (pI 6.12); bFGF-specific protein of M_r 73,000 (pI 5.43) (see also Table I). Reproduced from Zheng et al., 1990, by permission from the editor, *Growth Factors*.)

SGF, aFGF, and bFGF on TM_3 Leydig and TM_4 Sertoli cells on the synthesis and secretion of [^{35}S]methionine-labeled proteins (Zheng et al., 1990). Based on the resulting fluorograms, it is evident that both cell lines, after stimulation for 12 hr with one of the three peptides, show specific patterns of secreted proteins.

TM_3 Leydig cells secrete three qualitatively distinct proteins in response to SGF, another to aFGF, and one to bFGF (Fig. 6; Table I). Other secreted proteins exhibit quantitative changes, being either up or down regulated in response to an HBGF. Five of these proteins are regulated in the same direction by all three HBGFs, whereas two others are up regulated by SGF and down regulated by aFGF and bFGF (Table I).

TM_4 Sertoli cells secrete five distinct [^{35}S]methionine-labeled proteins in response to SGF, whereas no qualitative changes are evident on stimulating with aFGF or bFGF (Fig. 7; Table II). Three interesting quantitative differences are apparent. One secreted protein is up regulated by both SGF and bFGF, one by SGF and aFGF, and another by aFGF and bFGF (Table

TABLE I
Growth Factor-Induced Changes in [^{35}S]Methionine-Labeled Proteins Secreted by TM_3 Cells[a]

K_d	pI	aFGF	SGF	bFGF
Qualitative				
24	6.1	+	0	0
28	6.5	0	+	0
32	5.8	0	+	0
49	5.8	0	+	0
73	5.4	0	0	+
Quantitative				
28	5.4	↓↓	↓↓↓	↓
28	6.5	↓	↑	↓
33	6.3	↑↑↑	↑	↑↑
41	5.4	↑↑	↑↑↑	↑
54	5.4	↑	↑	↑
190	6.2	↓	↑↑	↓
~180	6.1	↑↑↑	↑	↑↑↑

[a] Proteins labeled with [^{35}S]methionine and secreted by TM_3 cells were subjected to SDS/PAGE and fluorography (see Fig. 6). The data are tabulated to indicate the relative expression of individual proteins. +, Positive expression; 0, no expression; ↑, up regulation; ↓, down regulation.

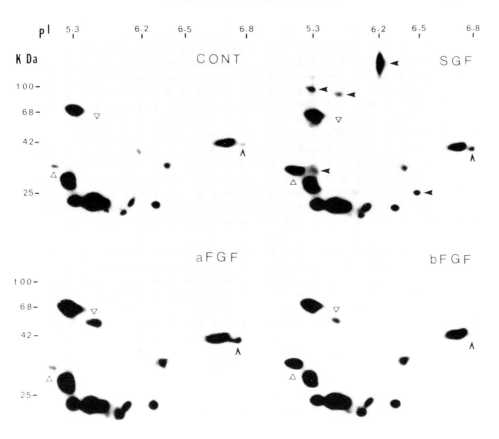

Fig. 7. IEF/SDS–PAGE fluorograms of [^{35}S]methionine-labeled proteins secreted by TM_4 Sertoli cells in response to SGF, aFGF, or bFGF. Cells were treated as described in the legend to Fig. 6. (Solid arrowheads) Five proteins induced only by SGF: M_r ~180,000 (pI 6.25), 90,000 (pI 5.39), 84,000 (pI 5.80), 34,000 (pI 5.39), and 28,000 (pI 6.51); (open arrowheads) proteins up regulated by SGF and aFGF (M_r 42,000; pI 6.82), SGF and bFGF (M_r 34,000; pI 5.24), and aFGF and bFGF (M_r 53,000, pI 5.79) (see also Table II). (Reproduced from Zheng et al., 1990, with permission from the editor, Growth Factors.)

II). These effects are analogous to the actions of SGF and bFGF on the extracellular accumulation of SGP-1.

These qualitative and quantitative differences in the expression of proteins secreted by TM_3 Leydig and TM_4 Sertoli cells substantiate the premise that the three HBGFs are capable of inducing different pleiotypic responses. Most importantly, SGF stimulates both cell lines to secrete distinct proteins and induces quantitative differences in the expressions of other proteins.

TABLE II
Growth Factor-Induced Changes in [^{35}S]Methionine-Labeled
Proteins Secreted by TM$_4$ Cells[a]

		Peptides		
K_d	pI	aFGF	SGF	bFGF
Qualitative				
28	6.5	0	+	0
34	5.3	0	+	0
84	5.8	0	+	0
90	5.3	0	+	0
180	6.2	0	+	0
Quantitative				
34	5.2	−	↑	↑
42	6.8	↑	↑	−
53	5.7	↑ ↑	−	↑

[a] Procedures were identical to those described in Table I. Data are derived from Fig. 7. +, Positive expression; 0, no expression; ↑, up regulation; ↓, down regulaton; −, no change.

The three HBGFs usually act to change protein secretion in the same direction, but in other cases the effect occurs in the opposite direction. These data therefore suggest that the three HBGFs act singly, in concert, and in opposite directions to regulate multiple cell functions on the two testis-derived cell lines.

IMMUNOLOGICAL PROPERTIES OF HEPARIN-BINDING GROWTH FACTORS

SGF induces TM$_3$ Leydig and TM$_4$ Sertoli cells to express distinct pleiotypic responses, and yet the peptide shares certain physiochemical properties with aFGF, bFGF, and FGF-5 (Feig et al., 1980, 1983; Thomas et al., 1984; Esch et al., 1985; Ueno et al., 1987; Zhan et al., 1988; Braunhut et al., 1990). This suggests the four HBGFs may be closely related and therefore could share sequence identity. This premise can be tested by using polyclonal antibodies (Abs) directed against peptide sequences unique to aFGF, bFGF, or FGF-5. On SDS–PAGE immunoblots, five Abs directed against the peptides aFGF$_{59-90}$, bFGF$_{1-12}$, bFGF$_{33-43}$, bFGF$_{135-145}$, or FGF-5$_{174-184}$ bound avidly to 50 ng of purified intact aFGF and bFGF or 50 ng of TrpE–FGF-5 without any detectable cross-reaction with the other peptides, thereby confirming their specificity (Fig. 8) (Zheng et al., 1990).

Fig. 8. Immunoblots of SGF, aFGF, and bFGF and the TrpE–FGF-5 fusion protein with polyclonal antibodies (Abs) directed against the respective peptide-specific sequences. The proteins were subjected to SDS–PAGE, transferred to nitrocellulose, and reacted separately or jointly with anti-aFGF and anti-bFGF Abs, or with the anti-FGF-5 Ab. Bound IgG-ATPase was reacted with 5-bromo-4 chloro-3-indoyl-phosphate p-toludine (BCIP) and nitroblue tetrazolium (**NBT**). *Note:* SGF did not bind Abs directed against aFGF, bFGF, or FGF-5. Lane 1, 50 ng aFGF with anti-aFGF$_{59-90}$ Ab; 2, 100 ng SGF with all Abs against aFGF and bFGF; 3, 50 ng bFGF, anti-bFGF$_{1-12}$ Ab; 4, 50 ng bFGF, anti-bFGF$_{33-43}$ Ab; 5, 50 ng bFGF, anti-bFGF$_{135-145}$ Ab; 6, 100 ng SGF with anti-FGF-5$_{174-184}$ Ab; 7, 50 ng TrpE–FGF-5 fusion protein, anti-FGF-5$_{174-184}$ Ab. (Data from Zheng *et al.*, 1990, by permission of the editor, *Growth Factors.*)

However, 100 ng of purified SGF is not recognized by any of the Abs, whether the peptide is incubated with the Abs separately or jointly. It is concluded that SGF does not share at least five epitopes with aFGF, bFGF, or FGF-5. This infers that the primary structure of SGF differs from aFGF, bFGF, and FGF-5.

DISCUSSION

SGF is distinct from aFGF and bFGF as a biological entity. First, SGF does not contain epitopes known to be present in aFGF, bFGF, or FGF-5. Second, SGF is able to induce a different set of pleiotypic responses in TM_3

and TM_4 cells, two mouse testis-derived cell lines having characteristics of Leydig cells and Sertoli cells, respectively. The set of SGF-dependent effects include the following: (1) the proliferation of TM_3 and TM_4 cells, (2) the accumulation of extracellular SGP-1 by TM_4 cells, and (3) the secretion of specific [^{35}S]methionine-labeled proteins by both cell lines. In regard to the latter, SGF is able to induce TM_3 and TM_4 cells to secrete proteins that are not produced in response to aFGF and/or bFGF. Further, SGF can up and down regulate the expression of other proteins in the same and occasionally in the opposite direction to those changes induced by aFGF and/or bFGF. These observations delineate several additional concepts. SGF can cause secretion of different proteins from a mesenchymal- versus an epithelial-derived cell type, both of which originate from mesoderm. SGF, aFGF, and bFGF are able to exert different pleiotypic effects on the same cell type. Last, and importantly, SGF is likely to be encoded by a distinct gene.

SGF can be classified as a member of the growing HBGF family. This peptide, in common with aFGF and bFGF, binds to heparin and stimulates the proliferation of endothelial cells (Braunhut *et al.*, 1990). SGF has other properties that are comparable to those of aFGF. The two peptides have M_r values of 16,000 to 17,000 and p*I* values of approximately 5.2 (Feig *et al.*, 1983; Thomas *et al.*, 1984), are eluted from heparin with 1.0 to 1.3 *M* NaCl (Shin *et al.*, 1984; Braunhut *et al.*, 1990), and their effects on endothelial cell proliferation are potentiated by heparin (Braunhut *et al.*, 1990). These properties differ from those of bFGF, including the truncated $bFGF_{16-146}$ that has been purified from bovine testes and identified from a partial amino acid sequence (Ueno *et al.*, 1987). The four oncogene-related peptides *int-2* (Dickson and Peters, 1987), *hst*/KS3 (Delli Bovi *et al.*, 1987), FGF-5 (Zhan *et al.*, 1988), and FGF-6 (Marics *et al.*, 1989) also have different physiochemical properties. Based on peptide sequences, aFGF and bFGF share 53% homology with each other and lower levels with *int-2*, *hst*/KS3, and FGF-5. These levels of homology may be too limited to take advantage of a possible consensus sequence for eliciting antibodies against SGF. With the exception perhaps of *int-2*, these growth factors, whether anionic or cationic, characteristically bind to heparin-Sepharose.

The present evidence is consistent with SGF, aFGF, and bFGF binding to specific receptors on the mesenchyme- and epithelial-derived cell lines. All three HBGFs, on addition to TM_3 Leydig and TM_4 Sertoli cells at approximately 10^{-9} *M*, stimulate proliferation and/or the synthesis of specific secretory proteins. These actions are indicative of receptor-mediated mechanisms, although it will be necessary to verify this premise directly by identifying and characterizing the receptors. Neufeld and Gospodarowicz (1985) have shown that a single class of high affinity binding sites and with Mr values of 145,000 and 125,000 membrane receptor proteins exist on baby hamster kidney cells (BHK-21 cells). These data are consistent with aFGF and bFGF sharing the same kind of receptors, perhaps the protein recently purified and predicted from the sequence of the cloned, complementary cDNA (Lee *et al.*, 1989). However, aFGF and bFGF do not compete for binding to receptors on bovine aorta endothelial cells (Schreiber *et al.*, 1985), suggesting that the two HBGFs act through different receptors. This could also be the case for TM_3 and TM_4 cells. This suggests that the different pleiotypic responses caused by SGF, aFGF and bFGF probably arise from the respective receptor–ligand complexes inducing transmembrane signals via separate transduction mechanisms and mediating their propagation along diverging intracellular pathways.

Questions need to be addressed for the mechanistic action of SGF. The elevated levels of extracellular SGP-1 at 48 hr was preceded by a commensurate increase in the steady-state levels of SGP-1 mRNA relative to that of actin mRNA. Thus, SGF has a selective effect on gene transcription in TM_4 Sertoli cells. Additionally, the delay in the "secretion" of SGP-1 leaves open the possibility that SGF may also act at the level of translational efficiency, glycosylation, and/or secretional targeting of SGP-1. The mechanisms by which SGF promotes the earlier synthesis and secretion of distinct proteins by TM_3 and TM_4 cells have yet to be resolved. Regardless, the current findings have important implications for the role of these peptides in regulating the functions of Leydig and Sertoli cells *in vivo*. In this context, bovine bFGF has been found to increase the numbers of FSH receptors and to enhance production of plasminogen activator by porcine Sertoli cells (Jaillard *et al.*, 1987). Clearly, modulation of testicular function by growth factors is likely to be complex. A fine balance of concerted and opposing actions of multiple growth factors could be necessary to coordinate events essential for sustaining mammalian spermatogenesis.

ACKNOWLEDGMENTS

The preparation of this chapter was funded, in part, by BIOCOL, Inc.

REFERENCES

Bellvé, A. R. (1979). *Oxford Rev. Reprod. Biol.* **1,** 159–261.
Bellvé, A. R., and Feig, L. A. (1984). *Recent Prog. Horm. Res.* **40,** 531–567.
Bellvé, A. R., and Zheng, W. (1989a). *J. Reprod. Fertil.* **85,** 771–793.
Bellvé, A. R., and Zheng, W. (1989b). *Ann. N.Y. Acad. Sci.* **564,** 116–131.
Bellvé, A. R., Cavicchia, J. C., Millette, C. F., O'Brien, D. A., Bhatnagar, Y. M., and Dym, M. (1977a). *J. Cell Biol.* **74,** 68–85.

Bellvé, A. R., Millette, C. F., Bhatnagar, Y. M., and O'Brien, D. A. (1977b). *J. Histochem. Cytochem.* **25**, 480–494.
Braunhut, S. J., Rufo, G. A., Ernisee, B. J., Zheng, W., and Bellvé, A. R. (1990). *Biol. Reprod.* **42**, 639–648.
Chen, C. L. C., Mather, J. P., Morris, M. P., and Bardin, C. W. (1984). *Proc. Natl. Acad. Sci. U.S.A.* **81**, 5672–5675.
Clark, J. M., and Eddy, E. M. (1975). *Dev. Biol.* **47**, 136–155.
Collard, M. W., Sylvester, S. R., Tsuruta, J. K., and Griswold, M. D. (1988). *Biochemistry* **27**, 4557–4564.
Delli Bovi, P., Curatola, A. M., Kern, F. G., Greco, A., Ittmann, M., and Basilico, C. (1987). *Cell* **50**, 729–737.
Dewji, N. N., Wenger, D. A., and O'Brien, J. S. (1987). *Proc. Natl. Acad. Sci. U.S.A.* **84**, 8652–8656.
Dickson, C., and Peters, G. (1987). *Nature (London)* **326**, 833.
Esch, F., Baird, A., Ling, N., Ueno, N., Denoroy, L., Klepper, R., Gospodarowicz, D., Böhlen, P., and Guillemin, R. (1985). *Proc. Natl. Acad. Sci. U.S.A.* **82**, 6507–6511.
Feig, L. A., Bellvé, A. R., Horbach-Erickson, N., and Klagsbrun, M. (1980). *Proc. Natl. Acad. Sci. U.S.A.* **77**, 4774–4778.
Feig, L. A., Klagsbrun, M., and Bellvé, A. R. (1983). *J. Cell Biol.* **97**, 1435–1443.
Folkman, J., and Klagsbrun, M. (1987). *Science* **235**, 442–447.
Freshney, R. I. (1987). In "Culture of Animal Cells. A Manual of Basic Technique" (R. I. Freshney, ed.), 2nd Ed., pp. 1–397. Alan R. Liss, New York.
Gilbert, S. F., and Migeon, B. R. (1975). *Cell* **5**, 11–17.
Griswold, M. D., Morales, C., and Sylvester, S. R. (1989). *Oxford Rev. Reprod. Biol.* **10**, 124–161.
Jaillard, C., Chatelain, P. G., and Saez, J. M. (1987). *Biol. Reprod.* **37**, 665–674.
Kao, W. W., and Prockop, D. J. (1977). *Nature (London)* **266**, 63–64.
Klagsbrun, M., Langer, R., Levenson, R., Smith, S., and Lillehei, C. (1977). *Exp. Cell Res.* **105**, 99–108.
Lee, P. L., Johnson, D. E., Cousens, L. S., Fried, V. A., and Williams, L. T. (1989). *Science* **245**, 57–60.
Lee, W., Mason, A. J., Schwall, R., Szonyi, E., and Mather, J. P. (1989). *Science* **243**, 396–398.
Maggi, M., Morris, P. L., Kassis, S., and Rodbard, D. (1989). *Int. J. Androl.* **12**, 65–71.
Magre, S., and Jost, A. (1980). *Arch. Anat. Microsc. Morphol. Exp.* **69**, 297–318.
Marics, I., Adelaide, J., Raybaud, F., Mattei, M. G., Coulier, F., Planche, J., de Lapeyriere, O., and Birnbaum, D. (1989). *Oncogene* **4**, 335–340.
Mather, J. P. (1980). *Biol. Reprod.* **23**, 243–252.
Mather, J. P., and Phillips, D. M. (1984). *J. Ultrastruct. Res.* **87**, 263–274.
Mather, J. P., Zhuang, L.-Z., Perez-Infante, V., and Phillips, D. M. (1982). *Ann. N.Y. Acad. Sci.* **385**, 44–68.
Nagy, F. (1972). *J. Reprod. Fertil.* **28**, 389–395.
Neufeld, G., and Gospodarowicz, D. (1985). *J. Biol. Chem.* **260**, 13860–13868.
Orth, J. M. (1982). *Anat. Rec.* **203**, 485–492.
Reyes, J., Allen, J., Tanphaichitr, N., Bellvé, A. R., and Benos, D. J. (1984). *J. Biol. Chem.* **259**, 9607–9615.
Reyes, J., Tanphaichitr, N., Bellvé, A. R., and Benos, D. J. (1986). *Biol. Reprod.* **34**, 809–819.
Robinson, J. M., Tanphaichitr, N., Bellvé, A. R., and Benos, D. J. (1986). *Am. J. Pathol.* **185**, 484–492.
Schreiber, A. B., Kenney, J., Kowalski, J., Thomas, K. A., Gimenez-Galego, G., Rios-Candelore, M., Di Salvo, J., Barritault, D., Courty, J., Courtois, Y., Moenner, M., Loret, C.,

Burgess, W. H., Mehlman, T., Friesel, R., Johnson, W., and Maciag, T. (1985). *J. Cell Biol.* **101,** 1623–1626.

Shabanowitz, R. B., DePhilip, R. M., Crowell, J. A., Tres, L. L., and Kierszenbaum, A. L. (1986). *Biol. Reprod.* **35,** 745–760.

Shin, Y., Folkman, J., Sullivan, R., Butterfield, C., Murray, J., and Klagsbrun, M. (1984). *Science* **223,** 1296–1299.

Thomas, K. A., Rios-Candelore, M., and Fitzpatrick, S. (1984). *Proc. Natl. Acad. Sci. U.S.A.* **81,** 357–361.

Ueno, N., Baird, A., Esch, F., Ling, N., and Guillemin, R. (1987). *Mol. Cell. Endocrinol.* **49,** 189–194.

Vanha-Perttula, T., Mather, J. P., Bardin, C. W., Moss, S. B., and Bellvé, A. R. (1985). *Biol. Reprod.* **34,** 870–877.

Zamboni, L., and Upadhyay, S. (1982). *Am. J. Anat.* **165,** 339–356.

Zhan, X., Bates, B., Hu, X., and Goldfarb, M. (1988). *Mol. Cell. Biol.* **8,** 3487–3495.

Zheng, W., Butwell, T. W., Heckert, L., Griswold, M. D., and Bellvé, A. R. (1990). *Growth Factors* **3,** 73–82.

16
Regulation of Blood Vessel Growth and Differentiation

WERNER RISAU, HANNES DREXLER,
HANS-GÜNTER ZERWES,[1] HARALD SCHNÜRCH,
URSULA ALBRECHT, AND RUPERT HALLMANN[2]
Max-Planck-Institut für Psychiatrie
D-8033 Martinsried Federal Republic of Germany

INTRODUCTION

Endothelial cells (ECs) form the inner surface of all functional blood vessels. They are polarized cells; the luminal surface has nonthrombogenic properties, whereas abluminally the cells produce a basement membrane. The maintenance of this polarity is important for endothelial and organ function. Other cell types also participate in the structure of blood vessels. The adventitial cells of capillaries are called pericytes (or mural cells, Rouget cells) which are embedded in the vascular basement membrane and extend processes around capillary tubes (1). Larger vessels have several layers of smooth muscle cells and adventitial fibroblasts. In this chapter we discuss the growth, development, and differentiation of vascular ECs and the cells of the vascular wall.

GROWTH AND DEVELOPMENT

Angiogenesis, the formation of new blood vessels, does not occur in the normal adult male. In fact, the turnover of adult ECs is very low, in humans of the order of years (2). During ovulation and corpus luteum formation, as

---

1. Present address: San doz AG, Präklinische Forschung, CH-4002 Basel, Switzerland.
2. Present address: Department of Pathology, Stanford University, School of Medicine, 300 Pasteur Drive, Stanford, California 94305.

well as during pregnancy, angiogenesis can be observed in females. This normal angiogenesis is tightly regulated, whereas during pathological conditions such as tumor growth and retinopathies it appears to be disturbed (3).

Angiogenesis also occurs during embryonic development. However, in the embryo the development of the vascular system is much more complex. Two fundamentally different mechanisms are involved. The first is the development of blood vessels from ECs of blood islands differentiating *in situ*. We call this process vasculogenesis to distinguish it from angiogenesis, the second mechanism, which is defined as the sprouting of capillaries from preexisting vessels (see Fig. 1).

Blood Island Induction

All blood vessels develop from blood islands which differentiate in the very early embryo (before somitogenesis) from the splanchnopleuric mesoderm of the area opaca and area pellucida. Blood islands have also been called hemangioblasts to indicate the close relationship between blood cell precursor cells, which differentiate in the center of blood islands, and the endothelial precursors, which differentiate in the periphery of blood islands. The term hemangioblast has led to confusion, and it has been suggested to use the term angioblast for EC precursors only (4–6). Nevertheless, the simultaneous differentiation and close association of blood cells and ECs in blood islands might indicate that these cells have a common stem cell. The discovery of monoclonal antibodies reacting with ECs as well as with blood cells has been taken as evidence in favor of this hypothesis (7); however, direct evidence is lacking.

The induction of blood islands is the first step in the development of the vascular system (Fig. 1). The fact that splanchnopleuric mesoderm gives rise to blood islands, together with evidence indicating that endoderm participates in blood island induction (see references cited in Ref. 6), raises the possibility that endoderm–mesoderm interaction is a prerequisite for blood island differentiation. It would be very important to determine whether cell–cell interaction leads to blood island differentiation, and whether endoderm can induce blood island differentiation during organogenesis.

Vasculogenesis

The second step of vascular development is vasculogenesis, the development of blood vessels from *in situ* differentiating endothelial cells of blood islands (4–6). Blood islands grow and fuse, thereby forming a so-called primary capillary plexus (Fig. 1). Figure 2 shows such a primary vascular plexus in the extraembryonic yolk sac.

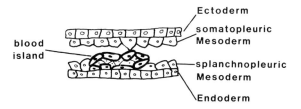

Fig. 1. Schematic delineation of mechanisms involved in the development of the vascular system.

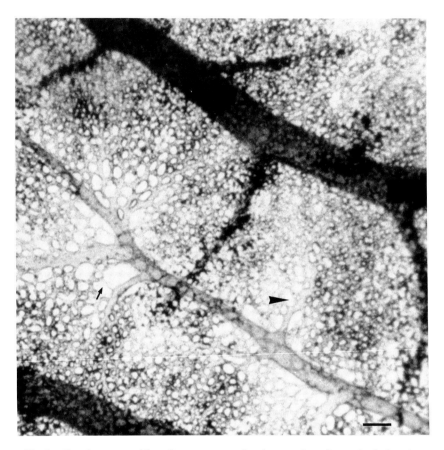

Fig. 2. Development and fate of a primary vascular plexus in the yolk sac of a chick embryo. A capillary plexus is present throughout the yolk sac. Some capillaries seem to fuse (arrowhead), thereby giving rise to larger vessels. Others seem to regress (arrow). Bar, 100 μm.

Since the early morphological studies of vascular development, it has been controversial whether vasculogenesis also occurs inside the embryo. The work of Reagan (8) and more recent experiments involving transplantation chimeras and species-specific monoclonal antibodies (9) indicate that blood vessels indeed differentiate *in situ* in the embryo. The paucity of intraembryonic blood islands, however, suggests that (a) the increase in EC number is achieved by a higher rate of EC proliferation rather than *in situ* differentiation, or (b) EC *in situ* differentiation in the embryo can occur without concurrent blood cell differentiation.

EMBRYONIC STEM CELLS AS AN EXPERIMENTAL MODEL SYSTEM FOR EARLY VASCULAR DEVELOPMENT

All studies concerning the early development of the vascular system have been descriptive so far. Experimental model systems are needed to manipulate blood island induction and vasculogenesis. We have taken advantage of *in vitro* differentiating embryonic stem cells (ESCs), which had been established directly from mouse blastocysts, to study early vascular development (10,11). In suspension culture ESCs spontaneously differentiate to blood island-containing cystic embryoid bodies in a simple culture medium. Apparently, ESCs produce all factors themselves to support blood island differentiation. More blood islands can be induced by factors present in human cord serum. Defined growth or hemopoietic factors tested so far were inactive (10,11). The nature of the factors present in human cord serum are unknown, but the ESC system promises to be a useful model system to characterize factors involved in blood island induction and differentiation.

Vasculogenesis, however, was observed only in ESCs grown intraperitoneally in syngeneic mice (11). These results suggest that factors supporting growth and fusion of blood islands are either limiting or absent in ESCs grown *in vitro*. It is possible that a functional vascular system is not necessary for—and may therefore not be induced in—ESC-derived cystic embryoid bodies, because sufficient nutrients may be supplied by diffusion in suspension culture.

Embryonic Angiogenesis

ESC aggregates and ESC-derived cystic embryoid bodies induce an angiogenic response when transplanted onto the quail chorioallantoic membrane (11). Quail ECs invade the transplants. In addition to vasculogenesis in the embryo proper, capillary invasion may therefore occur earlier than previously thought. Thus, embryonic angiogenesis may already occur at the embryonic stem cell stage. In addition, angiogenesis induced by ESCs may be necessary for implantation and placenta development.

Angiogenesis also occurs during organogenesis. Morphological evidence and data from chick–quail transplantation chimeras have shown that ECs invade the developing brain and kidney (12–15). Figure 3 shows a vascular sprout invading the neuroectoderm of a chick embryo. Capillaries penetrate the basement membrane surrounding the neural tube, radially invade and branch in the subependymal layer, but never penetrate through the ependyme. It is not yet clear whether ECs differentiate from organ-specific precursor cells in other organs. The vascular system of the liver, for example,

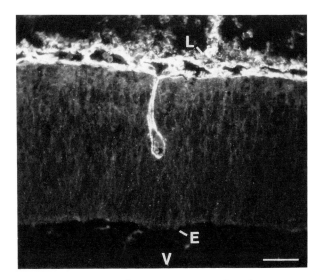

Fig. 3. Brain angiogenesis. A vascular sprout originating from the leptomeningeal vascular plexus (L) invades the neuroectoderm of a chick embryo (4 days old). The image was obtained by indirect immunofluorescence using monoclonal antifibronectin FN2B2 antibodies (27). E, Ependyme; V, ventricle. Bar, 50 μm.

shows a very complex developmental pattern. In addition to *in situ* differentiation of ECs and invasion of capillaries, both of which processes may occur in the embryonic liver, larger vessels are split up into an original vascular plexus.

Angiogenesis Factors

Diffusible factors that stimulate the growth of new capillaries were originally investigated in tumors, because a vast neovascularization occurs in tumor tissue. Tumor angiogenesis factors were characterized from different tumors with only limited success in the past. The development of novel *in vitro* and *in vivo* assays has allowed the recent purification to homogeneity of EC growth factors which are also angiogenic *in vivo* (for review, see Folkman and Klagsbrun, 3).

Prior to EC proliferation, chemotaxis, the directional migration of ECs, takes place during angiogenesis. Chemotactic factors for ECs, which are angiogenic *in vivo*, have also been purified and characterized. Table I summarizes the known properties of angiogenesis factors. There are factors that

TABLE I
Activities of Angiogenesis Factors[a,b]

Factor	EC proliferation	EC motility
Acidic FGF	+	+
Basic FGF	+	+
TGF-α	+	+
TNF	−	+
Adipocyte lipids	−	+
Wound fluid	0	+
Angiogenin	0	ND
Prostaglandins	0	ND
TGF-β	−	ND

[a] From ref. 3.
[b] FGF; Fibroblast growth factor; TGF, transforming growth factor; TNF, tumor necrosis factor; +, stimulation; −, inhibition; 0, no effect; ND, no date available.

have no direct effect on EC proliferation and motility. These may be considered "indirect" angiogenesis factors for the time being. It is unclear why there are so many different factors. Almost nothing is known about the regulation of their expression, secretion, and activity *in vivo*.

We have focused our attention on embryonic angiogenesis and its possible regulation by diffusible factors. Since ECs invade the developing brain and kidney, we have analyzed these embryonic organs for the presence of angiogenesis factors. We have characterized heparin-binding endothelial cell growth factors, which were also angiogenic *in vivo* (17,18). Both factors have biochemical characteristics similar to the fibroblast growth factors (FGFs). Amino-terminal sequence analysis of the embryonic chick brain-derived angiogenesis factor revealed its strong homology to human acidic FGF (19). Our results suggest that the FGFs are involved in the regulation of embryonic angiogenesis.

At present no information is available on the regulation of FGF expression during embryonic development. By using cDNA probes for human acidic FGF (gift of M. Jaye and T. Maciag) it may be possible to determine the expression and possible regulation of acidic FGF production in the chick brain. Our preliminary data show that the human gene for acidic FGF is conserved among vertebrates, because it hybridized under high stringency conditions to a single fragment of chick genomic DNA (Fig. 4). These data are consistent with our amino acid sequencing results (19).

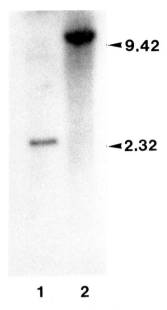

Fig. 4. Southern blot analysis of chick (lane 1) and human (lane 2) genomic DNA. A cDNA clone encoding human acidic FGF detects the typical 10-kilobase (kb) genomic EcoRI fragment in human DNA and a 2.4-kb EcoRI fragment in chick genomic DNA. Size markers (kb) are indicated at right.

Development of the Vascular Wall

As mentioned above, fusion of blood islands in the yolk sac gives rise to a primary capillary plexus. Major changes occur after the onset of circulation. Anastomoses disappear, some capillaries seem to fuse and give rise to arteries or veins, and the direction of blood flow may reverse several times (20). Other mechanisms may be involved in the development of larger vessels inside the embryo (e.g., dorsal aorta; 21). There is morphological evidence that the mechanical stress that ECs have to withstand after the onset of circulation is the predominant force responsible for the development of larger vessels (e.g., vitelline arteries and veins; 22,23).

To investigate the molecular basis for vascular wall development we have studied, as a first step, whether ECs *in vitro* produce chemotactic factors for vascular wall cells (i.e., pericytes, smooth muscle cells, adventitional fibroblasts). The rationale behind these experiments was (a) the observations by Clark and Clark (24) that adventitial cells seem to be attracted by ECs (see Fig. 5) and (b) evidence showing that the vascular wall cells belong to a different lineage than ECs (25). We found that ECs if cultured on nitro-

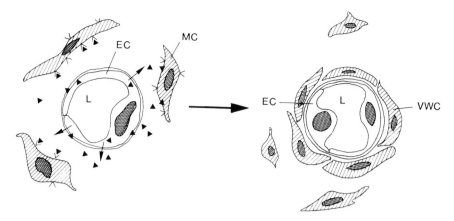

Fig. 5. Model of vascular wall development. Endothelial cells (EC) secrete a chemotactic factor (▲) abluminally. Upon binding of the factor to specific receptors on mesenchymal cells (MC) the cells migrate close to the capillary and eventually organize in layers forming the vascular wall cells (VWC). L, Lumen of the capillary.

cellulose membranes secreted a potent platelet-derived growth factor (PDGF)-like chemotactic factor almost exclusively into the basal (abluminal) compartment (26). This mode of polarized secretion would be expected for a factor involved in the development of the vascular wall (Fig. 5).

During embryonic vasculogenesis and angiogenesis changes also take place in the vascular extracellular matrix (ECM). We found that all blood vessels are surrounded by a fibronectin-rich ECM early in development (see Fig. 4), which is subsequently remodeled into a basement membrane-like matrix (27). However, laminin is expressed much earlier than an intact basement membrane has been observed by electron microscopic techniques. We therefore propose that laminin expression is an early marker for vascular maturation. Around larger vessels (e.g., dorsal aorta) which develop *in situ* in the embryo (vasculogenesis), several layers of fibronectin-expressing cells can be observed early on (Fig. 6). As for the capillary ECM, the ECM of the ECs as well as of the vascular wall cells of larger vessels is remodeled to a basement membrane-like matrix (27).

It is unknown how pericyte and smooth muscle cell differentiation is regulated. Cell–cell interaction with ECs or soluble factors released by ECs, or both mechanisms, could be envisaged. It should be noted that the differentiation of pericytes, for example, requires a remarkable transition of a motile mesenchymal cell to a stationary, almost quiescent cell type with many processes and smooth muscle actin-containing stress fibers.

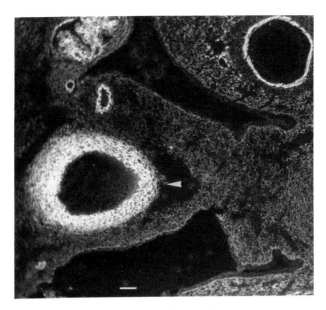

Fig. 6. Extracellular matrix of the vascular wall. Several layers of cells in the vascular wall of the dorsal aorta (6-day-old chick embryo) express fibronectin (arrowhead). The micrograph was obtained following indirect immunofluorescence using FN2B2 antibodies. Bar, 50 μm.

Regression of Blood Vessels

The phenomenon of vascular regression in the embryo has long been ignored. This process in the development of the vascular system is as important as the new formation of capillaries. Both events take place simultaneously. There seem to be different conditions, however, which consequently require the regression or death of the vasculature. These may be subdivided as follows: (1) changes in blood flow; (2) simultaneous regression of tissue or organ and its vasculature; (3) regression of blood vessels without surrounding tissue regression and without prior changes in blood flow. The disappearance of anastomoses of a capillary network and the regression of the sinus marginalis during limb bud outgrowth (28) are examples of the first type. There are many examples for the simultaneous regression of vasculature and tissues (e.g., yolk sac, interdigital tissue, 29) and organs (e.g., mesonephroi, corpus luteum; 30).

The third condition, of course, is the most interesting because it implies that factors might regulate the localized regression, in which only ECs are involved. Examples are the hyaloid vascular system (31) and the regression of capillaries in the skeletal elements of the limb bud (32; see Fig. 7). We

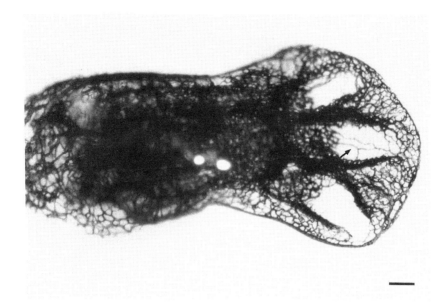

Fig. 7. Regression of capillaries during limb chondrogenesis. Dorsal view of a 6-day-old chick embryo leg bud. The vascular system was filled with India ink and the embryo subsequently cleared. The prospective areas of toes 3, 4, and 5 are avascular, with some capillaries remaining in toe 4 (arrow). Bar, 200 μm.

have studied the latter process in the chick limb bud and found that blood vessels regress before overt cartilage differentiation (33; Fig. 8). Thus, factors in the limb might control the proliferation of ECs—and may even induce the death of them (34)—in certain areas, thereby permitting cartilage differentiation. It is possible that chondrocyte precursor cells produce such factors.

ENDOTHELIAL CELL DIFFERENTIATION

All ECs throughout the body have common features; they are able to form a lumen, have a nonthrombogenic luminal surface, produce a basement membrane, and express common markers, such as Factor VIII-related antigen. Nevertheless, ECs are functionally and morphologically heterogeneous. Apart from the differences between small vessel and large vessel endothelium, the differences between capillary endothelial cells are most important for tissue and organ function. On the basis of morphology they have been subdivided into three groups: continuous, fenestrated, and dis-

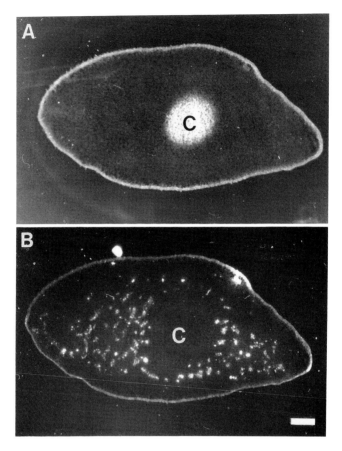

Fig. 8. Regression of capillaries precedes overt cartilage differentiation, as shown in a frozen section (10 μm) of the distal part of a 6-day-old chick embryo wing bud. The cartilage (C) of digit 3 is labeled by the monoclonal LIII antibody which reacts with chondrocytes (A). The vascular system of the same section is visualized by fluoresceinated acetylated low density lipoprotein which was taken up by endothelial cells. For details, see Ref. 33. An avascular zone is clearly seen in the area of digit 3 (B). It is larger than the cartilage (C). Bar, 100 μm.

continuous capillaries. As shown in Table II, differences in morphology reflect different functions. It is worth noting that the same organ (e.g., brain, kidney) can contain both fenestrated (choroid plexus, glomeruli) and continuous capillaries. There are basically two possible ways in which this heterogeneity may originate: (1) ECs differentiate from organ-specific precursor cells, or (2) ECs invade developing organs and differentiate to organ-specific ECs in response to the local environment. There is plenty of evidence that

TABLE II
Morphological and Functional Differences between Capillaries from Different Tissues and Organs

Capillary type	Example of tissue or organ	Properties	Function
Continuous	Muscle	High number of vesicles	Exchange/transport
	Central nervous system (CNS)	Very low number of vesicles, complex tight junctions	Blood–brain barrier
	Lymph node	High endothelial venules (HEV)	Lymphocyte homing
	Bone marrow	Marrow sinus	Hemopoiesis, delivery of blood cells
	Spleen	Splenic sinus of red pulp	Blood cell processing
Fenestrated	Gastrointestinal tract, endocrine glands, glomeruli/tubuli, choroid plexus	Fenestrae	Absorption, secretion, filtration
Discontinuous	Liver	Large gaps (sinusoids)	Rapid exchange of particles

the second mechanism is true for most tissues and organs. It is known that ECs invade the developing kidney and brain. At least in these organs ECs interact with the local environment to differentiate to the specialized endothelium. There are reports from other systems as well that ECs can be induced to differentiate in more than one direction (35,36).

We have focused our interest on the differentiation of blood–brain barrier (BBB) endothelium. BBB ECs are unique among ECs because they have complex tight junctions (37), which prevent the passage of molecules and provide a high electrical resistance (up to 2000 Ω/cm^2). Therefore, BBB endothelium has the characteristics of a tight epithelium, with which it shares the specialized transport systems. There is negligible if any pinocytotic transport.

To study the differentiation of brain ECs, we have correlated the expression of a number of proteins which were reported to be largely specific for brain ECs with the development of the BBB (38). In order to obtain more specific markers we have characterized a monoclonal antibody against a protein of chick BBB endothelium which is specific for endothelial cells in the brain (39; Fig. 9). This molecule is not present in ECs of the circumventricular organs (except for the subcommissural organ and the area

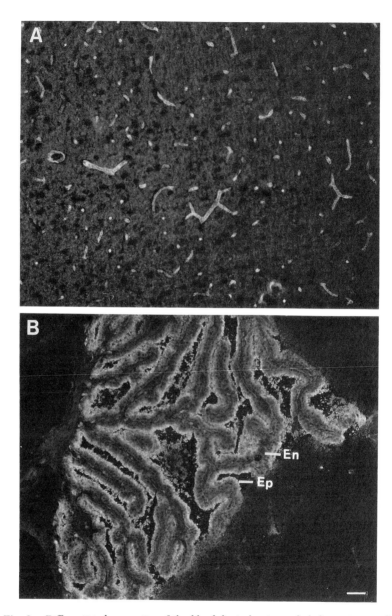

Fig. 9. Differentiated properties of the blood–brain barrier endothelium. A monoclonal antibody (HT7) against chick brain endothelial cells reacts with capillaries in the adult chick brain (A). In the choroid plexus the endothelium (En), which is fenestrated, does not express the antigen (B). However, choroid plexus epithelium (Ep) reacts with the antibodies. Bar, 50 μm.

postrema), which do not have a BBB (except for the subcommissural organ). Its additional expression on choroid plexus epithelium (Fig. 9), embryonic blood cells, and the basolateral membranes of kidney tubules suggests that it is involved in transport mechanisms at the BBB. More importantly, ECs derived from the chick chorioallantoic membrane which do not normally express this protein were induced to do so after they had invaded a brain transplant (39). This is consistent with earlier results by Stewart and Wiley (13). More recent direct evidence indicates that astrocytes which form the peculiar vascular endfeet can induce a permeability barrier to invading ECs (40). The differentiation of BBB endothelium is probably the best characterized system in which to look for factors necessary for the differentiation of this highly specialized endothelial cell.

CONCLUDING REMARKS

Endothelial cell growth is tightly regulated. It is out of control in a number of pathological conditions such as tumor growth. The endothelial cell is a very plastic cell type, able to differentiate into multiple endothelial phenotypes with different functional properties. The molecular mechanisms of endothelial cell growth control and differentiation are only marginally understood. Information about the regulation of normal endothelial cell growth and differentiation in the embryo might help us to understand and manipulate the same events in pathological conditions.

REFERENCES

1. Sims, D. E. (1986). *Tissue Cell* **18**, 153–174.
2. Denekamp, J. (1984). *Acta Radiol. Oncol.* **23**, 217–225.
3. Folkman, J., and Klagsbrun, M. (1987). *Science* **235**, 442–447.
4. Sabin, F. R. (1920). *Contrib. Embryol.* **9**, 215–262.
5. Gonzalez-Crussi, F. (1971). *Am. J. Anat.* **130**, 441–460.
6. Romanoff, A. L. (1960). Macmillan, New York.
7. Peault, B. M., Thiery, J. P., and LeDouarin, N. (1983). *Proc. Natl. Acad. Sci. U.S.A.* **80**, 2976–2980.
8. Reagan, F. P. (1915). *Anat. Rec.* **9**, 229–241.
9. Pardanaud, L., Altmann, C., Kitos, P., Dieterlen-Lievre, F., and Buck, C. A. (1987). *Development* **100**, 339–349.
10. Doetschman, T. C., Eistetter, H., Katz, M., Schmidt, W., and Kemler, R. (1985). *J. Embryol. Exp. Morphol.* **87**, 27–45.
11. Risau, W., Sariola, H., Zerwes, H.-G., Sasse, J., Ekblom, P., Kemler, R., and Doetschman, T. (1987). *Development* **102**, 471–478.
12. Bär, T. (1980). *Adv. Anat. Embryol. Cell Biol.* **59**, 1–62.

13. Stewart, P. A., and Wiley, M. J. (1981). *Dev. Biol.* **84,** 183–192.
14. Ekblom, P., Sariola, H., Karkinen, M., and Saxen, L. (1982). *Cell Differ.* **11,** 35–39.
15. Sariola, H., Ekblom, P., Lehtonen, E., and Saxen, L. (1983). *Dev. Biol.* **96,** 427–435.
16. Chen, W.-T., Chen, J.-M., and Mueller, S. C. (1986). *J. Cell Biol.* **103,** 1073–1090.
17. Risau, W. (1986). *Proc. Natl. Acad. Sci. U.S.A.* **83,** 3855–3859.
18. Risau, W., and Ekblom, P. (1986). *J. Cell Biol.* **103,** 1101–1107.
19. Risau, W., Gautschi-Sova, P., and Böhlen, P. (1988). *EMBO J.* **7,** 959–962.
20. Evans, H. M. (1909). *Anat. Rec.* **3,** 498–519.
21. Hirakow, R., and Hiruma, T. (1983). *Anat. Embryol.* **166,** 307–315.
22. Murphy, M. E., and Carlson, C. E. (1978). *Am. J. Anat.* **151,** 345–376.
23. Heine, U. I., Roberts, A. B., Munoz, E. F., Roche, N. S., and Sporn, M. B. (1985). *Virchows Arch. B:* **50,** 135–152.
24. Clark, E. R., and Clark, E. L. (1925). *Am. J. Anat.* **35,** 239–282.
25. Le Lievre, C. S., and Le Douarin, N. M. (1975). *J. Embryol. Exp. Morphol.* **34,** 125–154.
26. Zerwes, H.-G., and Risau, W. (1987). *J. Cell Biol.* **105,** 2037–2041.
27. Risau, W., and Lemmon, V. (1988). *Dev. Biol.* **125.**
28. Seichert, V., and Rychter, Z. (1971). *Folia Morphol. (Prague)* **19,** 367–377.
29. Hurle, J. M., Colvee, E., and Fernandez-Teran, M. A. (1985). *J. Embryol. Exp. Morphol.* **85,** 239–250.
30. O'Shea, J. D., Nightingale, M. G., and Chamley, W. A. (1977). *Biol. Reprod.* **17,** 162–177.
31. Latker, C. H., and Kuwabara, T. (1981). *Invest. Ophthalmol. Visual Sci.* **21,** 689–699.
32. Feinberg, R. N., Latker, C. H., and Beebe, D. C. (1986). *Anat. Rec.* **214,** 405–409.
33. Hallmann, R., Feinberg, R. N., Latker, C. H., Sasse, J., and Risau, W. (1987). *Differentiation* **34,** 98–105.
34. Latker, C. H., Feinberg, R. N., and Beebe, D. C. (1986). *Anat. Rec.* **214,** 410–417.
35. Hart, T. K., and Pino, R. M. (1986). *Lab. Invest.* **54,** 304–313.
36. Milici, A. J., Furie, M. B., and Carley, W. W. (1985). *Proc. Natl. Acad. Sci. U.S.A.* **82,** 6181–6185.
37. Stewart, P. A., and Hayakawa, E. M. (1987). *Dev. Brain Res.* **32,** 271–281.
38. Risau, W., Hallmann, R., and Albrecht, U. (1986). *Dev. Biol.* **117,** 537–545.
39. Risau, W., Hallmann, R., Albrecht, U., and Henke-Fahle, S. (1986). *EMBO J.* **5,** 3179–3183.
40. Janzer, R. C., and Raff, M. C. (1987). *Nature (London)* **325,** 253–257.

17

Vascular Endothelium: Role in Albumin Transport

G. E. PALADE[1] AND A. J. MILICI[2]
Yale University School of Medicine
Department of Cell Biology
New Haven, Connecticut 06510

MULTIPLICITY OF ENDOTHELIAL FUNCTIONS

The vascular endothelium of vertebrates performs a large variety of functions that have been studied in greater detail in man and laboratory mammals than in any other species. All the functions depend on the immediate contact of the circulating blood with the endothelium and on the interposition of the latter as a highly attenuated cellular partition between the blood plasma and the interstitial fluid.

The primary function of the endothelium is to mediate continuous, large-scale exchanges of water and solutes, both hydrophilic and hydrophobic, between the circulating blood plasma and the interstitial fluid. These exchanges are a critical prerequisite for keeping alive the vast majority of the cells of the organism which lives in the internal sea of the interstitial fluid. The development of an all-pervading vascular system, lined by a highly permeable endothelium, has been a key element in the evolution of higher vertebrates, man included. In the latter, the capacity of the system is used essentially to its limits; the margin of safety is extremely narrow, which explains the dominant position of cardiovascular diseases in human pathology.

Other endothelial functions include (a) hemostasis and control of blood coagulation; (b) participation in local defense reactions, as in inflammation or in immune reactions via antigen presentation; (c) regulation of blood flow via relaxing and constricting factors produced by endothelial cells; (d) regulation of cell–cell interactions within the circulatory system; and (e) participation in

---

1. Present address: Division of Cellular and Molecular Medicine, University of California, San Diego, School of Medicine, La Jolla, CA 92093.
2. Present address: Department of Immunology and Infectious Diseases, Pfizer Central Research, Groton, CT 06340.

angiogenesis via angiogenetic and growth factors as well as cell–cell interactions within the walls of the vessels. Notwithstanding their obvious importance, all these activities can be considered as subsidiary to the main function of the endothelium which, as already stated, is the mediation of blood plasma–interstitial fluid exchanges. These additional functions were apparently developed because of the strategic location of the endothelium, its basic organization, and its special environment.

GENERAL ORGANIZATION

As a general rule, the endothelium is organized as a highly attenuated cellular monolayer, the equivalent of a unicellular squamous epithelium, that lines without major interruptions the lumen of the entire vascular system. Some of its structural features are common to the entire endothelium irrespective of the size of the vessel and (to some extent) its function. Pronounced or extreme attenuation of the cell bodies (especially of their periphery), simplified intercellular junctions, and a large population of plasmalemmal vesicles are among these general features. But in detail endothelial structure is diversified, especially at the level of the microvasculature, from segment to segment, that is, from arterioles to capillaries to pericytic and muscular venules, as well as from one microvascular bed to another. It is at this last level that the most important structural variants appear (1,2). There are microvascular beds lined by a continuous endothelium in the limbs and body wall (skin, skeletal muscles, connective tissue), in visceral muscles (myocardium included), and in certain viscera (e.g., the lung). There are microvascular beds provided with a fenestrated endothelium as in the case of different mucosae and exocrine and endocrine glands. And there are also microvascular beds characterized by a discontinuous endothelial lining as in the sinusoids of the liver, spleen, and bone marrow. Finally there are highly specialized microvascular beds like those of the brain and of the renal glomeruli. In the former, the endothelium is continuous yet structurally and functionally different from any other microvascular endothelium in the body. In the latter, the endothelium is provided with large, diaphragm-free fenestrae, which means that the control of their permeability is relegated to their basement membrane.

Physiological data indicate even greater diversity than uncovered by structural findings, and the same seems to apply for the meager biochemical data so far available. There is considerable diversity within the endothelial lining of the vasculature, although the current tendency is to extend to the whole endothelium data obtained on restricted segments. The vascular system is not lined by a uniform endothelium, but by a wide variety of endothelia differentiated for locally relevant functions.

CONTINUOUS MICROVASCULAR ENDOTHELIUM

At present, the most substantial body of information concerns the continuous endothelium of the microvasculature of limbs (skeletal muscles included), myocardium, and the lung. Structurally, this type of continuous endothelium is characterized by a very large population of plasmalemmal vesicles of rather regular spherical shape and size (outer diameter 70 nm). Many of these vesicles are in continuity with the plasmalemma on either the blood front or the tissue front of the cells, and some of them form transendothelial channels (i.e., water-filled pathways uninterrupted by any hydrophobic barrier) by simultaneous fusion with one another and the plasmalemma (2,3). It has been known for some time that plasmalemmal vesicles can fuse to form chains (4) which eventually could become transendothelial channels (3). Work done in other laboratories (5,6) indicates, however, that in fixed specimens serially sectioned and examined by transmission electron microscopy, the plasmalemmal vesicles form dendritic, sessile structures by extensive vesicle to vesicle fusion. Free vesicles and transendothelial channels were only occasionally encountered (5,6). The physiological significance ascribed to these findings will be discussed later in this chapter. More recent studies have shown that the plasmalemmal vesicles have a characteristic surface structure on the cytoplasmic side of their membrane which consists of ridges disposed as meridians between two opposite poles (7).

Differentiated Microdomains on the Luminal Plasmalemma

The membrane of these vesicles appears to be different from the plasmalemma proper: it has a low surface density of anionic sites on its inner or luminal aspect, which contrasts with a high density of such sites on the luminal surface of the plasmalemma proper (8). The anionic sites were revealed by using cationized ferritin (pI 7.0–8.0) as a probe, and their nature was explored via enzyme perfusions: all sites were removed by broad specificity proteases, which indicates that they are provided by proteins, most likely sialoglycoproteins (10). These findings led to the conclusion that the luminal endothelial plasmalemma is provided with differentiated microdomains which include (a) plasmalemmal vesicles, the membranes of which do not bind anionic probes, but native (anionic) (4,8,11) or slightly cationized ferritin, pI 6.8–7.0, has access to their interior; (b) plasmalemma proper, which is heavily decorated by ferritin of pI 7.0–8.4; and (c) coated pits (few in number) that also bind cationized ferritin. The inquiry that led to the identification of differentiated microdomains added new elements to the spectrum of diversification of endothelia: additional microdomains were found in fenestrated endothelia associated with fenestral diaphragms (9,10), transendothelial channels,[1] and newly discovered structures called endothelial pockets (13). The luminal aspect of the fenestral diaphragms is heavily negatively charged in many but not all fenestrated endothelia (14), the negative charges being provided by heparan sulfate proteoglycans (10). The inquiry also revealed that plasmalemmal vesicles of similar or identical morphology differ in anionic site density from one endothelium to another, the density being much higher in the aortic endothelium than in that of the microvasculature (15).

Cell Junctions

The intercellular junctions of the continuous microvascular endothelium, studied on sectioned vessels and on freeze-cleaved specimens, consist of occluding junctions comparable to those of leaky epithelia and of poorly developed adhering zonules; desmosomes (adhering maculae) are generally absent. There are, however, segmental differentiations; communicating (gap) junctions are found in arteriolar endothelia associated with more elaborate occluding junctions, and rudimentary, occasionally open "tight" junctions appear in the endothelia of postcapillary venules (16,17).

PHYSIOLOGICAL DATA

Most of the information on capillary permeability comes from experiments carried out on microvascular beds provided with a continuous endothelium. The results obtained indicate a very high permeability to water and small hydrophilic solutes, two to three orders of magnitude higher than that of any other epithelium (18) in the animal body. This uniquely high permeability includes a convective component (hydraulic conductivity), which implies the existence of water-filled pores or channels across the endothelial layer, and a diffusive component that could be accounted for by either pores or vesicular transport. Such findings led early in the history of the field to the formulation of a pore theory of capillary permeability (19,20). Originally, a single population of small pores was postulated (19), but later on the theory was modified to accommodate the high permeability of the capillary endothelium to macromolecules. This was done by increasing the diameter of the small pores to approximately 11 nm and by including a small population of large pores or "leaks" of diameter 50–70 nm (20,21). From the beginning all pores were assumed to be located within intercellular junctions, primarily because

---

3. The transendothelial channels of the fenestrated endothelium are formed by single vesicles and are different in structure from their counterparts in the continuous endothelium (12).

no alternative was known at the time and because the existence of a permeable, intercellular "cement" had been postulated in earlier studies (22).

ATTEMPTS AT STRUCTURAL–FUNCTIONAL INTEGRATION

Early electron microscope studies of the vascular endothelium were essentially contemporary with the formulation of the pore theory of capillary permeability, which explains why attempts were made relatively early to identify and locate the pores by direct morphological studies or by the use of molecular tracers that, on account of their known size, could qualify as probes for either the large or small pores or both. Macromolecules of 10 nm or larger [ferritin, dextrans, and glycogens[2] (2)] were found exclusively in plasmalemmal vesicles while in transit through the endothelium. No tracer of such size appeared within intercellular junctions. The evidence was considered reliable because such tracer molecules could be identified and localized individually by electron microscopy.

Peroxidatic "Mass" Tracers

Smaller macromolecules (diameter <10 nm) that could qualify as probes for both pore systems (but could help locating primarily the small pores on account of their much higher surface density) were available but could be visualized only indirectly via a cytochemical peroxidatic reaction (24). They ranged in diameter from about 5 to 2 nm and included horseradish peroxidase (HRP), myoglobin, cytochrome c, and heme peptides derived from the latter (for review, see Ref. 2). The relevant cytochemical procedure involved a large amplification of the signal on account of the enzymatic activities of the probes but gave low resolution in space as well as in time because of the rapid diffusion of both probes and reaction products. Given these limitations, it proved impossible to arrive at a consensus in the interpretation of the experimental findings. Some investigators concluded that the tracers were leaving the blood plasma via intercellular junctions (25,26), as postulated by the pore theory; others maintained that tracer exit occurred primarily by vesicular shuttling (or transcytosis) (2,27) or by diffusion and convection along transendothelial channels (2,3). In time, it became clear that the experimental protocols had to be modified to obtain better spatial and temporal resolution and thereby improve the chances of achieving a reason-

---

4. Dextrans and glycogens were tested primarily on microvascular beds provided with a fenestrated endothelium and shown to exit the capillary lumina via fenestrae as well as plasmalemmal vesicles (23).

able consensus. Moreover, it seemed desirable to use probes of more direct physiological relevance than provided by the peroxidatic proteins already mentioned.

Do Plasmalemma Vesicles Transport Anionic Plasma Proteins?

In the meantime, the discovery of differentiated microdomains on the luminal plasmalemma of endothelial cells[3] provided a starting base for a new experimental approach. Plasmalemmal vesicles, already implicated in macromolecular transport by the work mentioned above (2), are the only differentiated microdomains endowed with a low concentration of anionic sites on their luminal (or inner) surface. In this respect, they are clearly different from other microdomains, for example, the coated pits—coated vesicle system, the plasmalemma proper, and the introits to intercellular junctions, which are found to be densely populated by anionic sites (8). Since plasma proteins are known to be, in their vast majority, anionic species (29), and since they are known to be transported to the interstitial fluid, we postulated (8–10) that plasmalemmal vesicles must be involved in this process, since access of anionic proteins to the vesicular contents is not expected to be hindered by electrostatic repulsion. Moreover, it was already known that native, anionic ferritin as well as modified, neutral ferritin have access to these vesicles (4,8).

Albumin (M_r ~68K, average diameter ~72 Å,[4] pI 4.5) appeared to be the test object of choice. In addition to many other important functions, it acts as a carrier of fatty acids, sterols, metal ions, and certain hormones and amino acids (29,30), that is, ingredients needed by all the cells living in the internal sea of the interstitial fluid. To perform this function the carrier itself must be transported or otherwise moved across the endothelium.

Albumin Used as Tracer

Albumin–Colloidal Gold Complexes

To test this hypothesis, two different experimental protocols were developed. In the first, the probe was an albumin–colloidal gold complex (A–Au), obtained by coating 5- to 7-nm colloidal gold particles (31) with bovine serum

---

5. On the tissue front of the plasmalemma, differentiated microdomains were found in fenestrated endothelia (28); the situation in capillaries with a continuous endothelium is still unknown.

6. The albumin molecule has the shape of a prolate ellipsoid whose large and small diameter measures 140 and 40 Å, respectively (30).

albumin (BSA). A–Au complexes can be directly visualized and reliably localized by conventional transmission electron microscopy (TEM) in thin sections of plastic-embedded tissue specimens, under conditions that ensure optimal preservation of endothelial structure. This signal advantage is counteracted in part by the fact that A–Au complexes are expected to behave not as monomeric albumin molecules, but as polymeric albumin or albumin aggregates. Moreover, because of their size, around 14 nm (see below), the complexes are expected to exit the blood plasma only through the large pore system or its equivalents. In addition, to maximize tracer–endothelium interactions, the protocol involves the exhaustive removal of endogenous (murine) albumin from the circulation, with full knowledge that the removal will increase microvascular permeability (see below).

In this first protocol, A–Au complexes suspended in glucose-supplemented phosphate-buffered saline (gPBS) were introduced by perfusion, in the systemic circulation of mice, after flushing the vasculature free of blood and, hence, of endogenous albumin. This step was followed by a short gPBS flush and then by fixation via perfusion with an aldehyde mixture. Specimens were collected primarily from the heart and lungs and processed for conventional TEM. The first results, published by Ghitescu et al. (32), showed that there was indeed extensive, preferential binding of the probe to the membrane of plasmalemmal vesicles, with limited binding to the plasmalemma proper and coated pits. In time (past 10 min), a rather small fraction of the A–Au complexes appeared in the pericapillary spaces, whereas another small fraction was diverted to endosomes and multivesicular bodies within endothelial cells. While in transit through the endothelium the distribution of the probe was strictly limited to plasmalemmal vesicles. No A–Au complexes were detected within intercellular junctions. Morphometric procedures were used to demonstrate that A–Au binding was saturable and competable by monomeric albumin (32).

The preferential binding of A–Au to plasmalemmal vesicles was confirmed by work done in our laboratory (33). The protocol we followed was essentially the same as previously (32), and the results obtained were comparable. We found rapid, extensive labeling of plasmalemma vesicles opened on the blood front of the cells, although there was concurrent but more limited labeling of the plasmalemma proper (Figs. 1a, 3). Occasionally the cytoplasmic aspect of the membrane of these vesicles (labeled or unlabeled by A–Au) was decorated with short radial spurs (Fig. 1, insets) that probably correspond to the meridional ridges described elsewhere (7). The extent of vesicular labeling could be easily assessed on grazing sections (Fig. 2) that showed that up to 75% of the luminal vesicles were A–Au labeled. In time (past 10 min) a small fraction of the probe appeared in the pericapillary spaces, often associated with or close to vesicles opened (and presumably

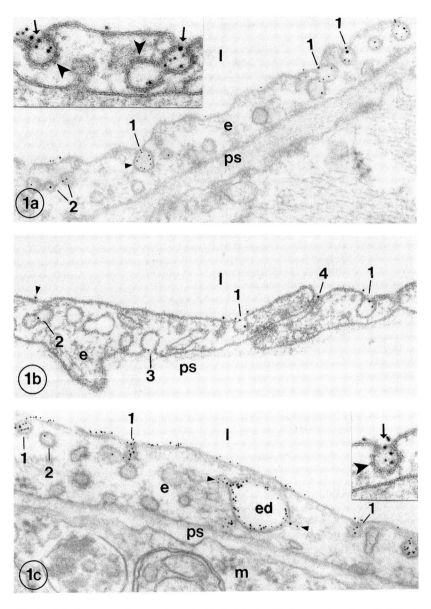

Fig. 1. Murine myocardium perfused *in situ* for 16 min with bovine albumin–colloidal gold complexes (A–Au), fixed by perfusion with 3% glutaraldehyde in 0.1 *M* sodium cacodylate buffer, pH 7.4, and then processed for transmission electron microscopy by standard procedures. (a) Segment of a normally sectioned capillary wall. All the plasmalemmal vesicles opened individually or as short chains on the blood front of the endothelial cell (e) are labeled by A–Au complexes (1). There is also labeling of vesicles apparently free in the cytoplasm (2), but there is no labeling of vesicles opened on the tissue front nor of the pericapillary spaces (ps).

discharging) on the tissue front of the cells (Figs. 1b, 3). Another small fraction was detected in endosomes, and multivesicular bodies. Most of these endocytic compartments were of usual vacuolar appearance (Figs. 1c, 4a), but occasionally complex tubular forms marked by the A–Au probes were encountered (Fig. 4b). The interaction of A–Au complexes with the plasmalemmal vesicle membrane and, in part, the plasmalemma proper were strong enough to resist a 2-min flushing of the vasculature (with gPBS) following the perfusion of the tracer. A–Au complexes were found occasionally tightly packed in introits leading from the lumina to intercellular junctions, but no tracer was detected within the junctions themselves or in intercellular spaces on the abluminal side of the junctions. No gradients of A–Au appeared in the pericapillary spaces centered on exits from endothelial junctions.

In a variant of this protocol, fixed frozen myocardium sections approximately 20 μm in thickness were also reacted with HRP-conjugated antialbumin. The results showed that, where accessible to the antibody, the albumin coat of the Au particles could be visualized by positive or negative staining as an approximately 5 nm thick layer to give an overall diameter of around 14 nm for the A–Au probe (Fig. 3). Complexes internalized in endosomes or multivesicular bodies had no stained coats or only lightly stained coats (Fig. 4a,b), and the same applied to probes found in coated vesicles (Fig. 5) and in plasmalemmal vesicles within the endothelial cytoplasm (Figs. 4, 6), even when the antibody had access to both sides of the endothelium (Figs. 4, 6). This means that not all vesicles are freely accessible to solutes as part of chains or "dendritic structures" opened on each front of the endothelium separately, without interlinking, as postulated previously (5,6).

On account of their size (>11 nm), A–Au complexes qualify, as already mentioned, for probes of the large pore system, assumed to be exclusively located at the level of intercellular junctions, according to the capillary physiology literature (34,35). Obviously our results are not in agreement with

---

Note the close association of A–Au complexes with the inner surface of the vesicle membrane at arrowhead. l, Lumen. Magnification 71,000×. (b) Segment of the wall of another capillary more sparsely labeled than the one in Fig. 1a. A–Au complexes label plasmalemmal vesicles opened on the blood front (1) apparently free in the cytoplasm (2), or open on the tissue front (3). There is limited labeling of the plasmalemma proper (arrowhead) and of the introit to an intercellular junction (4), but no labeling of the junction proper, intercellular space past the junction, nor pericapillary spaces (ps). l, Lumen; e, endothelium. Magnification 70,000×. (c) In addition to plasmalemmal vesicle labeling similar to that in Fig. 1a,b (1,2), A–Au complexes are found in an endosome (ed) and associated vesicles (arrowheads). l, Lumen; e, endothelium; ps, pericapillary spaces; m, macrophage. Magnification: 58,000×. (Insets) The cytoplasmic surface of plasmalemmal vesicles, open on the blood front (arrows) and marked by A–Au complexes, is provided with periodically distributed dense spurs (arrowheads). Magnification 113,000×.

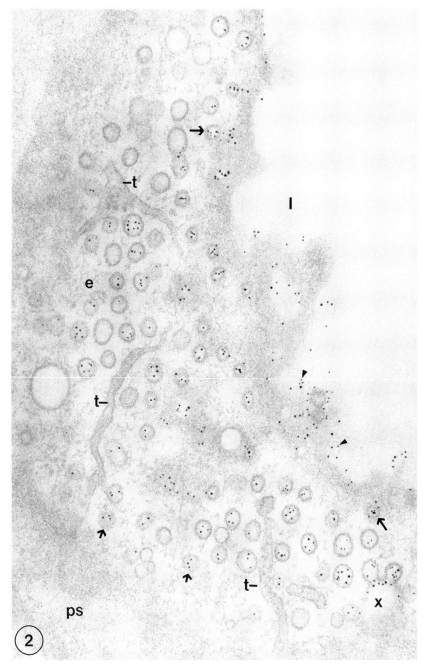

Fig. 2. Oblique section through a capillary endothelial cell in a mouse myocardium. The specimen was prepared and processed as in Fig. 1. The highly oblique orientation of the section reveals the extent of labeling of plasmalemmal vesicles across the entire thickness of the cell

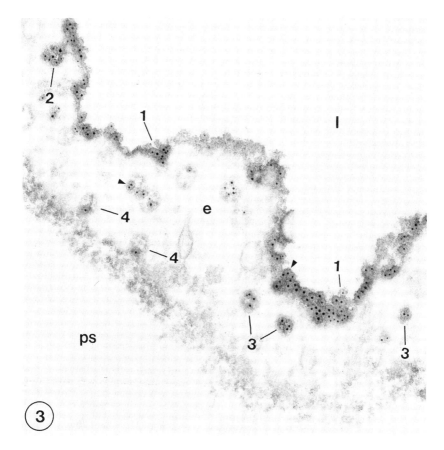

Fig. 3. Mouse myocardium perfused *in situ* with A–Au complexes. The specimen was fixed by perfusion and reacted (in frozen 20-μm sections) with HRP-conjugated antialbumin. A–Au complexes are found on the luminal plasmalemma (1) in relatively large amounts because of incomplete flushing of the tracer and in plasmalemmal vesicles either open on the luminal (blood) front (2), located apparently free in the cytoplasm (3), or open—apparently discharging—on the abluminal (tissue) front (4). The albumin shell on most complexes is rendered visible (arrowheads) by either positive or negative staining with HRP–antialbumin. l, Lumen; e, endothelium; ps, pericapillary spaces. Magnification: 90,000×.

from the luminal (long arrows) to the abluminal side (short arrows). A luminal surface dimple with a corona of labeled vesicles is marked x. A–Au complexes mark the plasmalemma proper (arrowheads) but not the pericapillary spaces. The tubular elements (t) running in between clusters of vesicles most probably belong to the endoplasmic reticulum (rather than the endosomal system). l, Lumen; e, endothelium; ps, pericapillary space. Magnification: 78,000×.

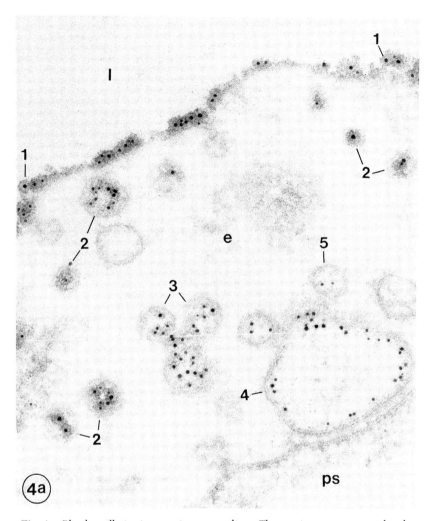

Fig. 4. Blood capillaries in a murine myocardium. The specimen was prepared and processed as in Fig. 3. (a) A–Au complexes label the luminal plasmalemma (1), plasmalemmal vesicles scattered apparently free, individually (2) or in clusters (3), in the cytoplasm, and an endosome (4). Note the wide variations in albumin shell staining from heavy (1 and 2) to light (3) to apparently missing (4 and 5). (b) In addition to the plasmalemma proper (1) and plasmalemmal vesicles either opened on the luminal front (2) or apparently free in the cytoplasm (3), A–Au complexes label heavily a large complex endosome (4) comprised of vacuolar and tubular elements. Note that A–Au complexes are not labeled by HRP–antialbumin in the endosome. l, Lumen; e, endothelium; ps, pericapillary spaces. Magnifications: a, 164,000×; b, 103,000×. (Figure continues.)

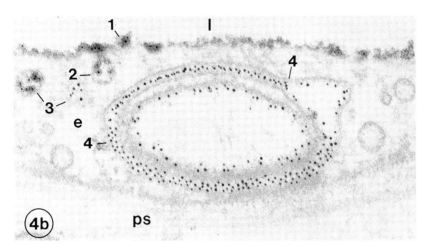

Fig. 4b (continued).

this assumption. A–Au complexes prepared with larger (20–25 nm) gold particles were often found stuck in vesicular introits (Fig. 7), apparently unable to pass through the narrow necks of many luminally open plasmalemmal vesicles. Extrapolating from measurements made possible by HRP-tagged antialbumin, it can be assumed that the diameter of these complexes

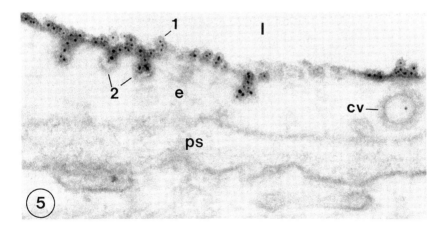

Fig. 5. Blood capillary in a murine myocardium. The specimen was prepared and processed as in Fig. 3. A–Au complexes with stained albumin shells appear on the plasmalemma proper (1) and in plasmalemmal vesicles open on the luminal front (2). The A–Au complex in the coated vesicle (cv) is not stained by HRP–antialbumin. l, lumen; e, endothelium; ps, pericapillary spaces. Magnification 117,000×.

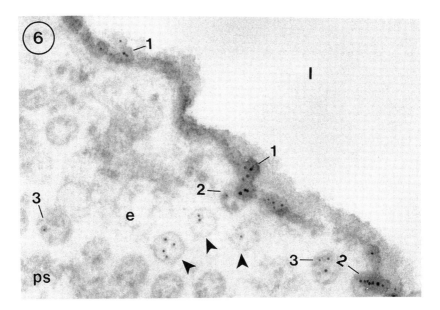

Fig. 6. Capillary in a murine myocardium. The specimen was processed as in Fig. 3. A–Au complexes with shells stained by HRP–antialbumin are seen on the plasmalemma proper (1) entering plasmalemmal vesicles open on the luminal front (2), and in plasmalemmal vesicles apparently free in the cytoplasm (3). Note that apparently internalized vesicles marked by A–Au complexes are not stained by the HRP–antialbumin (arrowheads). Note also that in this case the antibody had access to the pericapillary spaces (ps). l, Lumen; e, endothelium. Magnification: 122,000×.

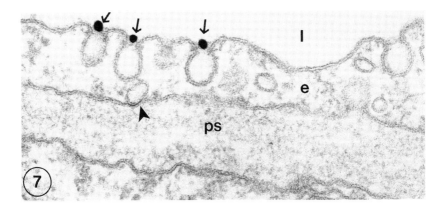

Fig. 7. Blood capillary in a murine myocardium. The specimen was processed as in Fig. 1, except that 20- to 25-nm gold particles were used to prepare the A–Au complexes. These large A–Au complexes are apparently stuck in the infundibula leading to plasmalemmal vesicles (arrows), unable to move through the neck of the vesicles. The arrowhead marks a vesicle fused to, but not open on, the abluminal plasmalemma. l, Lumen, e, endothelium; ps, pericapillary spaces. Magnification: 126,000×.

exceeds 30 nm. Such findings indicate that strictures at vesicular entries or along transendothelial channels may select according to size particles that otherwise have a common surface chemistry, imparted in this case by albumin coating.

The findings summarized above demonstrate (a) extensive, preferential binding of A–Au complexes to the membrane of plasmalemmal vesicles; (b) transcytosis of the probe at a slow rate; (c) no exit of the probe via intercellular junctions; and (d) its partial diversion to the endothelial endosome system. As already mentioned, some of these results may reflect the multifunctional character of the probe: A–Au complexes are expected to behave like polymeric albumin, and this may affect the rate of transport as well as the extent of diversion to endosomes.

Monomeric Albumin

The second protocol was developed in an attempt to circumvent such problems and to obtain evidence of more direct physiological relevance (33). To this intent, the probe used was monomeric exogenous albumin (bovine) administered as before by perfusion after flushing with gPBS the endogenous murine albumin from the circulation. The perfusion time was shortened progressively from 5 min to 15 sec and was followed by fixation, again by perfusion, with an aldehyde mixture with or without a short (1 min) preceding flush with gPBS. The specimens (heart, diaphragm) were processed through thin frozen sectioning, and the probe was detected by using a bovine albumin antibody (raised in rabbits) followed by a gold-tagged secondary antibody (raised in goats). Since the antialbumin used cross-reacted (to a limited extent) with murine albumin, the flushing of the vasculature was extended so as to remove the endogenous albumin not only from the vessel lumina but also from tissue interstitia. A 10-min perfusion gave satisfactory results. It reduced the background to a negligible level (Fig. 8).

The results obtained with this new approach were in part similar and in part different from those recorded in the case of A–Au complexes. Similar points included the following: (a) the tracer was restricted to plasmalemmal vesicles while transported across the endothelium, (b) no albumin was detected within intercellular junctions, and (c) no albumin gradients were found centered on the exit from intercellular spaces. Binding to the plasmalemma proper, including introits to intercellular junctions, also was limited. In part, however, the results were strikingly different because the transport of monomeric albumin was so much faster: the monomeric tracer was already detected in the pericapillary spaces after a 15- (Fig. 9) or 30-sec perfusion (Fig. 10); it was also much more effective: by 1 min (Fig. 11a,b) and 2 min the interstitia were flooded with exogenous albumin (Fig. 11). It is for this reason that the perfusion time of the tracer was gradually reduced so as to be able to detect the sites of earliest exit. At 15 sec most of the albumin

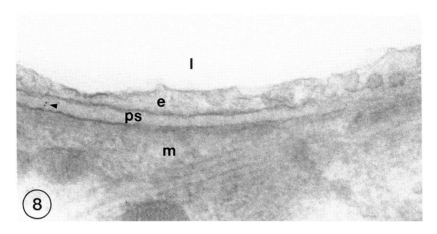

Fig. 8. Murine myocardium. This control specimen was flushed for 10 min to remove endogenous albumin, processed through thin frozen sectioning, and reacted with antialbumin (raised in rabbits) followed by gold-tagged anti-rabbit IgG (raised in goats). The arrowhead points to low background labeling. l, Lumen; e, endothelium; ps, pericapillary spaces; m, muscle fiber. Magnification: 68,000×.

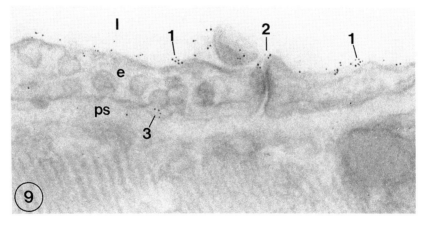

Fig. 9. Murine myocardium flushed for 10 min to remove endogenous albumin, perfused with bovine albumin (30 mg/ml) for 15 sec, processed through thin frozen sectioning, and reacted in succession with anti-bovine albumin (raised in rabbits) and gold-tagged anti-rabbit IgG (raised in goats). Monomeric albumin is localized on the plasmalemma proper (1), in the introit of an intercellular junction (2) but not in the junction proper nor the intercellular space past the junction, and in the pericapillary space at the exit from a plasmalemmal vesicle open on the abluminal front of the cell (3). l, Lumen; e, endothelium; ps, pericapillary space. Magnification: 73,000×.

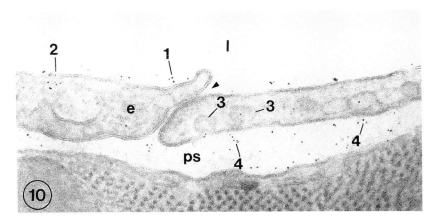

Fig. 10. The specimen was processed as in Fig. 9 except that perfusion with albumin lasted for 30 sec. Albumin is detected on the plasmalemma proper (1), in plasmalemmal vesicles opened on the luminal (2) or the abluminal front of the cell (3), and in the pericapillary space (4). No albumin is detected in the intercellular junction (arrowhead) nor in the intercellular space beyond the junction. l, Lumen; e, endothelium; ps, pericapillary space. Magnification: 80,000×.

in the pericapillary spaces appeared to be associated with discharging plasmalemmal vesicles (Fig. 9).

Still other marked differences concerned albumin binding. The association of monomeric albumin with plasmalemmal vesicles was less close than that of A–Au complexes. Most gold particles (used to tag antialbumin) were not tightly aligned along the vesicular membrane: they appeared randomly located at some distance (up to 40 nm) from either side of the membrane. This finding reflects at least in part the dimensions of our detecting system, which involves two antibodies and a gold particle, that is, a chain of approximately 40 nm. Hence, in our preparations the distance from the probe (albumin) to the reporting gold particle could vary from 0 to about 40 nm. The binding also appeared to be low affinity. The probe was easily removed from the plasmalemma proper as well as from vesicles opened on the luminal front of the endothelium by a 1-min flush with gPBS. Finally, no diversion of the tracer to endosomes (multivesicular bodies) was detected.

In some of these respects, the different results obtained with monomeric albumin approach physiological reality: the tight binding of A–Au to the membrane of plasmalemmal vesicles and the resistance of the binding to the flushing of the vasculature most probably reflect the polymeric character of the A–Au complexes. Monomeric albumin is expected to bind with low affinity to the endothelium (36). The diversion to endosomes and multi-

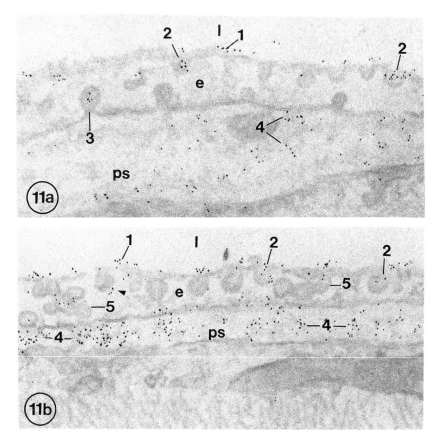

Fig. 11. Specimens were processed as in Fig. 9 except that perfusion with monomeric albumin was carried out for 1 min. Labeling for albumin is general but variable in extent from low (a) to high (b). Monomeric albumin is detected on the plasmalemma proper (1), in plasmalemmal vesicles opened on the luminal front (2) or on the abluminal front of the cell (3), in chains of fused vesicles (5), and in the pericapillary spaces (4). The arrowhead points to an area of spreading of the tagged antibody away from the target as discussed in the text. l, Lumen; e, endothelium; ps, pericapillary space. Magnification: 72,000×; b, 72,000×.

vesicular bodies is also ascribable to the polymeric character of the probe. Yet, in other respects, some of these results depart from physiological expectations: the extensive, nearly complete removal of endogenous albumin, which is part of our protocol, is expected to increase markedly the permeability of the microvasculature (37,38). Albumin binds to the endothelial surface, thereby generating a "fiber matrix" (39) that acts as an additional permselective barrier in front of the endothelium proper. Normal permeability is known to be reestablished at low concentrations of albumin in the perfusate (~1 mg/ml), but the process must require some time since the bolus of albumin (30 mg/ml) used in our experiments did not reduce immediately the increased permeability.

To obviate this problem, we modified the experimental protocol as follows: we raised in mice a polyclonal antibody to bovine serum albumin that does not cross-react (or cross-reacts minimally) with murine albumin. This new antibody gave us the possibility of retaining approximately 1 mg/ml murine albumin in the perfusate, thereby maintaining normal capillary permeability at the time when the probe was introduced as a bolus of 30 mg/ml. In the final preparations, bovine albumin was detected by the same procedures as described above. The exit sites were the same, but the rate of transport and the volume transported were considerably reduced. Albumin began to be detected in small amounts in the pericapillary spaces only after 1 to 2 min of perfusion.

Modified Albumins as Tracers

In more recent experiments, modified albumins were tested to find out whether the rates or even the pathways of exit vary as a function of given modifications. One of these variants involved glycated albumin, that is, the type of chemically modified albumin found in the circulation of diabetics. In this case, glucose is covalently linked to albumin amino groups, primarily ϵ-amino groups of lysine residues (40).

The microvasculature of diabetics is known to have higher than normal permeability to albumin, presumably on account of glycation (41), and experiments with freshly isolated endothelial cells have shown that they take up glycated albumin at a higher rate than its native counterpart (42). In recent experiments (43), glycated albumin (bovine)–colloidal gold complexes (gA–Au) were used as the probe, the rest of the protocol being the same as previously (32, 33). The general results were the same, except the gA–Au complexes decorated not only plasmalemmal vesicles but also the plasmalemma proper (Fig. 12a–c). Transcytosis was recorded with shorter exit times (<10 min) and higher efficiency (Figs. 12b,c) than in the case of A–Au complexes. Lack of exit along intercellular junctions (Figs. 12a, 13c) and diversion of part of the probe to endosomes (Fig. 13b) and microvesicular bodies (Fig. 13d) were again part of the general picture. The probe was differentially competable with glycated albumin, albumin, and glucose. Albumin removed the label from plasmalemmal vesicles; glucose was more efficient in competing the gA–Au complexes from the plasmalemma proper, whereas gA competed effectively for all binding sites (43).

Even more recently, fatty acid–albumin complexes (fA- -Au) were tested as a probe, again bound to colloidal gold complexes (44). The pattern of binding

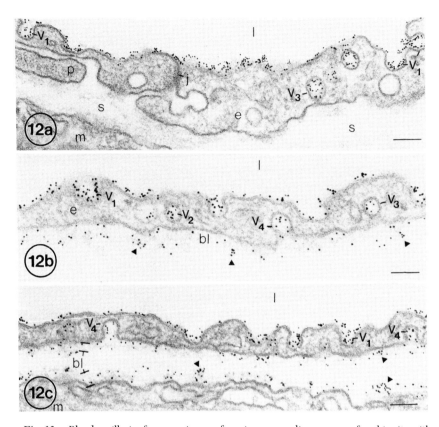

Fig. 12. Blood capillaries from specimens of murine myocardium were perfused *in situ* with glycated albumin–colloidal gold complexes (gA–Au), flushed (to remove excess tracer), fixed by perfusion with 2.5% formaldehyde–1.5% glutaraldehyde in 0.1 M cocadylate buffer (pH 7.2), and processed for TEM by standard procedures. (a) Blood capillary perfused for 3 min with gA–Au. The tracer labels the plasmalemma proper, all plasmalemmal vesicles opened on the blood front (v_1), and a vesicle apparently ready to open on the tissue front (v_3) The infundibulum leading to the junction (j) is heavily labeled, but the junction itself and the intercellular space past the junction are not labeled. (b,c) Blood capillaries perfused for 10 min with gA–Au. The tracer labels all plasmalemmal vesicles irrespective of their location, including vesicles apparently free in the cytoplasm (v_2) and opened on the tissue front (v_3, v_4) of the endothelial cell. Note the preferential association of gA–Au with the basement membrane (bl) and the extensive labeling of pericapillary spaces (arrowheads) in b and c. l, Lumen; e, endothelium; s, pericapillary space; p, pericyte; bl, basal lamina (or basement membrane); m, muscle. Bars, 0.1 μm.

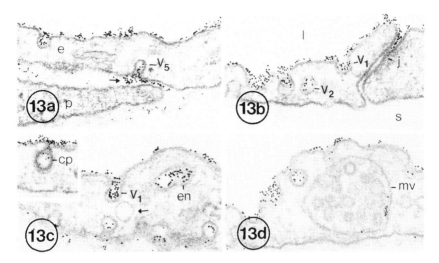

Fig. 13. Blood capillaries in myocardium specimens were perfused with gA–Au complexes for 10 min and processed as in Fig. 12. There is variable labeling of the plasmalemma proper and extensive labeling of plasmalemmal vesicles opened on the luminal front (v_1), apparently free in the cytoplasm (v_2), or discharging on the abluminal front (c). (1) The large accumulation (arrow) of gA–Au complexes associated with the vesicular profile v_5 opened on the abluminal front suggests that v_5 is the end of a transendothelial channel. The tracer gradient associated with it is probably due to slow diffusion in the narrow space between the endothelium (e) and a pericyte process (p). (b) The infundibulum leading to the junction (j) has accumulated gA–Au complexes, but the junction itself and the rest of the intercellular space is free of tracer. (c) A gA–Au-labeled endosome is marked en. The inset shows a labeled coated pit (cp) (A–Au and monomeric albumin do not bind to coated pits/coated vesicles). (d) gA–Au complexes appear in a multivesicular body associated closely to its membrane. l, Lumen; e, endothelium; s, pericapillary space; p, pericyte. Bar, 0.1 μm.

showed the same preference for plasmalemmal vesicles as A–Au complexes, but transcytosis, assessed by frequency and density of labeling of abluminal plasmalemmal vesicles, was more rapid and more efficient. Again, no evidence of exit along intercellular junctions was detected.

The series of experiments reviewed or reported above establishes, we believe, convincingly that transcytosis of monomeric and polymeric albumin operates via plasmalemmal vesicles in the case of the continuous microvascular endothelium of the myocardium and the lung.[5] None of the albumin–gold complexes, that is, A–Au, gA–Au, and fA–Au, that qualify as exclusive probes for the large pore system was found along intercellular junctions, where the large pores are assumed to be located, judging from the

7. Not all of the tracers mentioned were tested in the lung microvasculature.

current physiological literature (34,35). Tracer particles do accumulate on the luminal side of the junctions (Fig. 13c) but do not penetrate the intercellular spaces of the endothelium beyond the level of the occluding junctions. This is a remarkable negative finding that cannot be rationalized on the basis of strong convecting currents able to sweep away probes moving through intercellular junctions. In many of the latter there should be no such current, or (in the resorption part of the capillaries) the current should be reversed. All complexes were found instead in plasmalemmal vesicles that could represent either isolated shuttling vesicles or transendothelial channels. In fact, massive discharge from abluminal vesicular profiles (Fig. 13a) most probably marks exits from such channels.

General Discussion

The results obtained with monomeric albumin as a tracer deserve further comment. Albumin is an asymmetric molecule, a prolate ellipsoid, with long and short axes of 140 and 40 Å, respectively. Considering its dime sions alone, it is expected to exit the microvascular lumina through both small and large pores, but to mark primarily the former given their much higher endothelial surface density. Our results show that monomeric albumin, like its polymeric counterpart, exits primarily through plasmalemmal vesicles, not through intercellular junctions. What we identified by TEM as vesicular profiles may well correspond in three dimensions to transendothelial channels. The latter probably are transient structures, poorly preserved by our current specimen preparation procedures. They may well be more frequent *in situ* in the living functional endothelium. Our findings could be reconciled with current physiological data if we can assume that the large convective component in albumin transport (45) represents mostly the contribution of transendothelial channels.

The discrepancy between the large number of vesicles labeled by either monomeric or polymeric albumin and the postulated low frequency of small and large pores per unit endothelial surface (18,20) could be explained by the relatively low efficiency of vesicular transport or transendothelial channel formation by vesicular fusion. Finally, the differences recorded in transport efficiency in the absence or presence of 1 mg/ml albumin in perfusates suggest that the sites where the fibrillar matrix controls macromolecular permeability are not the intercellular spaces of the endothelium but the entry into either plasmalemmal vesicles or transendothelial channels. Results recently published by other investigators (11) are compatible with this interpretation.

What still remains to be understood is the lack of agreement between our findings and the conclusions drawn by Bundgaard, Frokjaer-Jensen, and collaborators (5,6) from tridimensional reconstructions of plasmalemmal vesicles from thin sections of different microvascular endothelia. As already mentioned, they found a marginal fraction of vesicles free in the cytoplasm of endothelial cells; they assume that a substantial fraction must exist to operate a vesicular shuttling mechanism and, on this ground, conclude that plasmalemmal vesicles cannot function in any transport process, transcytosis of macromolecules included. The conclusion rests essentially on morphological findings that can reflect events occurring in the twilight period between function still going on and function arrested as a result of general protein cross-linking by the fixatives used. The sessile chains of vesicles may represent the minimal free energy state of the system, the state to which vesicles come to rest when resistance to their movements increases and energy input becomes limiting. The advantage of our functional approach that follows a detectable tracer through the transport process is that it affords a look at the system while still in operation before it assumes a frozen, nonfunctional state as a result of fixation. We can hope that further studies, using other physiologically relevant macromolecules as tracers and perhaps more refined preparation procedures, will elucidate remaining areas of lack of agreement.

The intercellular junctions of the capillary endothelium are structurally highly reminiscent of those of leaky epithelia. Therefore, it can be reasonably assumed that a paracellular pathway functions in continuous microvascular endothelia as it does, for instance, in the epithelium of the proximal convolution of the nephron (46). In the case of the capillary endothelium; an upper size limit of approximately 20 Å diameter is suggested for permeant molecules by tracer studies (47). However, further work is needed to define this parameter as well as the importance of the paracellular pathway relative to that of transendothelial channels.

PERSPECTIVES

Although we have advanced toward an understanding of the structural basis of permeability in microvasculatures provided with a continuous endothelium, there is still a long way to go before achieving a reasonable consensus. Conflicting evidence regarding main pathways remains to be satisfactorily reconciled, and the contribution of other pathways, for example, across the cell or along membrane bilayers (see Ref. 45 for a discussion) remains to be more firmly assessed. Finally, the structural basis of the many other functions of the endothelium (mentioned in the introduction) is still to be established. As already stated, the vascular endothelium is not structurally and functionally uniform. What is true for microvascular beds with a continuous endothelium is not necessarily true for microvascular beds provided with different types of endothelia and perforce not true for the endothelium of large vessels. This means that for each type of endothelium we may expect a somehow different functional–structural type of correlation. By night, they say in French, all cats are grey. By TEM, all plasmalemmal vesicles, for instance, appear uniform in shape and size. But, as for cats, there are undoubtedly wide differences among their many other characteristics.

Recent work indicates that albumin binds with specificity to the surface of endothelial cells (36). Moreover, a number of albumin-binding proteins has been identified *in situ* as well as in culture (48–50), and some of them appear to be luminally exposed glycoproteins. There is as yet no full agreement as to their number and sizes, and their location, relative to plasmalemmal differentiated microdomains, remains unknown. But their apparent diversity should not be surprising: some of these proteins may act as partners involved in the generation of the fibrous matrix, and some others may function as receptors for albumin transcytosis as postulated previously (32).

REFERENCES

1. Bennett, H. S., Luft, J. H., and Hampton, J. C. (1959). *Am. J. Physiol.* **196**, 381–390.
2. Palade, G. E., Simionescu, M,. and Simionescu, N. (1979). *Acta Physiol. Scand. Suppl.* **463**, 11–32.
3. Simionescu, N., Simionescu, M., and Palade, G. E. (1975). *J. Cell Biol.* **64**, 586–607.
4. Bruns, R. R., and Palade, G. E. (1968). *J. Cell Biol.* **37**, 244–276, 277–299.
5. Bungaard, M. (1980). *Annu. Rev. Physiol.* **42**, 325–336.
6. Frokjaer-Jensen, J. (1980). *J. Ultrastruct. Res.* **73**, 9–20.
7. Peters, K.-R., Carley, W. W., and Palade, G. E. (1985). *J. Cell Biol.* **101**, 2233–2238.
8. Simionescu, M., Simionescu, N., Santoro, F., and Palade, G. E. (1985). *J. Cell Biol.* **100**, 1397–1407.
9. Simionescu, N., Simionescu, M., and Palade, G. E. (1981). *J. Cell Biol.* **90**, 605–613.
10. Simionescu, M., Simionescu, N., Silbert, J., and Palade, G. E. (1981). *J. Cell Biol.* **90**, 614–621.
11. Schneeburger, E. E., and Hamelin, M. (1984). *Am. J. Physiol.* **247**, H206–H217.
12. Milici, A. J., L'Hernault, N., and Palade, G. E. (1985). *Circ. Res.* **56**, 709–717.
13. Milici, A. J., Peters, K.-R., and Palade, G. E. (1986). *Cell Tissue Res.* **244**, 493–499.
14. Bankston, P. W., and Milici, A. J. (1983). *Microvasc. Res.* **26**, 36–48.
15. Baldwin, A. L., and Chien, S. (1985). *Atherosclerosis* **55**, 233–245.
16. Simionescu, M., Simionescu, N., and Palade, G. E. (1975). *J. Cell Biol.* **67**, 863–885.
17. Simionescu, N., Simionescu, M., and Palade, G. E. (1978). *J. Cell Biol.* **79**, 27–44.
18. Renkin, E. M., and Curry, F. E. (1978). *In* "Membrane Transport in Biology" (G. Giebisch, D. C. Tosteson, and H. H. Ussing, eds.), Vol. 4, pp. 1–45. Springer-Verlag, Berlin.
19. Pappenheimer, J. R., Renkin, E. M., and Borrero, L. M. (1951). *Am. J. Physiol.* **167**, 13–46.

20. Landis, E. M., and Pappenheimer, J. R. (1963). In "Handbook of Physiology, Volume II: Circulation" (W. F. Hamilton and P. Dow, eds.), pp. 1035–1073. American Physiology Society, Washington, D.C.
21. Grotte, G. (1956). *Acta Chir. Scand. Suppl.* **211**, 1–84.
22. Chambers, R., and Zweifach, B. W. (1947). *Physiol. Rev.* **27**, 436–463.
23. Simionescu, N., Simionescu, M., and Palade, G. E. (1972). *J. Cell Biol.* **53**, 365–392.
24. Graham, R. C., and Karnovsky, M. J. (1966). *J. Histochem. Cytochem.* **14**, 291–302.
25. Karnovsky, M. J. (1970). In "Capillary Permeability, Alfred Benzon Symposium II" (C. Crone and N. A. Lassen, eds.), p. 341. Academic Press, New York.
26. Wissig, S. L., and Williams, M. C. (1978). *J. Cell Biol.* **76**, 341.
27. Simionescu, N. (1981). In "International Cell Biology, 1980–81" (H. G. Schweiger, ed.), pp. 657–672. Springer-Verlag, Berlin.
28. Simionescu, M., Simionescu, N., and Palade, G. E. (1982). *J. Cell Biol.* **95**, 425–434.
29. Andersson, L.-O. (1979). In "Plasma Proteins" (B. Blombäck and L. A. Hanson, eds.), pp. 43–71. Wiley, New York.
30. Peters, T. J. (1970). *Adv. Clin. Chem.* **13**, 37–111.
31. Slot, J., and Geuze, H. (1981). *J. Cell Biol.* **90**, 533–536.
32. Ghitescu, L., Fixman, A., Simionescu, M., and Simionescu, N. (1986). *J. Cell Biol.* **102**, 1304–1311.
33. Milici, A. J., Watrous, N. E., Stukenbrok, H., and Palade, G. E. (1987). *J. Cell Biol.* **105**, 2603–2612.
34. Rippe, B., Kamya, A., and Folkow, B. (1979). *Acta Physiol. Scand.* **105**, 171–187.
35. Haraldsson, B. (1986). *Acta Physiol. Scand.* **128** (Suppl. 553), 1–40.
36. Schnitzer, J. E., Carley, W. W., and Palade, G. E. (1988). *Am. J. Physiol.* **254**, H425–H437.
37. Danielli, J. F. (1940). *J. Physiol. (London)* **98**, 109–129.
38. Mason, J. E., Currey, F. E., and Michel, C. C. (1977). *Microvasc. Res.* **13**, 185–204.
39. Curry, F. E., and Michel, C. C. (1980). *Microvasc. Res.* **20**, 96–99.
40. Garlick, R. L., and Mazer, J. S. (1983). *J. Biol. Chem.* **258**, 6142–6146.
41. Feld-Rasmussen, B. (1986). *Diabetologia* **29**, 282–286.
42. Williams, S. K., Devenny, J. J., and Bitensky, M. W. (1981). *Proc. Natl. Acad. U.S.A.* **78**, 2393–2397.
43. Predescu, D., Simionescu, M., Simionescu, N., and Palade, G. E. (1988). *J. Cell Biol.* **107**, 1729–1738.
44. Galis, Z., Ghitescu, L., and Simionescu, M. (1988). *Eur. J. Cell. Biol.* **47**, 358–365.
45. Renkin, E. M. (1988). In "Endothelial Cell Biology" (N. Simionescu and M. Simionescu, eds.), pp. 51–68. Plenum, New York and London.
46. Claude, P., and Goodenough, D. (1973). *J. Cell Biol.* **56**, 390–400.
47. Simionescu, N., Simionescu, M., and Palade, G. E. (1978). *Microvasc. Res.* **15**, 17–36.
48. Schnitzer, J. E., Carley, W. W., and Palade, G. E. (1988). *Proc. Natl. Acad. Sci. U.S.A.* **85**, 6773–6777.
49. Ghinea, N., Fixman, A. Alexandru, D., Popov, D., Hasu, M., Ghitescu, L., Eskenasy, M., Simionescu, M., and Simionescu, N. (1988). *J. Cell Biol.* **107**, 231–239.
50. Ghinea, N., Eskenasy, M., Simionescu, M., and Simionescu, N. (1989). *J. Biol. Chem.* **264**, 4755–4758.

18

A Growth Factor Homologous Gene Controlling Pattern Formation in *Drosophila*

WILLIAM M. GELBART
Department of Cellular and Developmental Biology
Harvard University
Cambridge, Massachusetts 02138–2097

INTRODUCTION

Numerous gene products which play controlling roles in early *Drosophila* development have now been identified. Many have properties suggesting that they act as sequence-specific DNA-binding proteins (see the chapters in this volume by Kornberg *et al.*, Lewis, and Hoey *et al.*). These contribute to a cascade of cell-intrinisic events which specify cell fate. Another group of early acting functions contributes to pattern formation by mediating cell–cell interactions. Only a few examples of this latter type of controlling genes have been characterized molecularly as yet, but functions of the latter group have already been identified among genes controlling the establishment and/or maintenance of cell fate along both the anterior–posterior (*wingless*, 1,2; *patched*, 3) and dorsal–ventral (*decapentaplegic*, 4,5) axes of the *Drosophila* embryo. In this chapter, we summarize our progress in understanding the contribution of one of these genes, *decapentaplegic*, to the specification of cell fate.

OVERVIEW OF THE ORGANIZATION OF THE *decapentaplegic* GENE

The *decapentaplegic* gene (hereafter referred to as *dpp*) is a complex genetic unit contributing to numerous developmental events, including dorsal–ventral patterning of the embryo, undefined processes necessary for survival and development of the larva, and proximal–distal outgrowth of the adult appendages that derive from the imaginal disks of the larva (6,7). It has been possible to subdivide the *dpp* gene according to phenotypes elicited by mutations in a given region (Fig. 1). Thus, the *shv* region contains the necessary cis-regulatory information for those aspects of *dpp* expression necessary for proper development through the larval period (8). Similarly, the *disk* region contains the cis-regulatory apparatus necessary for proximal–distal development of the adult appendages (9). Finally, the centrally located *Hin* region, on which I focus in this report, contains the cis-regulatory information necessary for proper dorsal–ventral patterning of the embryo.

There is also transcriptional complexity at *dpp*, with the identification of at least five different transcript species representing different promoters and different 5' exons (7). In each case, however, these alternative 5' exons are spliced to exactly the same middle and 3' exons. Four of the alternative 5' exons reside within the *shv* region, whereas one resides within *Hin*. The middle and 3' exons both reside entirely within *Hin*. Each transcript exhibits its own temporal pattern of expression, as determined by developmental Northern blots. While we assume that these transcripts will also prove to differ (at least partially) in their spatial patterns of expression, we have as yet no direct information on this point.

This transcriptional diversity is not reflected in heterogeneity of protein products. Rather, the alternative 5' exons all appear to be nontranslated, and the entire open reading frame of the transcript is contained within the middle and 3' exons common to all of the transcripts. Thus, it appears that all of the cis-regulatory regions located within approximately 50 kilobases (kb) of the *dpp* gene drive the expression of the same polypeptide product of the gene (10; 27). I discuss the nature of this polypeptide product below.

CONTRIBUTION OF THE *Hin* REGION TO *dpp* FUNCTION

The preceding discussion points to the fact that the *Hin* region contributes two types of information to the *dpp* gene: cis-regulatory information for proper embryonic development along the dorsal–ventral axis and the entire coding information for the *dpp* polypeptide. Virtually all mutations within the *dpp* region were selected on the basis of functional defects in aspects of *dpp* expression required for normal adult appendage development (4). It is only when these mutations were examined in dpp^{Hin}/dpp^+ heterozygotes or in dpp^{Hin} homozygotes that we realized that they were defective in a function required for embryonic development. [dpp^{Hin} refers to a null mutation

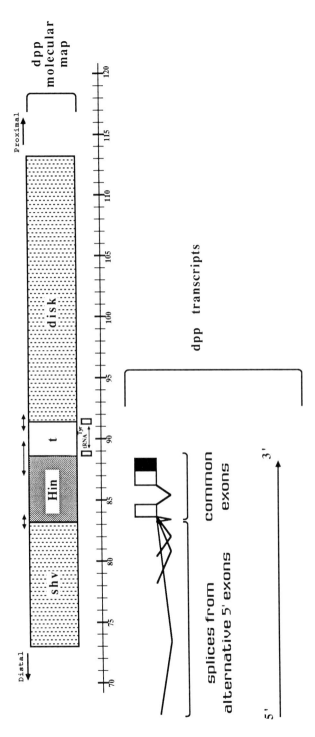

Fig. 1. Molecular structure of the *dpp* gene of *Drosophila melanogaster*. Distal and proximal orient the *dpp* map relative to the telomere and centromere, respectively, of chromosome arm 2L. At the top of the map are listed the major regions of the *dpp* gene, including *shv* (shortvein), *Hin* (haploinsufficient), and *disk* (imaginal disk-specific). Two tRNA^{Tyr} genes are located at the juncture of the *Hin* and *disk* regions. It is unclear at present if the region between the two tRNA^{Tyr} genes should be considered part of the *disk* region or if it represents a fourth major domain of the *dpp* gene. For this reason, it is currently designated the *t* (tRNA) region. Double-headed arrows above the boundaries separating these different regions represent uncertainties in the positions of the boundary lines. The scale represents the molecular coordinates of the map, in kilobases. The transcript map below the scale indicates the approximate positions of splice donor sites from the several 5' exons and the positions of the two common exons shared by all transcripts. The unshaded portions of the rectangles representing the two common exons delineate the extent of the *dpp* open reading frame. This figure summarizes work described in Ref. 7.

in the *Hin* region; three of these mutations have been shown to delete portions of the *dpp* open reading frame (7).] Because the dominant lethality of dpp^{Hin}/dpp^+ individuals is rescued by the addition of another copy of dpp^{Hin+} to the genome, we conclude that the dominant lethality is due to insufficient *dpp* gene product in the monosomic; hence, we term this region of *dpp* the *Hin* or haplo-insufficient region.

Genetic limits on the size of the *Hin* region have been established in two different ways. First, by molecular mapping of *dpp* translocation or inversion breakpoints in which the *Hin* region embryonic functions remain normal, we infer that *Hin* must lie between 82.7 and 90.6 on the molecular map, and thus that its molecular size is at most 7.9 kb (7). This localization is confirmed by the fact that an 8.2-kb segment extending from 82.6 to 90.8, when introduced into the germ line by P element-mediated germ-line transformation, completely rescues the embryonic functions of *dpp* (11). These results indicate that the one transcription unit within the *Hin* region is sufficiently regulated by *Hin* region sequences to fulfill the embryonic functions of *dpp*.

Irish and Gelbart (4) described *dpp* embryonic phenotypes in detail. dpp^{Hin}/dpp^+ heterozygotes die as embryos, displaying cephalic and caudal defects suggestive of a weak dorsolateral to ventrolateral transformation. This phenotypic transformation is much more pronounced in dpp^{Hin} homozygous embryos. In such embryos, there is a complex loss of normal dorsal ectodermal structures (derived from the amnioserosa and dorsal epidermal regions of the fate map, schematically depicted in Fig. 2), and in their place there is an expanded set of ventral ectodermal derivatives. Mutant embryos are abnormal as early as 4 hr in development, exhibiting defects in several of the morphogenetic movements characteristic of early gastrulation.

The mutant embryonic phenotypes of *dpp* are entirely dependent on the genotype of the embryo itself (4). Homozygous dpp^{Hin} embryos derived from germ-line cells which are also homozygous for such dpp^{Hin} null alleles are phenotypically indistinguishable from null embryos derived from mothers carrying two copies of dpp^+. *dpp* is thus one of the earliest acting zygotically expressed genes affecting dorsal–ventral patterning in the embryo. Indeed, dpp^{Hin} alleles are unique among known zygotically expressed genes in causing the virtually complete transformation of dorsal into ventral ectoderm. Based on all of these observations, we have concluded that the gene is required to determine dorsal ectoderm in the *Drosophila* embryo.

This conclusion is bolstered by an examination of the *in situ dpp* transcript accumulation pattern (5) (Fig. 3). Using probes homologous to the exons shared by all *dpp* transcripts, embryonic RNA accumulation can be detected as early as the 12th zygotic cleavage. Already at this time, *dpp* expression occurs in a swatch along the dorsal 40–45% of the embryo and extends around the anterior and posterior poles as well. This pattern remains

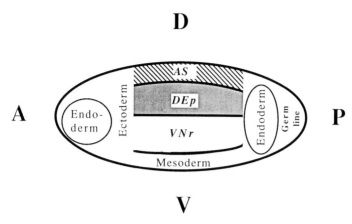

Fig. 2. Schematized fate map of the *Drosophila* embryo. This diagram depicts the general location of the major germ layer primordia on the blastoderm stage embryo (12). The ectoderm is divided into three domains, the ventrolaterally located ventral neurogenic region (VNr), the dorsolaterally positioned dorsal epidermal region (DEp), and the dorsal amnioserosa (AS). A and P represent anterior and posterior, respectively; D and V similarly connote dorsal and ventral locations on this diagram.

through the cellular blastoderm stage. As gastrulation begins, and throughout the germband extension stage, expression is extinguished in the dorsalmost region of the embryo, in the presumptive amnioserosa, and at the poles. The transcript accumulation pattern which remains is in a region coincident with the expected fate map position of the dorsal epidermis (as seen in a comparison of Fig. 2 and 3).

The final change in the embryonic ectodermal expression pattern occurs at or around the transition from germband extension to germband retraction. Much or all of the expression in the dorsal epidermal region is extinguished and is replaced by two thin stripes of expression. One stripe is clearly in the dorsalmost cells of the dorsal epidermis, just ventral to the amnioserosa. The other is in cells at or near the boundary between the dorsal epidermal region (which is dorsolateral on the embryonic fate map) and the ventral neurogenic region (which is ventrolateral). How these different stages in the *dpp* expression pattern relate in detail to the various embryonic phenotypes associated with the inactivation of the *dpp* gene remains to be determined. Given the early time of initial expression of *dpp*, and the absence of other zygotically expressed genes with similar or more severe effects on the dorsal epidermal region of the fate map, we deem it likely that the early *dpp* transcripts are regulated directly by the maternally expressed dorsal–ventral gene products (13).

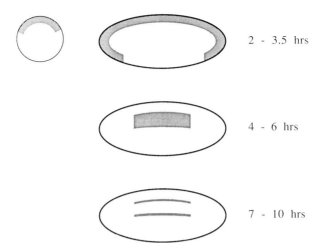

Fig. 3. Schematized description of the *dpp* transcript accumulation pattern in the embryonic ectoderm, summarizing the results of St. Johnston and Gelbart (5). The first pattern of *dpp* expression is seen in blastoderm stage embryos, ranging from 2 to 3.5 hr postfertilization. The shaded area represents the region of transcript accumulation. The elliptical diagram at right represents a side view of an embryo, with the dorsal aspect at the top. The circular diagram at left represents the expression pattern as viewed in cross section. The 4–6 hr and 7–10 hr side views depict the patterns observed during germband elongation and germband shortening, respectively. However, for simplicity, these representations depict the pattern in terms of the location of the primordia of the expressing tissue patches on the blastoderm fate map and ignore the shape changes caused by the movements of the germband during these stages.

NATURE OF THE *dpp* POLYPEPTIDE

We have been able to infer a great deal about the possible role of *dpp* from the sequence of *dpp* cDNAs (10; R. W. Padgett *et al.*, in preparation). The predicted 588 amino acid long *dpp* polypeptide is a member of the transforming growth factor-β (TGF-β) family of secreted factors. This family comprises several polypeptides identified in vertebrate systems, including several closely related TGF-β molecules (14–16), the inhibin subunits (17,18), Müllerian inhibiting substance (19), Vgl (20), Vgr-1 (21), and the bone morphogenesis proteins (22). Many of these factos have been implicated in the control of determination or differentiation events.

For each family member for which the protein product has been isolated, it is a dimer of the processed C-terminal fragment of approximately 110 amino acids. The conserved sequences among the family members lie in these C-terminal fragments. The most similar domains are centered around a set of 7 completely positioned cysteine residues, suggesting that aspects of the tertiary structure of the molecules are being maintained within the family (10). This C-terminal fragment derives from much longer precursor polypeptides, ranging from 300 to 600 amino acids in length. The precursor polypeptides share several other features, including N-terminal signal secretion sequences, N-glycosylation sites, and dibasic serine protease cleavage sites which are utilized to release the mature C-terminal fragment. The *dpp* polypeptide shares all of these structural features (10).

The *dpp* polypeptide is especially closely related to two recently described members of the TGF-β family: BMP-2A and BMP-2B (Fig. 4). These two bone morphogenesis proteins show approximately 75% amino acid identity to *dpp* in their respective C-terminal domains (22). Indeed, while the precursor regions of the polypeptides of most family members do not show any significant sequence similarity to one another, there is about 25–30% sequence identity in the precursor regions of *dpp* and the BMP-2 polypeptides. Based on these observations, it is tantalizing to speculate that *dpp* and the BMP-2 proteins may represent the descendants of a common ancestral gene in the arthropod and vertebrate lineages, respectively.

Based on the membership of *dpp* in the TGF-β family, we have suggested that the *dpp* product is a secreted molecule which acts in the intercellular communication of cell specification (10). By analogy with other members of the family, we presume that the secreted *dpp* protein acts on target cells by binding to a membrane-bound receptor protein (23,24). Little is known at present about the nature of these receptors, nor is there much information about signal transduction systems utilized by this ligand–receptor system. No other loci in *Drosophila* are known to have phenotypes similar to those of *dpp*; thus, no obvious candidates for *dpp* receptor genes are apparent.

The nature of *dpp* as a signaling molecule is consistent with several developmental observations. Based on single cell transplantation experiments, Technau and Campos-Ortega (25) showed that, even during gastrulation, cells of the dorsal epidermal region of the embryonic fate map were able to adopt a ventral neurogenic fate when transplanted in ventrolateral regions of the embryo. This implies that such dorsal epidermal cells depend on proper communication with their neighbors to maintain their dorsolateral fate. The secreted *dpp* product may be an important component of this signaling system. The nonautonomy of *dpp*, as determined by analyzing genetically mosaic adult appendages, is another feature anticipated for such a signaling molecule (26). We hope that by pursuing an intensive genetic, embryological, and molecular analysis of *dpp* and genes which interact with it, we shall be able to understand the cellular basis of this signaling system and how it contributes to the specification of cell fate.

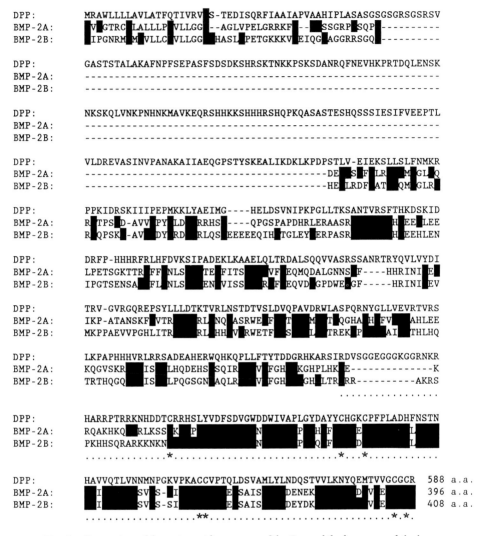

Fig. 4. Comparison of the amino acid sequences of the *Drosophila dpp* gene and the human BMP-2 genes. The alignment of these sequences is diagrammed to indicate regions of similarity. Amino acids of BMP-2A and/or BMP-2B which are identical to the *dpp* amino acid at a given position are depicted by black rectangles. Gaps are shown by dashes. The approximate extent of the mature C-terminal fragment hypothesized for each of these proteins is shown by the dotted line. Asterisks indicate the positions of the conserved cysteine residues. Note that the most conserved region of the protein begins with the first of the conserved cysteines. Considerable conservation is also noted in the precursor regions. Also note that the *dpp* polypeptide is almost 200 amino acids longer than the BMP-2 proteins. The *dpp* sequence is from Ref. 10, and the BMP-2 sequences are from Ref. 22.

ACKNOWLEDGMENTS

This work summarizes the excellent and invaluable intellectual and technical contributions of many investigators over a 10-year period, most notably (in alphabetical order) Ronald Blackman, F. Michael Hoffmann, Vivian Irish, Richard W. Padgett, Leila Posakony, Laurel Raftery, Daniel Segal, Forrest Spencer, and R. Daniel St. Johnston. This work has been supported by research grants from the U.S. Public Health Service and from the American Cancer Society.

REFERENCES

1. Cabrera, C. V., Alonso, M. C., Johnston, P., Phillips, R. G., and Lawrence, P. A. (1987). *Cell (Cambridge, Mass.)* **50**, 659–663.
2. Rijsewijk, F., Schuermann, M., Wagenaar, E., Parren, P., Weigel, D., and Nusse, R. (1987). *Cell (Cambridge, Mass.)* **50**, 649–657.
3. Nakano, Y., Guerrero, I., Hidalgo, A., Taylor, A., Whittle, J. R. S., and Ingham, P. W. (1989). *Nature (London)* **341**, 508–513.
4. Irish, V. F., and Gelbart, W. M. (1987). *Genes & Dev.* **1**, 868–879.
5. St. Johnston, R. D., and Gelbart, W. M. (1987). *EMBO J.* **6**, 2785–2791.
6. Gelbart, W. M., Irish, V. F., St. Johnston, R. D., Hoffmann, F. M., Blackman, R. K., Segal, D., Posakony L. M., and Grimaila, R. (1985). *Cold Spring Harbor Symp. Quant. Biol.* **50**, 119–125.
7. St. Johnston, R. D., Hoffmann, F. M., Blackman, R. K., Segal, D., Grimaila, R., Padgett, R. W., Irick, H. A., and Gelbart, W. M. (1990). *Genes & Dev.* **4**, 1114–1127.
8. Segal, D., and Gelbart, W. M. (1985). *Genetics* **109**, 119–143.
9. Spencer, F. A., Hoffmann, F. M., and Gelbart, W. M. (1982). *Cell (Cambridge, Mass.)* **28**, 451–461.
10. Padgett, R. W., St. Johnston, R. D., and Gelbart, W. M. (1987). *Nature (London)* **325**, 81–84.
11. Hoffmann, F. M., and Goodman, W. (1987). *Genes & Dev.* **1**, 615–625.
12. Hartenstein, V., Technau, G. M., and Campos-Ortega, J. A. (1985). *Wilhelm Roux's Arch. Dev. Biol.* **194**, 213–216.
13. Anderson, K. V. (1987). *Trends Genet.* **3**, 91–97.
14. Derynck, R., Jarrett, J. A., Chen, E. Y., Eaton, D. H., Bell, J. R., Assoian, R. K., Roberts, A. B., Sporn, M. B., and Goeddel, D. V. (1985). *Nature (London)* **316**, 701–705.
15. Dijke, P. T., Hansen, P., Iwata, K. K., Pieler, C., and Foulkes, J. G. (1988). *Proc. Natl. Acad. Sci. U.S.A.* **85**, 4715–4719.
16. Marquardt, H., Lioubin, M. N., and Ikeda, T. (1987). *J. Biol. Chem.* **262**, 12127–12131.
17. Mason, A. J., Hayflick, J. S., Ling, N., Esch, F., Ueno, N., Ying, S.-Y., Guillemin, R., Naill, H., and Seeburg, R. H. (1985). *Nature (London)* **318**, 659–663.
18. Mayo, K. E., Cerelli, G. M., Spiess, J., Rivier, J., Rosenfeld, M. G., Evand, R. M., and Vale, W. (1986). *Proc. Natl. Acad. Sci. U.S.A.* **83**, 5849–5833.
19. Cate, R. L., Mattaliano, R. J., Hession, C., Tizard, R., Farber, N. M., Cheung, A., Ninfa, E. G., Frey, A. Z., Gash, D. J., Chow, E. P., Fisher, R. A., Bertonis, J. M., Torres, G., Wallner, B. P., Ramachandran, K. L., Ragin, R. C., Managanaro, T. F., MacLaughlin, D. T., and Donahoe, P. K. (1986). *Cell (Cambridge, Mass.)* **45**, 685–698.
20. Weeks, D. L., and Melton, D. A. (1987). *Cell (Cambridge, Mass.)* **51**, 861–867.

21. Lyons, K., Graycar, J. L., Lee, A., Hashmi, S., Lindquist, P. B., Chen, E. Y., Hogan, B. L. M., and Derynck, R. (1989). *Proc. Natl. Acad. Sci. U.S.A.* **86**, 4554–4558.
22. Wozney, J. M., Rosen, V., Celeste, A. J., Mitsock, L. M., Whitters, M. J., Kriz, R. W., Hewick, R. M., and Wang, E. A. (1988). *Science* **242**, 1528–1534.
23. Boyd, F. T., and Massagué, J. (1989). *J. Biol. Chem.* **264**, 2272–2278.
24. Cheifetz, S., Weatherbee, J. A., Tsang, M. L.-S., Anderson, J. K., Mole, J. E., Lucas, R., and Massagué, J. (1987). *Cell (Cambridge, Mass.)* **48**, 409–415.
25. Technau, G. M., and Campos-Ortega, J. A. (1986). *Wilhelm Roux's Arch. Dev. Biol.* **195**, 445–454.
26. Posakony, L. M., Raftery, L. A., and Gelbart, W. M. (1990). *Mech. Devel.* (in press).
27. Padgett, R. W., Findley, S., Taylor, M., de Cuevas, M., and Gelbart, W. M. (In preparation).

PART IV

SEQUENCE-SPECIFIC DNA-BINDING PROTEINS: ROLE IN DIFFERENTIATION AND DEVELOPMENT

19
Role of Master Regulatory Genes in Controlling Development

E. B. LEWIS
Division of Biology
California Institute of Technology
Pasadena, California 91125

The chapters in this section have as their central theme the role of clusters of master regulatory genes in programming the development of an organism. The discovery of such genes has a curious history. They are sometimes called homeotic genes because when mutated they tend to result in one part of the adult organism transforming toward a corresponding homologous part. Thus, in the first such case in *Drosophila*, namely, bithorax (bx), the halteres of the third thoracic (T3) segment transform toward wings, the homologous appendages of the corresponding second (T2) segment (Bridges and Morgan, 1923). A genetic fine structure study of this mutant and closely related ones showed that they were not simply multiple alleles of a single gene but rather were part of a gene cluster that had presumably evolved by tandem gene duplication (Lewis, 1951). This bithorax gene complex (BX-C) has now been extensively studied molecularly, commencing with the pioneering work of Hogness and colleagues. Their analysis has confirmed the existence of a gene cluster and has shown that it spans nearly 400 kilobases of DNA (Bender *et al.*, 1983; Karch *et al.*, 1985).

Another striking example of a homeotic mutant is the transformation of the antenna toward a leglike structure. The first case of this kind in *D. melanogaster* was reported by S. Yu in 1949, who named it Antennapedia (Antp) after a mutant of this type in *D. affinis* (Sturtevant, 1940). Le Calves reported in 1948 a dominant allele of aristapedia (a homeotic mutant type in which the antenna transforms toward only the tarsal portion of the leg); on cytological grounds it is now assumed that this was an Antp allele [see

Lindsley and Grell (1968) for references to the work of Yu and Le Calves]. Studies (Kaufman *et al.*, 1980) of other homeotic mutants mapping to the vicinity of Antp showed that they, too, are part of a gene cluster, the Antp gene complex (ANTP-C or ANT-C), that is separate from but lies in the same chromosome arm as that of the BX-C. Molecularly, the ANT-C is also very large, spanning over 400 kilobases (see review by Scott and O'Farrell, 1986). Furthermore, certain regions of the BX-C and ANT-C share a remarkable degree of homology at the level of DNA base sequence in the form of the homeobox (Hox), a highly conserved sequence coding for a protein domain of 60 amino acids (McGinnis *et al.*, 1984a; Scott and Weiner, 1984); five Hox-containing genes are in the ANT-C and three in the BX-C (McGinnis *et al.*, 1985). The strong conservation of sequences not only among the Hox regions of each complex but between those of the ANT-C and those of the BX-C provides suggestive evidence that at least parts of those complexes have evolved from a common ancestral gene(s) by tandem gene duplication.

Hox sequences that are relatively highly conserved with those found in the ANT-C and BX-C were soon found in the frog, earthworm, beetle, chicken, mouse, and human genomes (McGinnis *et al.*, 1984b; Carrascos *et al.*, 1984; Levine *et al.*, 1984). Moreover, in mice and human beings they are known to exist in clustered groups (e.g., Ruddle *et al.*, 1985; Boncinelli *et al.*, 1985).

Mutants producing homeotic transformations in the adult organism were generally only partial inactivations of the relevant genes. As deletions for BX-C became available it was possible to deduce the nature of the wild-type gene functions within the complex (Lewis, 1978). Thus, when the organism lacks all genes of the complex, not only T3 but more posterior segments up to and including A7 are transformed toward T2. Addition of wild-type genes to this null genotype allowed us to establish functions for many of the individual genes. Such functions can be expressed in a formal sense in terms of their controlling certain levels (L) of segmental development, as summarized in Table I. Thus, the wild-type BX-C functions appear to specify the entire pattern of differentiation of all of the segments commencing with T3 and extending to the terminal body segments A8 and possibly A9. Although many of the genes of the BX-C have some functions in common, each gene seems to have unique functions as well, and many of these need not be homeotic at least in the strict sense of the word. Thus, the BX-C genes seem to play a master regulatory role in directing much of the thoracic and abdominal development of the organism.

What began as a study of mutants which produced flies with four wings has revealed a vast cluster of genes having far more profound effects on the organism than had been conceived at the start. A parallel conclusion follows from the studies of the mutants converting antennae to legs. Here, too, the

TABLE I
Functions of BX-C Genes in Terms
of Transforming Segments from One Level
of Development to Another[a]

Gene	First strongly expressed in segment	Wild-type function	COE effect of mutants
Ubx	T3	LT2→LT3	None
bxd	A1	LT3→LA1	pT2→pT3
iab-2	A2	LA1→LA2	None
iab-3	A3	LA2→LA3	A1→A2
iab-4	A4	LA3→LA4	A2→A3
iab-5	A5	LA4→LA5	A3→A4
iab-6	A6	LA5→LA6	A4→A5
iab-7	A7	LA6→LA7	None
iab-8	A8	LA7→LA8	A6→A7

[a] L, Level of development; T, thoracic; A, abdominal; p, posterior; Ubx, Ultrabithorax; bxd, bithoraxoid, iab, infra-abdominal. The last column shows the cis-overexpression (COE) effects produced by certain chromosomal rearrangements having breakage points within the corresponding gene in the first column on the next most proximal gene.

mutant effects on the adult are clearly homeotic, but the wild-type genes again are acting as master regulators of the ultimate differentiation of the head and anterior thorax of the organism.

Since the Hox regions code for domains that are expected to have DNA-binding properties (Shepard *et al.*, 1984; Laughon and Scott, 1984), the molecular studies support the idea, deduced on genetic grounds, that the BX-C genes (and presumably the ANT-C genes, as well) function as regulators of still other "downstream" genes. In the case of the BX-C it was assumed that the genes act as repressors of systems of cellular differentiation, thereby allowing other systems to come into play (Lewis, 1964). This idea now seems too restrictive since such genes may equally well act directly as activators or repressors of downstream genes. Garcia-Bellido (1975) has suggested a model by which the BX-C genes might control downstream or "realisator" genes.

A third and especially relevant example is the history of the study of the *engrailed* (*en*) gene in *Drosophila*. The original mutant gene was found by Evang (Eker, 1929). It has a striking homeotic effect in that the posterior wing margin transforms toward a mirror image of the anterior wing margin. This mutant has not only played an important role in the development of the compartment hypothesis (Garcia-Bellido et al., 1973), but the genetic and molecular analysis of the *en* gene (Kornberg, 1981; Kornberg et al., 1983) has shown that it plays a profound role in the earliest stages of embryonic development as discussed in the next chapter by Kornberg et al. Here, too, instead of a single gene, at least two genes each with Hox sequences have been identified (reviewed by Joyner et al., 1985). In this case the Hox sequences although related to those of the ANT-C and the BX-C show less conservation to those sequences than do the Hox sequences of the ANT-C to those of the BX-C.

Probes for detecting the RNA and protein products of the Hox-containing genes of the ANT-C, BX-C, and the *en* region have been used to follow the distribution of these products during embryonic and later stages of development (see reviews by Scott and O'Farrell, 1986; Gehring and Hiromi, 1986; Akam, 1987). These and related studies on other genes have provided evidence that the ANT-C and BX-C genes are regulated in trans by such early segmentation genes as *en*, possibly through their Hox regions. Thus, the Hox sequences are implicated in a role during ontogeny that may be related to their phylogenetic origin.

In the BX-C there are, in addition to the three known Hox-containing genes *Ultrabithorax* (*Ubx*), *infra abdominal-2* (*iab-2*), and *iab-7*, the latter also being known as *abd-A* and *Abd-B* (reviewed by Duncan, 1987), other elements which we designate as genes since they appear to be transcription units if the *bithoraxoid* (*bxd*) case studied by Lipshitz and Hogness (1987) is a typical one. These non-Hox-containing genes act as transcription regulatory regions (Peifer et al., 1987). They have unusual genetic properties of which only two are briefly mentioned here. Most noteworthy is the finding that in the case of the *bxd* region the more extreme the loss of function in the *bxd* gene the more proximal is the break in the *bxd* gene (Bender et al., 1985). Since that gene is known to be transcribed from a distal to a proximal direction, it is surprising that breaks near the 5' end have only very slight bxd mutant effects; indeed, the results are more consistent with the antisense strand having the determining role in the *bxd* function. Alternatively, the *bxd* region may function primarily as an enhancer region as assumed by Lipshitz and Hogness (1987). A second genetic property that thus far characterizes the non-Hox-containing genes of the BX-C is their ability to produce a cis-overexpression (COE) effect. Such an effect is seen when the complex is disrupted by chromosomal rearrangements. In many but not all such cases there is a COE effect on the next most proximal gene to that sustaining a rearrangement breakage point. The cis-overexpression of this latter gene is seen one segment anterior in the body to the segment where that gene is normally fully expressed. Thus, several breakage points that have an inactivation of the *iab-4* gene show an overexpression of the *iab-3* wild-type function in A2 which is then transformed toward A3 (Lewis, 1985). As shown in Table I, such COE effects have not been found associated with rearrangements having breakage points within the Hox-containing genes of the complex, suggesting that there is some fundamental difference between those genes and the remaining regions of the complex.

The long-awaited synthesis of genetics, embryology, and molecular biology is clearly at hand and promises to tell us much about the ontogeny and phylogeny of living organisms.

REFERENCES

Akam, M. (1987). *Development* **101**, 1–22.
Bender, W., *et al.* (1983). *Science* **221**, 23–29.
Boncinelli, *et al.* (1985). *Cold Spring Harbor Symp. Quant Biol.* **50**, 301–306.
Bridges, C. B., and Morgan, T. H. (1923). *Carnegie Inst. Washington Publ.* **327**.
Carrasco, A. E., *et al.* (1984). *Cell (Cambridge, Mass.)* **37**, 409–414.
Duncan, I. (1982). *Genetics* **102**, 49–70.
Duncan, I. (1987). *Annu. Rev. Genet.* **21**, 285–319.
Eker, A. (1929). *Hereditas* **12**, 217–222.
Garcia-Bellido, A. (1975). *Ciba Found. Symp.* **29**, 161–182.
Gehring, W., and Hiromi, Y. (1986). *Annu. Rev. Genet.* **20**, 147–173.
Joyner, A., *et al.* (1985). *Cold Spring Harbor Symp. Quant. Biol.* **50**, 291–300.
Karch, F., *et al.* (1985). *Cell (Cambridge, Mass.)* **43**, 81–96.
Kaufman, T., *et al.* (1980). *Genetics* **94**, 115–133.
Kornberg, T. (1981). *Proc. Natl. Acad. Sci. U.S.A.* **78**, 1095.
Kornberg, T., *et al.* (1983). *Cell (Cambridge, Mass.)* **40**, 45.
Laughon, A., and Scott, M. P. (1984). *Nature (London)* **310**, 25–31.
Levine, M. *et al.* (1984). *Cell (Cambridge, Mass.)* **38**, 667–673.
Lewis, E. B. (1951). *Cold Spring Harbor Symp. Quant. Biol.* **16**, 159–174.
Lewis, E. B. (1964). *In* "The Role of Genes in Development" (M. Locke, ed.), pp. 231–252. Academic Press, New York.
Lewis, E. B. (1978). *Nature (London)* **276**, 565–570.
Lewis, E. B. (1985). *Cold Spring Harbor Symp. Quant. Biol.* **50**, 155–164.
Lindsley, D., and Grell, E. H. (1968). *Carnegie Inst. Washington Publ.* **627**.
Lipshitz, H., and Hogness, D. (1987). *Genes & Dev.* **1**, 307–322.
McGinnis, W. *et al.* (1984a). *Nature (London)* **308**, 428–433.
McGinnis, W., *et al.* (1984b). *Cell (Cambridge, Mass.)* **38**, 675–680.
Martinez-Arias, A., and Lawrence, P. (1985). *Nature (London)* **313**, 639–642.
Peifer, M., *et al.* (1987). *Genes & Dev.* **1**, 891–898.
Ruddle, F., *et al.* (1985). *Cold Spring Harbor Symp. Quant. Biol.* **50**, 277–284.
Scott, M. P., and O'Farrell, P. H. (1986). *Annu. Rev. Cell Biol.* **2**, 49–80.
Scott, M., and Weiner, A. J. (1984). *Proc. Natl. Acad. Sci. U.S.A.* **81**, 4115–4119.
Shepherd, J. C. W., *et al. Nature (London)* **310**, 70–71.
Sturtevant, A. H. (1940). *Genetics* **25**, 337–353.

20
engrailed, A Gene for All Segments

THOMAS KORNBERG,[1] CHIHIRO HAMA,[2] NICHOLAS J. GAY,[3] AND STEPHEN J. POOLE[4]

1. Department of Biochemistry and Biophysics
University of California
San Francisco, California 94143
and
2. National Institute of Neuroscience
NCNP 4-1-1 Ogawahigashi, Kodaira
Tokyo 187, Japan
and
3. Department of Biochemistry
University of Cambridge
Cambridge CB2 1QW
England
and
4. Department of Biology
University of California
Santa Barbara, California 93106

INTRODUCTION

Even casual inspection of an insect reveals a fundamental organizational principle: throughout their life cycle, insects are subdivided into a linear series of segments. Originating during early embryogenesis as reiterated units that are identical in appearance, segments persist and differentiate through to the adult stage. In *Drosophila melanogaster,* homeotic mutants are known in which segmental appendages have been transformed, for instance replacing antennae with legs (the *Antennapedia* mutants), the proboscis with legs (the *proboscopedia* mutants), or the halteres with wings (the *bithorax* mutants). Such mutant phenotypes suggest that the physical subdivisions, the segments, reflect an underlying genetic organization that governs insect development.

Not obvious morphologically are the subdivisions of the segments themselves. Each segment is composed of one anterior and one posterior compartment (Garcia-Bellido et al., 1973; Kornberg, 1981). Compartments are distinguished by the lineage of their consituent cells (Garcia-Bellido, 1975). Each compartment originates as a small group of founder cells whose only known relationship to each other is their physical proximity. Once assigned to a compartment, the constituent cells grow as a developmental unit, confined to a defined area but not specifically restricted to a differentiated cell type.

The compartments and segments are first defined during the cellular blastoderm stage of the early *Drosophila* embryo (Hafen et al., 1984). Although the cells of cellular blastoderm embryos are morphologically indistinguishable, they express numerous genes in patterns that reflect the partitioning of the embryo (Akam, 1983; Hafen et al., 1984; Kornberg et al., 1985; Harding et al., 1986; Bopp et al., 1986; Macdonald et al., 1986). For instance, among the primordia of the metameric segments are the anlagen of the posterior compartments, arranged as single cell-wide annuli regularly spaced along the anterior–posterior axis of the embryo. The *engrailed* gene is expressed only in these posterior compartment cells (Kornberg et al., 1985).

The importance of the *engrailed* gene to embryonic and subsequent development is revealed by the effects of *engrailed* mutations. Embryos die in the absence of *engrailed* function, their segments fused together, altering the epidermal morphology as well as the organization of the nervous system and other internal organs (Kornberg, 1981). *engrailed* mutations have similar effects in adults. Cells of the adult posterior compartments that have developed in the absence of *engrailed* function grow to produce abnormal patterns without obeying the compartmental or segmental borders which would normally restrict their growth (Lawrence and Morata, 1976). In order to understand how the *engrailed* gene is regulated so precisely such that it is expressed in only the posterior compartment cells of the embryo and later developmental stages, and to understand how the product of the *engrailed* gene functions to direct normal posterior compartment development, we have isolated the *engrailed* gene in recombinant form (Kuner et al., 1985). In the following, we describe an analysis of the *engrailed* promoter and a study of the *engrailed* protein.

THE *engrailed* GENE PROMOTER

The *engrailed* gene has a highly unusual structure. Although *engrailed* mutations that have been localized are spaced over 70 kilobases (kb) in the

48A region of the *Drosophila* second chromosome, the transcription unit is limited to a small portion of this region, approximately 4 kb. More than 45 and 15 kb of transcriptionally silent upstream and downstream sequence, respectively, flank this transcription unit (Drees *et al.*, 1987). Changes in the pattern of *engrailed* expression in mutants that have chromosome breaks in these flanking regions suggest that these outlying sequences are involved the spatial regulation of *engrailed* transcription (Weir and Kornberg, 1985).

To identify the functional elements of the *engrailed* promoter, we have constructed promoter fusions in which an 8-kb portion of the *engrailed* promoter was joined to *lacZ* coding sequences. These constructs were transferred to the germ line of *Drosophila* by P element-mediated transformation, and animals bearing these chimeric genes were analyzed for β-galactosidase activity. Among strains transformed with the promoter fusion were several that expressed *lacZ* specifically in the cells of the posterior compartments. Genetic tests revealed that these strains, although initially containing a wild-type *engrailed* allele, had acquired *engrailed* mutations upon transformation. Furthermore, as indicated by hybridization of insert-specific probes to mutant polytene chromosomes *in situ* and to appropriate Southern blots, the promoter–*lacZ* fusion DNA had inserted into the *engrailed* gene in the 48A region of the second chromosome. This is the first report of directed mutagenesis in *Drosophila*.

Analysis of the spatial disposition of β-galactosidase activity in tissues from transformed strains revealed the posterior compartment cells of the embryo, larva, and adult. Patterns of β-galactosidase activity in gastrulating embryos (Fig. 1) were not distinguishable from patterns of *engrailed* RNA or protein previously described for embryos of this developmental stage (Kornberg *et al.*, 1985; Weir and Kornberg, 1985; DiNardo *et al.*, 1985). Of particular interest, *lacZ* was also found to be expressed in the cuticle of the larva (Fig. 2). The cells of the posterior compartments form single-layered rings that encircle the embryo in the posterior portion of each segment. The existence of posterior compartments in the larval epidermis had not been known previously. *lacZ* expression in the tissues of the adult revealed the posterior compartment cells with startling clarity. For illustration, the wing imaginal disk, the larval organ which will produce the adult wing blade, and the wing blade of the adult are shown (Fig. 3). In both specimens, β-galactosidase activity is restricted to the cells of the posterior compartments.

THE *engrailed* PROTEIN IS A PHOSPHOPROTEIN

The primary structure of the *engrailed* protein is unusual. It has stretches of polyalanine and polyglutamine, and toward the C-terminal end there is a

Fig. 1. Expression of *lac*Z in a gastrulating embryo. Wild-type embryos were dechorionated, and β-galactosidase protein was localized with anti-β-galactosidase antiserum and horseradish peroxidase-coupled secondary antibody.

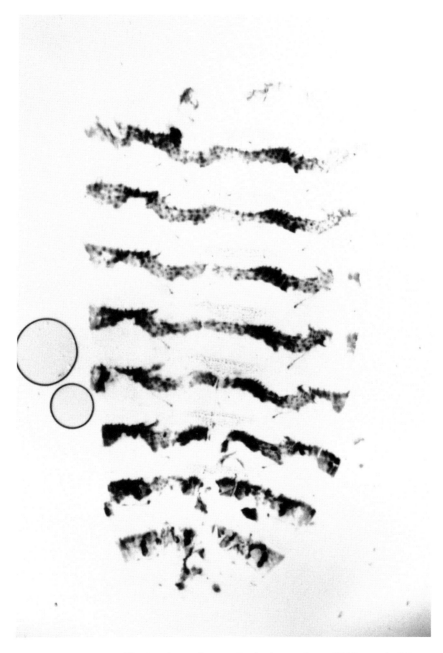

Fig. 2. Expression of *lacZ* in the epidermis of a third instar larva. Wild-type third instar larvae were decapitated, and the cuticle was inverted, scraped clean of adhering tissue, fixed with glutaraldehyde, and incubated with 0.2% 5-bromo-4-chloro-3-indolyl β-D-galactopyranoside (X-gal) for 12 hr at 37°C.

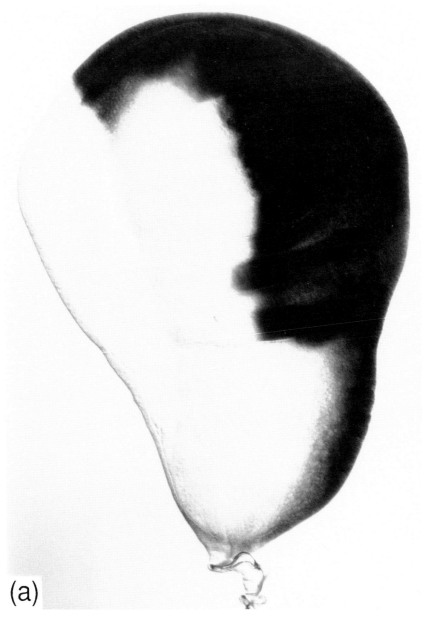

Fig. 3. Expression of *lacZ* in the wing imaginal disk (a) and in the wing blade (b). Wing imaginal disks were isolated from crawling third instar larvae and the wing blades from freshly eclosed adults. The specimens were fixed in glutaraldehyde, then incubated in a solution of X-gal for 12 hr at 37°C. (*Figure continues.*)

Fig. 3 (*continued*).

highly conserved region, the homeobox domain, also found in several other *Drosophila* proteins involved in segmentation (Poole et al., 1985). Homeoboxes are also encoded in the genomes of vertebrate species, including humans (Hart et al., 1985). The *engrailed* protein mediates sequence-specific DNA binding *in vitro* to numerous sites in both *Drosophila* and bacteriophage λ DNA (Desplan et al., 1985). The biochemical properties of these interactions remain poorly characterized and their functional significance unproved. Yet, these data, taken together with genetic evidence (Garcia-Bellido and Santamaria, 1972; Garcia-Bellido, 1975) and the observed nuclear localization of *engrailed* protein (DiNardo et al., 1985), suggest that the *engrailed* gene directly orchestrates a subset of other genes whose role is to elaborate the developmental program of a cell.

Inasmuch as the *engrailed* protein is expressed at very low levels in embryos, we constructed a *Drosophila* cell line, HS-EN, in which *engrailed* protein expression can be induced to high levels by heat-shock treatment. On induction at 37°C, *engrailed* protein was synthesized at high levels and accumulated in the cell nucleus (Fig. 4). In order to determine whether the *engrailed* protein produced by these cells is modified by the action of protein kinases, HS-EN cells were incubated in $^{32}PO_4$. Aside from a few minor differences, the HS-EN cell line produced an array of phosphorylated proteins similar to that of the untransfected parental cell line, Schneider 2 (Fig. 5a,b). Immunoprecipitation of the ^{32}P-labeled HS-EN cell extract with antibodies directed against the *engrailed* protein produced a single labeled species (Fig. 5d). This protein had a mobility in sodium dodecyl sulfate–polyacrylamide gel electrophoresis SDS–PAGE indistinguishable from that of *engrailed* protein produced in *Escherichia coli*. Immunoprecipitation of the parental Schneider 2 cell extract yielded no such phosphoprotein (Fig. 5c).

The phospho-*engrailed* protein from HS-EN cells was subjected to amino acid analysis. The phosphorylated amino acids of the *engrailed* protein proved to be mostly phosphoserine, with a trace of phosphothreonine. Phosphotyrosine was not detected.

The *engrailed* protein contains 79 serine residues (Poole et al., 1985), 64 of which are conserved in the *engrailed* protein of a related fruit fly, *Drosophila virilis* (Kassis et al., 1986). The homeobox domain contains 5 serines, one of which, Ser-462, is conserved or substituted with threonine in all known homeobox sequences. Several groups of serine/threonine phosphorylating protein kinases have been characterized in mammalian cells, and for two such groups of kinases, a primary amino acid sequence specificity has been recognized (Krebs and Beavo, 1979). In the case of cAMP- and cGMP-dependent protein kinases, a basic dipeptide two or three residues to the amino side of the phosphorylated amino acid is required, whereas the Ca^{2+}-

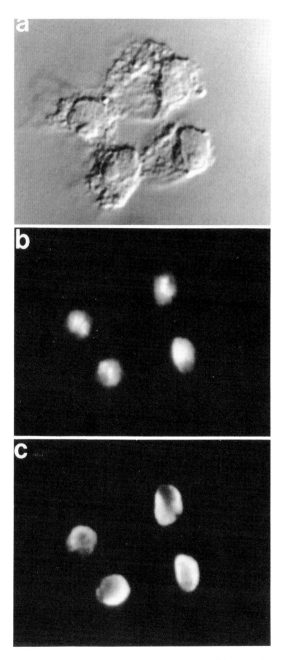

Fig. 4. Expression of *engrailed* protein in HS-EN cells. A stable line of Schneider 2 cells transfected with a plasmid containing the *engrailed* coding sequence under the control of the HSP-70 promoter was heat-shocked to induce expression of the *engrailed* protein, fixed, and stained for indirect immunofluorescence. Shown are representative HS-EN cells after heat-shock treatment that have been visualized by (a) differential interference contrast optics, (b) fluorescence optics in the presence of the DNA-binding dye 4,6-diamidino-2-phenylindole (DAPI) to localize nuclei, and (c) fluorescence optics after binding of anti-*engrailed* antibody and rhodamine-coupled secondary antibody to localize *engrailed* protein.

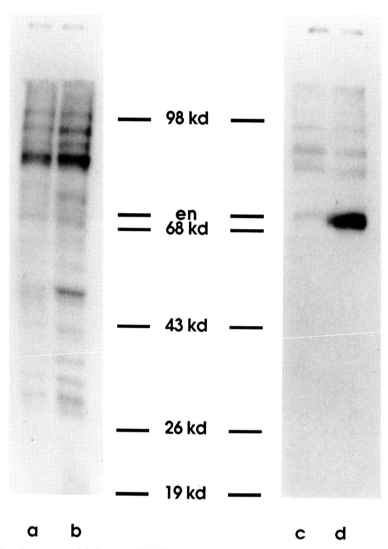

Fig. 5. *In vivo* labeling of HS-EN and Schneider 2 cells and purification of phospho-*engrailed* protein by immunoprecipitation. Cells were labeled with $^{32}PO_4$, the labeled protein was separated in a 12.5% polyacrylamide gel in the presence of SDS, and the gel was dried and autoradiographed (6 hr). (a) Soluble protein extract of Schneider 2 cells. (b) Soluble protein extract of HS-EN cells. (c) Products of immunoprecipitation of a Schneider 2 cell extract with anti-*engrailed* antibodies. (d) Products of immunoprecipitation of an HS-EN cell extract with anti-*engrailed* antibodies. Protein molecular mass standards were run in adjacent lanes and detected by staining with Serva G. en, *E. coli* produced *engrailed* protein.

dependent group exemplified by phosphorylase kinase requires an arginine residue two amino acids on the C-terminal side of the modified amino acid. In the *engrailed* protein, no sequence that conforms to the former consensus is found, but Ser-182, Ser-367, and Ser-432 have an appropriate C-neighboring arginine residue. None of these is in the homeodomain, but all three are conserved in the *D. virilis engrailed* protein. Serine 432 is also conserved in the *D. melanogaster invected* protein and in the *engrailed*-related mouse proteins (Joyner and Martin, 1987; Coleman *et al.*, 1987).

Based on other examples, phosphorylation of the *engrailed* protein is likely to be of functional significance by causing a direct or allosteric activation, inactivation, or modification of *engrailed* protein function. Histone H1 becomes modified when the cell is committed to mitosis (Bradbury *et al.*, 1973); the degree of phosphorylation of the large T antigen of SV40, a nuclear phosphoprotein involved in the regulation of viral transcription and initiation of viral replication, is correlated with an increase in site-specific DNA-binding affinity (Montenarh and Hennig, 1980), and phosphorylation of the NS phosphoprotein of vesicular stomatitis virus potentiates its role as an activator of viral RNA synthesis (Chattopadhyay and Banerjee, 1987).

The reversible activation of substrate-specific serine/threonine protein kinases is known to mediate the transduction of a variety of extracellular signals. One example is the G protein-mediated activation of phospholipase C by agonist receptor interactions, resulting in the production of two second messengers by the hydrolysis of phosphatidylinositol 4,5-bisphosphate to 1,2-(*sn*)-diacylglycerol and inositol 1,4,5-trisphosphate. Diacylglycerol directly activates the serine/threonine-specific protein kinase C, and inositol 1,4,5-trisphosphate mobilizes intracellular Ca^{2+}, a second activator of protein kinase C enzymes (Nishizuka, 1986). These are a family of closely related kinases, ubiquitous in animal cells, and are implicated in the mitogenic effects mediated by many effector molecules (Stiles *et al.*, 1979; Berridge *et al.*, 1984; Farrar and Anderson, 1985; Müller *et al.*, 1984). Similarly, the receptor-mediated activation of adenylate cyclases results in the synthesis of cAMP, an activator of the cAMP-dependent protein kinases. The *cdc2* gene product of *Schizosaccharomyces pombe*, which is required to commit a cell to the mitotic cycle (Lee and Nurse, 1987), is related to these mammalian cAMP-dependent enzymes. Thus, phosphorylation of the *engrailed* protein may modify its function in response to external stimuli, such as embryonically specified positional information. Alternatively, the stimuli may be mitogenic, and the *engrailed* protein function could be modulated in a cell cycle-dependent manner. In this regard, a protein kinase C homolog which has recently been isolated from *Drosophila* contains a metal/DNA-binding domain very similiar to that found in *Xenopus* TFIIIA and several other DNA-binding proteins (Rosenthal *et al.*, 1987). Sequence-specific DNA binding of the kinase homolog might facilitate phosphorylations of adjacent bound factors, such as *engrailed* protein.

REFERENCES

Akam, M. (1983). *EMBO J.* **2**, 2075–2084.
Berridge, M., Berridge, M. J., Heslop, J. P., Irvine, R. F., and Brown, K. D. (1984). *Biochem J.* **222**, 195.
Bopp, D., Burri, M., Baumgartner, S., Frigerio, G., and Noll, M. (1986). *Cell (Cambridge, Mass.)* **47**, 1033–1040.
Bradbury, E., Inglis, R., and Matthews, H. (1973). *Nature (London)* **247**, 257.
Chattopadhyay, D., and Banerjee, A. (1987). *Cell (Cambridge, Mass.)* **49**, 407–414.
Coleman, K., Poole, S., Weir, M., Soeller, W., and Kornberg, T. (1987). *Genes & Dev.* **1**, 19.
Desplan, C., Theis, J., and O'Farrell, P. (1985). *Nature (London)* **318**, 630.
DiNardo, S., Kuner, J., Theis, J., and O'Farrell, P. (1985). *Cell (Cambridge, Mass.)* **43**, 59.
Drees, B., Ali, Z., Soeller, W., Coleman, K., Poole, S., and Kornberg, T. (1987). *EMBO J.* **6**, 2803.
Farrar, W., and Anderson, W. (1985). *Nature (London)* **315**, 233.
Garcia-Bellido, A. (1975). *Ciba Found. Symp.* **29**, 161.
Garcia-Bellido, A., and Santamaria, P. (1972). *Genetics* **72**, 87.
Garcia-Bellido, A., Ripoll, P., and Morata, G. (1973). *Nature (London) New Biol.* **245**, 251–253.
Hafen, E., Kuroiwa, A., and Gehring, W. (1984). *Cell (Cambridge, Mass.)* **37**, 833–841.
Harding, K., Rushlow, C., Doyle, H., Hoey, T., and Levine, M. (1986). *Science* **233**, 953–959.
Hart, C., Awgulewitch, A., Fainsod, A., McGinnis, W., and Ruddle, F. (1985). *Cell (Cambridge, Mass.)* **43**, 9.
Joyner, A., and Martin, G. (1987). *Genes & Dev.* **1**, 29.
Kassis, J., Poole, S., Wright, D., and O'Farrell, P. (1986). *EMBO J.* **5**, 3583.
Kornberg, T. (1981). *Proc. Natl. Acad. Sci U.S.A.* **78**, 1095.
Kornberg, T., Siden, I., O'Farrell, P., and Simon, M. (1985). *Cell (Cambridge, Mass.)* **40**, 45.
Krebs, E., and Beavo, J. (1979). *Annu. Rev. Biochem.* **48**, 923.
Kuner, J., Nakanishi, M., Ali, Z., Drees, B., Gustavson, E., Theis, J., Kauvar, L., Kornberg, T., and O'Farrell, P. (1985). *Cell (Cambridge, Mass.)* **42**, 309–3116.
Lawrence, P., and Morata, G. (1976). *Dev. Biol.* **56**, 321–337.
Lee, M., and Nurse, P. (1987). *Nature (London)* **327**, 31.
MacDonald, P., Ingham, P., and Struhl, G. (1986). *Cell (Cambridge, Mass.)* **47**, 721–734.
Montenarh, M., and Hennig, R. (1980). *FEBS Lett.* **114**, 107.
Müller, R., Bravo, R., and Burkhardt, J. (1984). *Nature (London)* **312**, 716.
Nishizuka, Y. (1986). *Science* **233**, 305.
Poole, S., Kauvar, L., Drees, B., and Kornberg, T. (1985). *Cell (Cambridge, Mass.)* **40**, 37.
Rosenthal, A., Rhee, L., Yadegari, R., Paro, R., Ulbrich, A., and Goeddel, D. (1987). *EMBO J.* **6**, 433–441.
Stiles, C., Capone, G., Scher, C., Antoniades, H., Van Wyk, J., and Pledger, W. (1979). *Proc. Natl. Acad. Sci. U.S.A.* **76**, 1279–1283.
Weir, M., and Kornberg, T. (1985). *Nature (London)* **318**, 433–435.

21
Sequence-Specific DNA-Binding Activities of *Even-Skipped* and Other Homeo Box Proteins in *Drosophila*

TIMOTHY HOEY, MANFRED FRASCH,
AND MICHAEL LEVINE
*Department of Biological Sciences
Fairchild Center
Columbia University
New York, New York 10027*

The homeo box gene *even-skipped* (*eve*) plays a crucial role in the segmentation of the *Drosophila* embryo. There is evidence that the *eve* protein controls morphogenesis by regulating the expression of the segmentation gene *engrailed* (*en*) and by autoregulating its own expression. Here we show that a full-length *eve* protein binds with high affinity to specific sequences associated with both the *eve* and *en* transcription units. Many of the binding sites contain at least one copy of a 10-base pair (bp) consensus sequence: T-C-A-A-T-T-A-A-A-T. Several of the *eve* binding sites are also recognized by other homeo box proteins, lending support to the proposal that cross-regulatory interactions among homeo box genes involve a competition of different homeo box proteins for similar cis-regulatory sequences.

INTRODUCTION

There are at least 30 homeo box genes in *Drosophila*, of which 18 have been cloned and characterized (McGinnis *et al.*, 1984; Scott and Weiner, 1984; reviewed in Levine and Harding, 1987). Most of these genes (16 of 18) control the morphogenesis of pattern elements along the anterior–posterior body axis, and mutations in these genes usually disrupt the characteristic segmentation pattern of the embryo. In situ localization studies of homeo box transcripts and proteins indicate that each homeo box gene is expressed within a unique subset of embryonic cells during early development (e.g., Levine et al., 1983; McGinnis et al., 1984; Hafen et al., 1984; Akam and Martinez-Arias, 1985; Fjose et al., 1985; Kornberg et al., 1985; Carroll and Scott, 1985; DiNardo et al., 1985). Consequently, it appears that virtually every embryonic cell contains a unique combination of active and inactive homeo box genes. It is thought that embryonic cells follow different pathways of morphogenesis as a result of these unique permutations of gene activity (Garcia-Bellido, 1977; Scott and O'Farrell, 1987). A key question regarding the homeo box gene control of development is a problem of regulation: how do the different homeo box genes come to be expressed in specific regions of the developing embryo?

There is evidence that selective patterns of homeo box gene expression involve cross-regulatory interactions, whereby one homeo box gene can influence the expression of other homeo box genes (Hafen et al., 1984; Harding et al., 1985, 1986; Ingham and Martinez-Arias, 1986). Since the homeo box protein domain appears to contain a sequence-specific DNA-binding activity, it has been proposed that these cross-regulatory interactions occur at the level of transcription (Desplan et al., 1985). In order to further investigate this possibility we have examined the DNA-binding activity of the *even-skipped* (*eve*) protein.

eve mutations cause the most severe developmental defects among the known zygotically active homeo box genes in *Drosophila*, in that eve^- embryos completely lack segmentation (Nusslein-Volhard et al., 1985). Thus, *eve* is a good candidate for examining regulatory interactions among homeo box genes since it would appear that *eve* should influence the expression of other genes that control morphogenesis. In an effort to identify possible "target" genes that might be regulated by the *eve* protein, the expression patterns of approximately 20 different segmentation and homeo box genes were examined in eve^- embryos (Harding et al., 1986; Macdonald et al., 1986; Rushlow et al., 1987b; K. Harding, and M. Levine, unpublished results). Five of these genes show abnormal patterns of expression in eve^-, including the pair-rule genes *eve*, *fushi tarazu (ftz)*, *hairy (h)*, and *engrailed (en)*, as well as the homeotic gene *Deformed (Dfd)*. Based on the timing and sites of the altered patterns that were observed, *eve*, *en*, and *Dfd* are the most likely targets for eve^+ gene activity.

As a first step toward determining whether *eve* functions as a specific transcription factor for the regulation of either *eve* or *en* expression we have performed DNA-binding studies using a full-length *eve* protein synthesized in *Escherichia coli*. We showed that this *eve* protein binds with high affinity to several sites near the *eve* and *en* coding regions. In addition, the binding activity of the *eve* protein was compared with the activities of three other full-length homeo box proteins. The four proteins that were examined contain highly divergent homeo domains, but nonetheless display some similarities in their binding activities. This supports the proposal that cross-regulatory interactions among homeo box genes involve a competition of different homeo box proteins for similar cis-regulatory sequences (Harding *et al.*, 1985; Desplan *et al.*, 1985).

RESULTS

Patterns of *eve* and *en* Expression

eve encodes a 1.4-kilobase (kb) mRNA that specifies a 42-kDa protein composed of 376 amino acid residues (Macdonald *et al.*, 1986; Frasch *et al.*, 1987). Anti-*eve* antibodies have been used to localize this protein in whole-mount preparations of developing embryos Frasch *et al.*, 1987). The *eve* protein is distributed within a series of 7 transverse stripes during early stages of development, and at later stages 7 stripes of expression also appear, giving a pattern of 14 stripes (Fig. 1a). Each of the strong stripes is about 3 cells in width, whereas the weaker stripes span about 2 cells apiece. This pattern is very similar to that observed for *en* (DiNardo *et al.*, 1985). *en* proteins first appear within several minutes after a periodic *eve* pattern is established during cellularization. *en* proteins are distributed in a series of 14 stripes, each of which is somewhat narrower than the *eve* stripes (Fig. 1b). Double staining experiments done with a mixture of anti-*eve* and anti-*en* antibodies show that the *eve* and *en* stripes are virtually coincident (compare Fig. 1a,b). The marked cells in Fig. 1a correspond to those cells that express *en*, indicating that the *en* stripes occur at the anterior border of each *eve* stripe.

The similarities in the spatial and temporal limits of *eve* and *en* expression are consistent with the possibility that regulatory interactions between these genes are direct. It has been previously shown that *en* transcripts are not observed above background levels in the middle body region of eve^- embryos, which suggests that *eve* is required for the activation of *en* expression (Harding *et al.*, 1986; Macdonald *et al.*, 1986). Further support for this proposal is indicated by the *en* protein patterns in wild-type (Fig. 1c) versus eve^- embryos (Fig. 1d). As can be seen, *en* proteins do not occur within the region of eve^- embryos that would normally express *eve*. The only *en* proteins that are detected occur in an anterior region that is ouside the normal domain of *eve* expression.

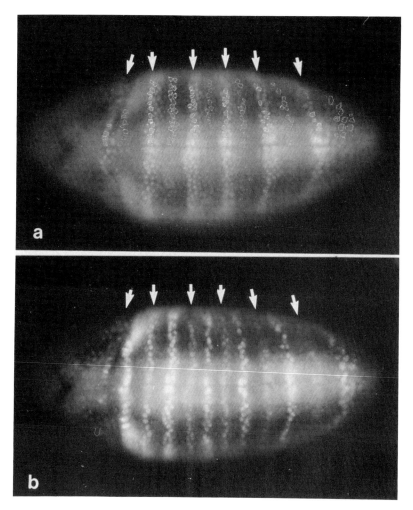

Fig. 1. Patterns of *eve* and *en* expression in wild-type and mutant embryos. Embryos were stained with rabbit anti-*eve* antibodies or a monoclonal mouse anti-*en* antibody. All embryos are oriented so that anterior is to the left. Visualization was done with either fluorescein (a) or rhodamine (b–d). (a) and (b) are different micrographs of the same embryo after staining with a mixture of anti-*eve* and anti-*en* antibodies. (a) Wild-type gastrulating embryo showing *eve* staining. The arrows mark the anterior margins of each of the original 7 stripes that appear during cellularization. By this stage weaker stripes can be seen between the arrows. (b) Same embryo as in (a) showing *en* fluorescence. The arrows show the positions of the anterior margins of each of the strong *eve* stripes. The marked cells in the top half of the embryo in (a) show those that are also stained by both *eve* and *en* antibodies. (c) Lateral view of a wild-type embryo undergoing germ band elongation after staining with anti-*en* antibody. (d) *eve*⁻ embryo at a similar stage as that shown in (c) after staining with anti-*en*. Note that the only *en* stripe observed is anterior to the normal limits of *eve* expression in the wild type.

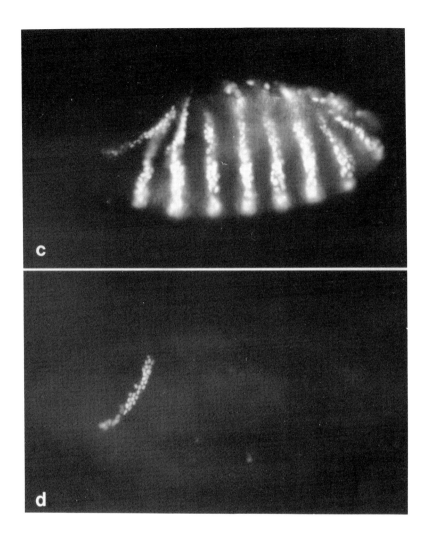

Expression of Full-Length Homeo Box Proteins in *Escherichia coli*

Homeo box proteins used for DNA-binding studies were synthesized in *E. coli* using a T7 expression system developed by Studier and Moffatt (1986). This system involves the insertion of a foreign coding sequence downstream from a T7 promoter and its overexpression in a bacterial strain that contains the T7 polymerase gene (gene *1*) under the control of an isopropyl-β-D-thiogalactopyranoside (IPTG)-inducible *lac* UV5 promoter. In order to prepare full-length proteins that do not contain extraneous amino acid residues from the vector, *in vitro* mutagenesis was done (Zoller and Smith, 1984) to create an *Nde*I restriction site at the initiating AUG for each of the four coding sequences that were used (*eve*, *en*, *zerknullt*, and *Dfd*). There is a unique *Nde*I site in the T7 plasmid, which is located just downstream from a Shine–Dalgarno ribosome-binding sequence. The "mutagenized" coding sequences were cloned into this *Nde*I site, thereby allowing for the efficient

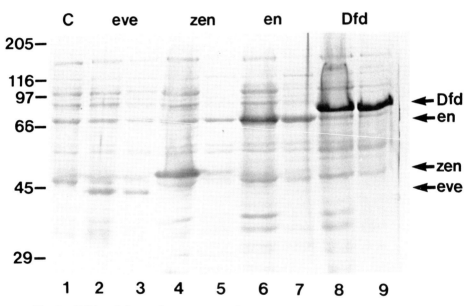

Fig. 2. Full-length homeo box proteins used for DNA-binding studies. Extracts prepared from IPTG-induced cultures containing each of the four homeo box coding sequences were electrophoresed in a 9% polyacrylamide/sodium dodecyl sulfate (SDS) gel and stained with Coomassie blue. Approximately 5 μg of total protein extract was loaded in each lane. Molecular weight markers ($\times 10^{-3}$) are indicated to the left of the gel. Lane 1, IPTG-induced total extract of BL21(DE3) cells that contain the T7 expression vector without insert. Lanes 2, 4, 6, and 8 were loaded with total induced cultures that contain the *eve*, *zen*, *en*, and *Dfd* expression vectors, respectively. Lanes 3, 5, 7, and 9 show the corresponding guanidine-HCl extracts that were used for the DNA-binding experiments.

expression of homeo box proteins that possess amino acid sequences identical to the native *Drosophila* proteins.

Figure 2 shows the protein preparations that were obtained with the T7 expression system. Lanes 1 and 2 compare total protein extracts from IPTG-induced bacterial cultures containing the T7 expression vector with no insert (lane 1) and cultures that contain the vector with the inserted *eve* coding sequence (lane 2). The *eve* protein migrates with an apparent molecular weight of 42,000, which is close to the predicted size of 39,000 based on nucleotide sequence analysis (Macdonald *et al.*, 1986; Frasch *et al.*, 1987). At least 10–20% of the total induced protein corresponds to *eve*.

The *en* Transcription Unit Contains High-Affinity *eve* Binding Sites

In order to initially identify any *eve* binding sites near *en*, immunoprecipitation assays (McKay, 1981) were performed with a 5-kb *Eco*RI fragment that includes the first exon and intron of *en* as well as 2.5 kb of 5′ flanking sequences (Kuner *et al.*, 1985). A 10-kb recombinant plasmid that contains this *en* fragment was digested with a mixture of *Eco*RI, *Cla*I, and *Hin*fI, yielding about 20 smaller fragments with an average size of approximately 500 bp. After end-labeling with ^{32}P, the fragments were incubated with a preparation of the full-length *eve* protein (shown in Fig. 2, lane 3), and protein–DNA complexes were immunoprecipitated with anti-*eve* antibodies. Four of the *en* fragments specifically bind to the *eve* protein (Fig. 3a), which are designated C, H, F, and K according to the nomenclature of Desplan *et al.* (1985). Fragment K appears to bind with the highest affinity, with fragments F, H, and C showing progressively weaker binding. These correspond to the same fragments that are bound by *en*–β-galactosidase fusion proteins (Desplan *et al.*, 1985). The locations of these fragments relative to the *en* gene are summarized in Fig. 3b. Also shown are the results of immunoprecipitation assays using cloned DNAs from the 3′ region of *en* (data not shown).

The *eve* binding sites present in fragments K and F were further characterized by DNase I footprint experiments (Galas and Schmitz, 1978). Fragment K is 240 bp in length and located between −950 and −710 bp upstream from the *en* transcription start site. Fragment K was ^{32}P-labeled at a *Cla*I site on either the coding strand or the noncoding strand, incubated with increasing amounts of *eve* protein extract, and then partially digested with DNase I (Fig. 4). Three different regions of protection are observed, designated k1, k2, and k3. Each of these sites is 20 to 30 nucleotides in length, and there are some similarities shared among these sequences (summarized in Fig. 8). A fourth binding site, k0, is also located within fragment K. Protection of k0 is not observed in this experiment since its 5′ border is

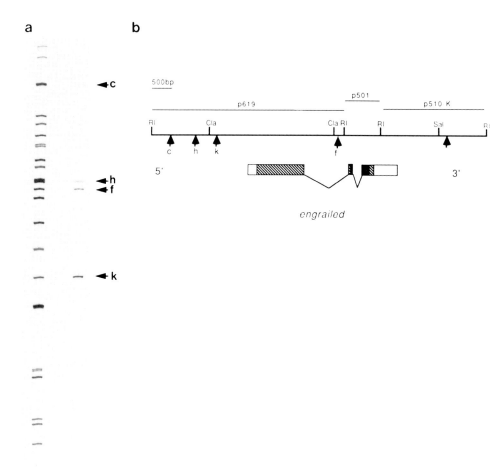

1 2 3 4

Fig. 3. Immunoprecipitation assays using *eve* protein and cloned *en* DNAs. (a) Autoradiogram of a polyacrylamide gel that shows *en* DNAs before and after immunoprecipitation with *eve* protein. Lane 1, *en* plasmid p619, which contains the first intron and exon of *en* as well as 5' flanking sequences, after digestion with a mixture of *Hin*fI/*Cla*I/*Eco*RI and end-labeling with ^{32}P. Over 20 bands with an average molecular weight corresponding to 500 bp are observed. Lane 2, the *en* fragments shown in lane 1 were incubated with 1 μg of a protein extract from BL21(DE3) cells that carry the T7 expression vector without the *eve* insert. None of the *en* fragments are precipitated after addition of anti-*eve* antibodies. Lane 3, *en* DNA digest was incubated with addition of *eve* protein extract and then with Pansorbin without anti-*eve* antibodies. Lane 4, *en* DNA digest was incubated with 1 μg of total protein extract from cells carrying the T7 vector plus *eve* coding sequences, then immunoprecipitated with anti-*eve*

contained within the *Cla*I recognition sequence used for end-labeling (data not shown; the sequence of k0 is shown in Fig. 8). The boundaries of the k3 binding site were further established by labeling the DNA at an *Eag*I site, near the 3' end of fragment K (data not shown).

Fragment F is located within the first intron, from +2.1 to +2.4 kb downstream from the *en* transcription start site (see Fig. 3b). Fragment F contains a cluster of at least three *eve* binding sites, designated f1, f2, and f3, which are located within an interval of approximately 100 bp (Fig. 5). The f1 and f3 binding sites are quite divergent from the sequences of the other binding sites associated with the *en* and *eve* genes (see below; Fig. 8), and are protected only at the highest concentrations of the *eve* protein that were assayed. There is a region of hypersensitivity located between the f1 and f2 binding sites, which becomes progressively more pronounced as the two sites become "filled" by *eve* protein (Fig. 5b).

High-Affinity *eve* Binding Sites Near the *eve* Gene

Based on examining the distribution of the *eve* protein and transcript in various *eve* mutants, it appears that *eve* might autoregulate its own expression (M. Frasch, unpublished results; Warrior *et al.*, in preparation). Weak *eve* mutants (i.e., hypomorphs) prevent the narrowing of the *eve* stripes during gastrulation, so that only 7, not 14, *eve* stripes are observed at advanced stages of development. Strong *eve* mutants alter the establishment of the *eve* pattern during cellularization. In order to determine whether this apparent autoregulation involves the binding of *eve* protein to specific regions near the *eve* gene, we initially performed immunoprecipitation assays with cloned *eve* DNAs. A total of about 6.5 kb of genomic DNA was tested for binding, including the entire *eve* coding region as well as approximately 3.5 kb of 5' and 1.5 kb of 3' flanking sequence. The only high-affinity binding sites that were identified map at about −3 kb upstream from the transcrip-

---

antibodies. Four fragments are selectively immunoprecipitated, which are designated C, H, F, and K according to the nomenclature of Desplan *et al.* (1985). The location of these fragments relative to the *en* transcription unit are shown in (b). (b) Map of the *en* gene showing the location of the DNA fragments used for immunoprecipitation assays. The *en* gene is 2.7 kb in length and is composed of three exons that are interrupted by two introns (Poole *et al.*, 1985). Open boxes indicate the untranslated sequences present in the mature *en* mRNA; the hatched regions indicate protein coding sequences. The *en* homeo domain is indicated by the black box. The lines connecting the boxes represent the introns. The arrows indicate DNA fragments that are selectively immunoprecipitated by anti-*eve* antibodies. The results of the immunoprecipitation assays using the p501 and p510-K plasmids are not shown. There is a high-affinity *eve* binding site located in p510-K, which maps about 1.5 kb downstream from the 3' end of *en*.

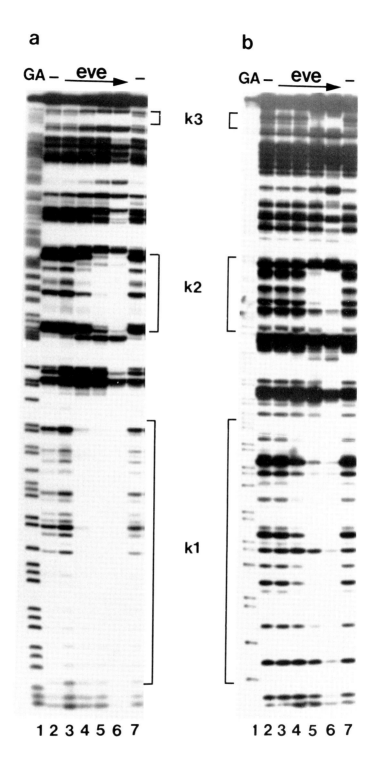

tion start site (summarized in Fig. 6). The *eve* binding sites were further characterized by DNase I protection assays (Fig. 7). Three sites, designated e1, e2, and e3, were identified. These binding sites are clustered within an interval of approximately 200 bp, and they possess affinities that are comparable to those observed for *en* fragment K.

The proportion of the total *E. coli* extract corresponding to the *eve* protein was determined by densitometric scanning of the polyacrylamide/SDS gel shown in Fig. 2 (lane 3). On this basis *eve* represents about 15% of the total extract. Assuming that 100% of the *eve* protein present in this extract has retained its native DNA-binding activity, it is possible to measure the binding constants for each of the *eve* binding sites present in the *en* fragments K and F, as well as those present in the 5′ region of *eve*. There is a broad range of binding affinities. The *en* k1 sequence shows the highest affinity ($K_D \sim 3 \times 10^{-9}$ M), whereas the *en* f1 and f3 sites show the weakest binding (each possesses about a 30-fold lower affinity than the k1 site). The highest affinity binding sites include at least one copy of the following 10-bp consensus sequence: T-C-A-A-T-T-A-A-A-T (see Fig. 8).

Comparison of the DNA-Binding Activities of Four Different Homeo Box Proteins

We have compared the binding activities of the *eve* protein with full-length proteins encoded by *en*, the homeotic gene Deformed (Dfd) (Regulski *et al.*, 1987; Chadwick and McGinnis, 1987), and the dorsal–ventral gene *zerknullt* (*zen*) (Wakimoto *et al.*, 1984; Doyle *et al.*, 1986). Full-length proteins were prepared in *E. coli* using the T7 expression system, as described above. The protein extracts that were used for the binding studies are shown in Fig. 2.

As shown above, the 5′ *en* fragment K was found to contain the highest affinity sites among the *eve* and *en* DNA fragments that bind to the *eve* protein. We therefore selected fragment K for comparative binding studies with the different homeo box proteins. The 5′ end of fragment K was ^{32}P-

---

Fig. 4. Footprint assays using *eve* protein and *en* fragment K. Fragment K was labeled with ^{32}P at the *Cla*I site on either the coding strand (a) or the noncoding strand (b). In this and subsequent figures, lanes labeled with a dash (—) indicate DNase I digestion of fragment K without added protein, and lanes labeled GA indicate the deoxyguanosine plus deoxyadenosine sequence of the fragment. Lanes 3 through 6 were run with increasing amounts of *eve* protein: 0.04, 0.2, 1, and 5 μg. There are three regions that are protected from digestion with DNase I, labeled k1, k2, and k3. The *eve* protein extract used for this and subsequent experiments is shown in Fig. 2, lane 3.

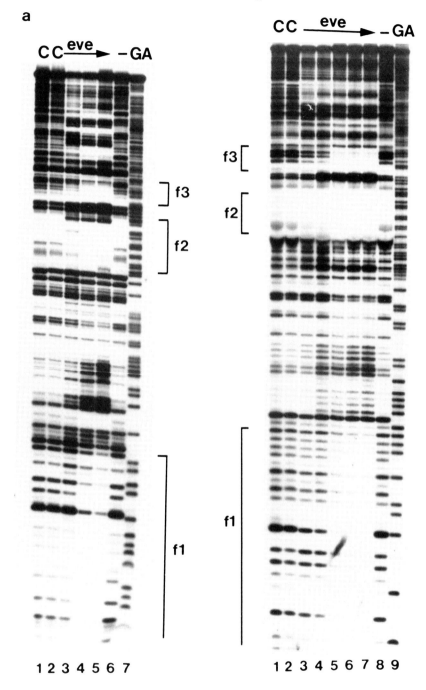

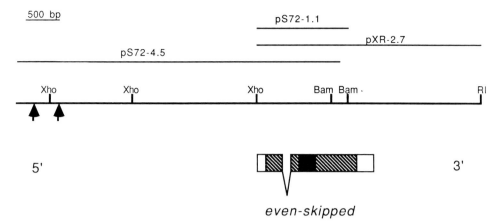

Fig. 6. Location of high-affinity *eve* binding sites near the *eve* gene. The *eve* gene is 1.5 kb in length and contains two exons separated by a single intron located near the 5' end. The boxes indicate the transcribed sequences that are contained in the mature *eve* mRNA. Hatching indicates the protein coding sequence, and the homeo domain is represented by the black box. Three plasmids that encompass a total of approximately 6.5 kb of genomic DNA around *eve* were tested for binding to the *eve* protein by immunoprecipitation. Only one of these, pS72-4.5, was found to contain high-affinity binding sites. Two clusters of binding sites were found within an interval of about 200 bp that map approximately −3 kb upstream from the *eve* transcription start site. Fragments that contain these two clusters were used for the footprint experiments shown in Fig. 7.

labeled on either the coding strand (Fig. 9a) or the noncoding strand (Fig. 9b), separately incubated with each of the four protein extracts, and partially digested with DNase I. As shown in Fig. 4, the *eve* protein protects three different regions of fragment K, called k1, k2, and k3. Each of these binding sites is also protected by the *zen*, *en*, and *Dfd* proteins. None of the proteins bind to additional regions within fragment K that are not protected by the *eve* protein.

There are both qualitative and quantitative differences in the way these homeo box proteins bind to the *en* k1, k2, and k3 sites. As can be seen, the *eve* and *zen* proteins show similar patterns and affinities of binding to k1,

Fig. 5. Footprint assays using *eve* protein and *en* fragment F. Fragment F was labeled at a *Cla*I site on either the coding strand (a) or the noncoding strand (b). The lanes labeled "C" (lanes 1 and 2) were run with 1 and 5 μg, respectively, of total protein from induced BL21(DE3) cells that contain the T7 expression vector without *eve* insert. No protected regions are observed using such protein extracts. In (a) lanes 3, 4, and 5 were run with increasing amounts of *eve* protein extracts: 1, 5, and 10 μg. In (b) lanes 3–7 were run with 1, 2, 5, 7.5, and 10 μg of *eve* protein extract. Three regions of protection are observed when *eve* extracts are used, labeled f1, f2, and f3.

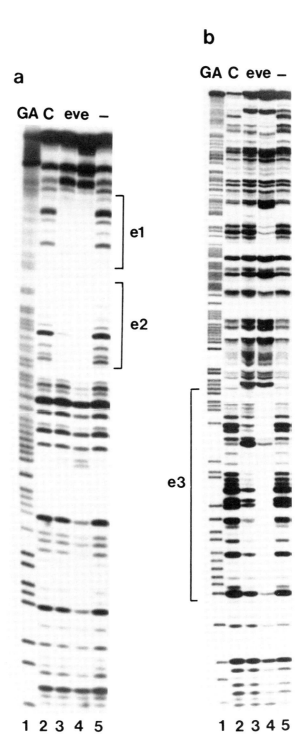

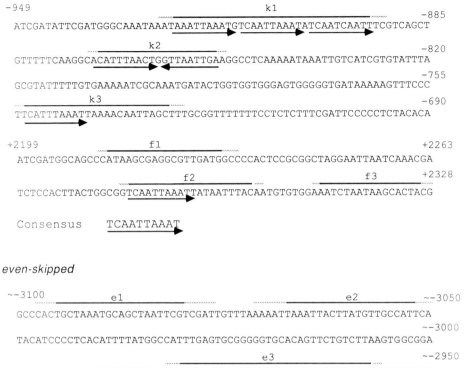

engrailed

```
-949                          k1                              -885
ATCGATATTCGATGGGCAAATAAATAAATTAAATGTCAATTAAATATCAATCAATTTCGTCAGCT
                   k2                                         -820
GTTTTTCAAGGCACATTTAACTGGTTAATTGAAGGCCTCAAAAATAAATTGTCATCGTGTATTTA
                                                              -755
GCGTATTTTTGTGAAAAATCGCAAATGATACTGGTGGTGGGAGTGGGGGTGATAAAAAGTTTCCC
              k3                                              -690
TTCATTTAAATTAAAACAATTAGCTTTGCGGTTTTTTTCCTCTCTTTCGATTCCCCCTCTACACA

+2199                f1                                      +2263
ATCGATGGCAGCCCATAAGCGAGGCGTTGATGGCCCCACTCCGCGGCTAGGAATTAATCAAACGA
                 f2                         f3              +2328
TCTCCACTTACTGGCGGTCAATTAAATTATAATTTACAATGTGTGGAAATCTAATAAGCACTACG

Consensus     TCAATTAAAT
```

even-skipped

```
~-3100              e1                              e2        ~-3050
GCCCACTGCTAAATGCAGCTAATTCGTCGATTGTTTAAAAATTAAATTACTTATGTTGCCATTCA
                                                              ~-3000
TACATCCCCTCACATTTTATGGCCATTTGAGTGCGGGGGTGCACAGTTCTGTCTTAAGTGGCGGA
                          e3                                 ~-2950
TGGTTTCCACCAGATTTCTCGAGGGATGATGTGCTCTAATATCTCCTCATCAAATGGGATGGTTT
```

Fig. 8. Nucleotide sequences of the *eve* binding sites. These sequences represent the maximum limits for the protein–DNA contact points. The highest affinity sites show a conserved 10-bp sequence: T-C-A-A-T-T-A-A-A-T. The k1 site from the 5′ *en* fragment K contains three tandem copies of this sequence, which are oriented as direct repeats. k2 contains two copies arranged as inverted repeats.

Fig. 7. Footprint assays using *eve* protein and *eve* DNA fragments. (a) A 150-bp *eve* fragment located at approximately -3 kb from the transcription start site (see Fig. 6) was ^{32}P-labeled on the noncoding strand at an *Afl*II site. Lanes 3 and 4 were run with 1 and 5 μg of *eve* protein extract, respectively. A large region (~50 bp) is protected from DNase I digestion. There might be two adjacent binding sites, e1 and e2, since a faint hypersensitive site appears in the center of this protected region upon addition of increasing amounts of *eve* protein. (b) A 400-bp *eve* fragment was ^{32}P-labeled on the coding strand at the same *Afl*II site as in (a). Lanes 3 and 4 were run with 1 and 5 μg of *eve* protein extract. There is a single region of protection that is about 30 bp in length. The lanes labeled C in (a) and (b) are controls that were run with 5 μg of protein extract from cells which contain the T7 expression vector without the *eve* insert. The three binding sites are labeled e1, e2, and e3 according to their relative 5′ to 3′ positions with respect to the direction of *eve* transcription.

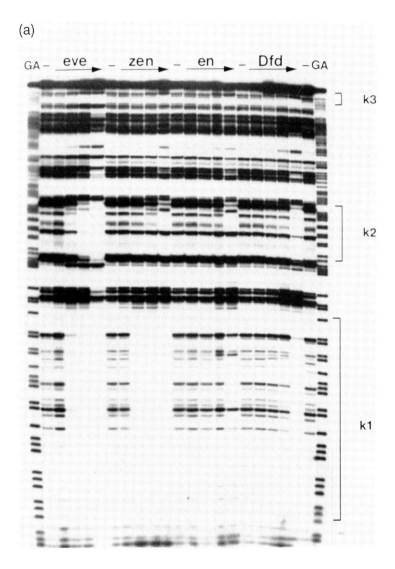

Fig. 9. Comparison of the binding activities of four different homeo box proteins. The *en* fragment K was ^{32}P-labeled on either the coding strand (a) or noncoding strand (b) and incubated with increasing amounts of *eve*, *zen*, *en*, and *Dfd* proteins. The horizontal arrows at top indicate increasing quantities of each protein. Each set of assays involved the use of four different amounts of protein extract. For *eve* and *Dfd*, 0.04, 0.2, 1, and 5 μg were used. For *en* and *zen*, 0.08, 0.4, 2, and 10 μg were used. Each of the four proteins bind to the k1, k2, and k3 sites, although there are both qualitative and quantitative differences in the footprints observed.

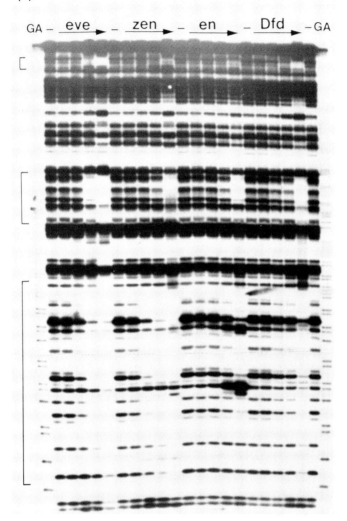

whereas the *en* and *Dfd* proteins bind with about 30–40 times lower affinity. There are differences in the pattern of DNase I hypersensitivity among the different proteins. Progressive binding of the *en* protein to the k1 site results in the appearance of several hypersensitive bands that are protected by the *eve* protein. Similar differences in the binding behaviors are also seen for the k2 region. *eve* shows a substantially higher affinity to k2 as compared with *en*, *zen*, and *Dfd*. Furthermore, *eve* and *Dfd* protect an approximately 20-bp sequence in the k2 region, while *zen* and *en* protect a sequence of only about 10 bp. It is possible that these differences in binding to k2 reflect protein–protein associations. The k2 sequence that is protected by *eve* and *Dfd* is an inverted repeat of two 10-bp sequences (see Fig. 8), suggesting that these proteins might be binding as dimers. In contrast, *zen* and *en* protect only one-half of this dyad, suggesting that they might bind as monomers.

DISCUSSION

We have shown that full-length *eve*, *zen*, *en*, and *Dfd* proteins produced in *E. coli* bind with high affinity to specific sequences near the *en* transcription initiation site. This observation is consistent with previous proposals that the homeo box protein domain contains a sequence-specific DNA-binding activity (Laughon and Scott, 1984; Desplan et al., 1985) and that homeo box proteins might control development by the trans-regulation of batteries of "target" genes (Garcia-Bellido, 1977). The binding of the *eve* protein to specific regions of *en* and *eve* is consistent with genetic evidence that *eve*$^+$ activity influences its own expression (Warrior et al., in preparation) and regulates the expression of *en* (Harding et al., 1986; Macdonald et al., 1986). The affinities that were determined for the binding of *eve* protein to *en* and *eve* sequences are comparable to estimates for the binding of bacterial proteins to their operator sequences (Ptashne et al., 1980; Fried and Crothers, 1984). The demonstration that the *eve*, *en*, *zen*, and *Dfd* proteins can bind to similar DNA sequences is consistent with the proposal that cross-regulatory interactions among homeo box genes involve a competition of different homeo box proteins for similar cis-regulatory sequences (Harding et al., 1985; Desplan et al., 1985).

Regulation of *eve* and *en* by the *eve* Protein

eve$^+$ gene activity is required for the correct expression of *eve* and *en*. There is a failure to correctly initiate *en*$^+$ expression in *eve*$^-$ embryos, suggesting that *eve* in some way exerts a positive effect on *en* (see Fig. 1). The timing and sites of *eve* and *en* expression during wild-type development are consistent with the possibility that this regulatory effect might be direct. There is only a short lag, less than 10–20 min, between the time eve^+ proteins first appear as a series of stripes in wild-type embryos and the time *en* products are first detected (DiNardo *et al.*, 1985; Frasch *et al.*, 1987). Furthermore, there is a close spatial linkage in the wild-type *eve* and *en* expression patterns, in that *en* products accumulate in cells which comprise the anterior boundary of each *eve* stripe (Fig. 1a,b). There is evidence that *eve* negatively regulates its own expression. Weak *eve* mutants (i.e., hypomorphs) overexpress *eve* products, in that they persist for longer periods during germ band elongation as compared with wild type (M. Frasch, unpublished results; Warrior *et al.*, in preparation).

The identification of high-affinity *eve* binding sites within cloned *en* and *eve* DNA fragments is consistent with the possibility that the *eve* protein functions as a specific transcription factor in the regulation of these genes. The affinities that are calculated for several of the *eve* binding sites, particularly those located in the 5' *en* fragment K, are similar to those observed for bacterial DNA-binding proteins (Ptashne *et al.*, 1980; Fried and Crothers, 1984). However, we have not yet established a direct link between the binding of *eve* protein to *eve* and *en* DNA sequences and the expression of these genes *in vivo*. For example, it is anticipated that an *eve* promoter which lacks the *eve* binding sites at -3 kb (see Fig. 6) would direct an abnormal *eve* expression pattern, similar to that observed in *eve* mutants. P element mediated-transformation studies are currently underway to test this, and other, predictions. Evidence that the *eve* binding sites observed *in vitro* might be important for the regulation of *en* expression *in vivo* stems from the observation that many of these binding sequences are evolutionarily conserved within the *en* transcription unit. For example, even though the first *en* intron shows strong sequence divergence between *D. melanogaster* and *D. virilis*, a 26-bp sequence which includes the f2 binding site is perfectly conserved (Kassis *et al.*, 1986) (see Figs. 3 and 5).

There are several potential problems with the estimates of the *eve* binding constants. First, it is possible that these values represent underestimates since enriched extracts, rather than pure proteins, were used for binding. In order to reduce background, it was necessary to use fairly stringent conditions of binding, including relatively high concentrations of salt and competitor DNA. Less stringent binding conditions would probably give higher binding values.

A second potential problem concerns the use of *eve* protein extracts that were denatured and then renatured prior to their use in the binding assays. Although the *eve* and *en* proteins are quite soluble and do not require extraction with denaturants, such treatment was necessary to obtain sufficient solubilization of *zen* and *Dfd* for use in the DNase I footprint assays (note that the immunoprecipitation assays shown in Figs. 3 and 6 involved the use of undenatured *eve* proteins). The footprint experiments presented here involved the use of *eve* and *en* proteins that were denatured so their binding behaviors could be directly compared with *zen* and *Dfd*. It is possible that the native binding activity of these proteins is not fully restored upon removal of the denaturant (see Materials and Methods section). To test this possibility, we compared the binding of undenatured *eve* extracts with the guanidine-treated extracts used in this study. Undenatured *eve* protein displays a binding activity with 5' *en* fragment K sequences that is not detectably different from the denatured–renatured extracts used in this study (data not shown). Thus, it does not appear that the use of denatured–renatured proteins has strongly biased our binding estimates.

A third potential problem involves the use of bacterial proteins. It is possible that the *eve* protein produced in *E. coli* does not function in a manner analogous to the native protein in *Drosophila*. For example, the native protein might possess modifications (e.g., phosphate groups) that are not elaborated in bacterial hosts. It is unclear to what extent such modifications might influence the binding activities of homeo box proteins *in vivo*.

Mechanism of Binding

The *eve* protein was found to bind to a broad spectrum of sequences (see Fig. 8). Nonetheless, sequences that display the highest binding affinities contain at least one copy of a 10-bp consensus sequence: T-C-A-A-T-T-A-A-A-T. A similar consensus binding sequence has been determined for *en*-β-galactosidase fusion proteins (C. Desplan and P. O'Farrell, personal communication). The *en* k1 site contains 3 tandem copies of this sequence and shows the tightest binding with the *eve* protein. The close linkage of these binding sites is consistent with the possibility that the *eve* protein binds as a multimer. The *en* f1 and f3 sites do not contain well-conserved copies of the consensus sequence, and each shows only one-thirtieth the affinity for the *eve* protein as compared with the *en* k1 site. Some of the binding sites show a dyad of symmetry in that they contain a copy of the consensus sequence next to an inverted repeat. This provides further support for the possibility that the most efficient binding of *eve* involves a mulimeric form of the protein. It has been proposed that the homeo domain mediates binding through an α-helix–turn–α-helix mechanism (Laughon and Scott, 1984), which is observed for a number of bacterial DNA-binding proteins such as cro and λ repressor (Anderson *et al.*, 1981, 1982; Pabo and Lewis, 1982). In general, such bacterial proteins bind as dimers, and their target sequences contain two copies of an inverted repeat (Johnson et al., 1981). A similar mechanism of binding has been proposed for the yeast mating type protein α_2 (Miller et al., 1985; Johnson and Herskowitz, 1985; Hall and Johnson, 1987), which shares substantial homology with the carboxy-terminal one-third of the homeo domain (Laughon and Scott, 1984; Shepherd et al., 1984) and includes the putative recognition helix. There is evidence that α_2 might bind as a multimer (Miller et al., 1985).

The occurrence of DNase I hypersensitivity between adjacent eve binding sites lends further support to the possibility that eve binding might involve protein–protein interactions. Such hypersensitive sites are best seen in the en intron fragment F (Fig. 5). The sequence of the f1 binding site is quite divergent from the other eve binding sequences (see Fig. 8) and is protected only by high concentrations of the eve protein. In contrast, the f2 binding sequence is similar to high-affinity binding sequences present in fragment K and in the 5' region of eve. It is possible that binding to f1 involves a cooperative interaction with eve protein at the f2 site. There are 44 bp between the borders of f1 and f2, which corresponds to four helical turns of DNA. This region becomes hypersensitive as the f1 and f2 sites become occupied by eve protein. This hypersensitivity might result from the looping-out of the DNA between the f1 and f2 sites. A similar type of hypersensitivity was seen when adjacent binding sites for λ repressor were separated by insertion of variable lengths of DNA (Hochschild and Ptashne, 1986).

Binding of *eve, zen, en,* **and** *Dfd* **Proteins**

The *eve, en, zen,* and *Dfd* proteins each bind to the k1, k2, and k3 sites of *en* (see map, Fig. 3b). These proteins are related only by virtue of homeo box homology, and they do not share additional homologies (Poole et al., 1985; Macdonald et al., 1986; Frasch et al., 1987; Regulski et al., 1987; Rushlow et al., 1987a). Thus, it would appear that the DNA-binding activities of these proteins are mediated by their homeo domains. However, the homeo domains of these proteins are highly divergent and represent different classes of homeo box genes in *Drosophila*. There is no clear relationship between the binding behavior of these different homeo box proteins to the 5' *en* fragment K and the primary sequence of their homeo domains. According to the helix–turn–helix model for DNA binding, the most critical region of the homeo domain that determines binding specificity should correspond to the 8 amino acid residues that comprise the putative recognition helix (residues 42–49 of the homeo domain). Among all pair-wise comparisons of the four proteins that we have examined, the *zen* and *Dfd* homeo domains are the most closely related within this region and share 7 of the 8 residues. However, *zen* and *Dfd* are no more similar in their binding activity to fragment K than they are to *eve* or *en*. For example, the binding of *zen* to k1 is both qualitatively and quantitatively more like that of *eve* than *Dfd*. *eve* and *zen* share only 5 of 8 residues in the putative recognition helix, yet bind with nearly the same affinity to k1; *Dfd* shows approximately 40 times lower affinity to k1 than either *zen* or *eve*. This suggests that additional regions of the homeo domain are important for determining binding specificity, and it calls into question the validity of a strict helix–turn–helix mechanism of DNA binding.

There is a possible correlation between the homeo box proteins that show the highest affinity for *en* fragment K *in vitro* and the regulatory interactions which are observed *in vivo*. Among the four proteins that were assayed for DNA binding, genetic studies on *in vivo* interactions suggest that *eve* is the most likely to exert a regulatory effect on *en* expression (Harding *et al.*, 1986; Macdonald *et al.*, 1986). Consistent with this prediction is the finding that the *eve* protein shows the strongest overall binding to fragment K (i.e., k1 + k2 + k3). There is no current evidence suggesting that *en* or *Dfd* regulate *en* expression *in vivo*, and these proteins were found to possess the lowest affinities for fragment K *in vitro*. The *zen* protein shows intermediate binding to fragment K, and there is some evidence that *zen* might influence *en* expression *in vivo*. en^+ products appear to be overexpressed in dorsal regions of zen^- embryos (C. Rushlow and M. Levine, unpublished results). Despite this possible correlation between *in vitro* binding affinities and genetic circuitry, it should be cautioned that other factors, in addition to absolute affinity, might determine whether a given homeo box protein will regulate a particular target gene *in vivo*.

There are at least two possible mechanisms for cross-regulatory interactions among homeo box genes. First, perhaps different homeo box proteins recognize qualitatively different DNA sequences. The "on/off" condition of a particular target gene could depend on what combination of different homeo box proteins is bound to the diverse cis-regulatory sequences associated with that gene. Alternatively, it is possible that all homeo box proteins bind to the same limited set of DNA-binding sequences with different affinities. In this latter model different homeo box proteins "compete" for binding to a specific cis-regulatory element, and the "on/off" condition of a target gene depends on which homeo box protein has the highest affinity (Harding *et al.*, 1985; Desplan *et al.*, 1985). In their extreme forms, both of these models are probably incorrect; it is possible that a combination of the two mechanisms is actually used *in vivo*. Nonetheless, the demonstration that *eve*, *en*, *zen*, and *Dfd* can all bind to similar sequences with diverse affinities favors the occurrence of a competition mechanism.

MATERIALS AND METHODS

Plasmid Constructions

The T7 expression vector pAr3040 contains the promoter, initiating methionine, and the first 11 codons of the T7 gene, $\phi10$ (Studier and Moffatt, 1986). There is a unique *Nde*I site at the T7 initiating methionine, which was used to insert the full-length *eve*, *zen*, *en*, and *Dfd* coding sequences. In vitro mutagenesis was done to create an *Nde*I site at the initiating methionine of each of these homeo box genes. Mutagenic oligonucleotides were prepared for converting the nucleotide sequences immediately 5′ to the initiating A-T-G to the *Nde*I recognition sequence C-A-T-A-T-G. Mutagenic primers of about 20 nucleotides were annealed to M13 DNA templates that contained one of the full-length homeo box protein coding sequences. For *en*, the M13 templates were prepared in an ung^-, dut^- mutant strain of *E. coli* in order to introduce several uracil residues in place of thymidine (Kunkel, 1985). After annealing each of the templates to a mutagenic primer, double-stranded DNA was prepared by extension with Klenow according to the method of Zoller and Smith (1984). RF DNA was prepared from mutant phage, and cloned into the pAR3040 expression vector.

Preparation of Protein Extracts

Homeo box proteins were expressed in the bacterial strain BL21(DE3), which contains a single copy of the T7 RNA polymerase gene under the control of the inducible *lac* UV5 promoter (Studier and Moffatt, 1986). Cells transformed with pAR3040/homeo box gene recombinant plasmids were grown at 37°C in 2× YT media supplemented with 200 μg/ml ampicillin and 0.4% glucose. The cells were grown to a density (A_{600}) of 0.5, IPTG was added to a final concentration of 0.4 mM, and the cells were grown for an additional 2 hr.

The following steps were done on ice or at 4°C. Induced cells were harvested by centrifugation and resuspended in 1/200 volume of buffer Z [100 mM KCl; 25 mM HEPES, pH 7.8; 12.5 mM $MgCl_2$; 1 mM dithiothreitol (DTT); 0.1% Nonidet P-40 (NP-40) (Sigma); 20% glycerol] plus protease inhibitors [1 mM phenylmethylsulfonyl fluoride (PMSF) (Sigma)); 2 mM benzamidine (Sigma); 5 μg/ml leupeptin (Sigma); 5 μg/ml pepstatin A (Sigma)] and 0.5 mg/ml lysozyme. Cells were incubated on ice for 15 min and then lysed by sonicating 2 times, 15 sec each, at setting 2 with a Branson sonifier. The lysate was centrifuged at 15,000 rpm for 10 min. The supernatant was recovered and stored at −70°C. The pellet was resuspended in buffer Z plus 4 M guanidine-HCl and incubated for 30 min. After solubilization of the pellet, the denaturant was removed by dialysis against buffer Z plus 1 M guanidine-HCl, followed by dialysis against buffer Z without guanidine-HCl. Insoluble material was removed by centrifugation at 15,000 rpm for 2–5 min. DNA-binding experiments were done with the original supernatant obtained from induced cells, and with the guanidine-HCl extracted protein. Similar results were obtained with these two protein preparations.

Immunoprecipitation Assays

Immunoprecipitation reactions were done essentially as described by Desplan et al. (1985). Binding reactions were done with 40 ng of ^{32}P-labeled DNA and 1 µg of induced protein extract (undenatured) in a final volume of 50 µl for 30 min on ice. The final concentrations of binding buffer components were 170 mM NaCl, 20 mM Tris (pH 7.6), 250 µM EDTA, 1 mM DTT, 10% glycerol, and 100 µg/ml sonicated calf thymus DNA. After binding, DNA–protein complexes were immunoprecipitated by addition of *Staphylococcus aureus* cells (Johnson and Herskowitz, 1985) coupled to anti-*eve* antibody (see below). The mixture was incubated for an additional 30 min on ice. The *S. aureus* cells were pelleted by centrifugation and washed 2 times with binding buffer plus 0.5% NP-40. The *S. aureus* cells were resuspended in 200 µl TE (pH 7.5), and the DNA was recovered by phenol extraction and ethanol precipitation. The samples were electrophoresed in 5% polyacrylamide/7.5 M urea gels.

Staphylococcus aureus cells were prepared for immunoprecipitation by washing 200 µl of the extract provided by the manufacturer (Pansorbin, CalBiochem) in 500 µl of binding buffer. The cells were mixed with 1 ml of rabbit serum containing anti-*eve* antibody and incubated at 4°C with gentle shaking for 1 hr. The cells were pelleted, washed in 500 µl of binding buffer, and then resuspended in 200 µl of binding buffer.

Footprint Assays

DNase I protection assays were done essentially as described by Heberlein et al. (1985). Binding reactions were done with 2 to 5 ng of ^{32}P-labeled DNA and 5 µg of sonicated calf thymus DNA in 50 µl for 30–45 min on ice. The concentrations of binding buffer components were 110 mM KCl, 47.5 mM HEPES (pH 7.8), 13.75 mM MgCl$_2$, 50 µM EDTA, 1 mM DTT, 17% glycerol, and 0.05% NP-40. After binding, 50 µl of 10 mM MgCl$_2$, 5 mM CaCl$_2$ was added to the reaction, followed by 5 µl of freshly diluted DNase I (Worthington) at a final concentration of 10 µg/ml. DNase I digestion was done for 5 min on ice. The reaction was stopped by addition of 90 μl of 1% SDS, 20 mM EDTA, 200 mM KCl, 250 μg/ml yeast tRNA. The samples were extracted 2 times with phenol–chloroform (1:1), ethanol precipitated, and electrophoresed in 6–10% polyacrylamide/7.5 M urea gels.

DNA fragments used for footprint assays were prepared by labeling with [γ-^{32}P]ATP and polynucleotide kinase or with [α-^{32}P]dCTP and Klenow polymerase. Labeled fragments were purified after fractionation in 5% polyacrylamide gels. Aliquots of the labeled DNA were used for the binding experiments or for sequencing by the chemical cleavage method (Maxam and Gilbert, 1980).

ACKNOWLEDGMENTS

We thank Mike Regulski and Bill McGinnis for the *Dfd*/T7 plasmid, and for allowing us to use the *Dfd* protein for binding assays. We thank Chris Doe and Tom Kornberg for the monoclonal anti-*en* antibody, and Claude Desplan and Pat O'Farrell for cloned *en* DNAs. We also thank C. Desplan for advice and encouragement. We are grateful to Dr. L. Chasin for his help with the densitometric scans. M.F. was supported by a fellowship from the Deutsche Forschungsgemeinschaft. This work was funded by grants from the National Institutes of Health and the Searle Scholars Program.

REFERENCES

Akam, M. E., and Martinez-Arias, A. (1985). *EMBO J.* **4**, 1689–1700.
Anderson, W. F., Ohlendorf, D. H., Takeda, Y., and Matthews, B. W. (1981). *Nature (London)* **290**, 754–758.
Anderson, W. F., Takeda, Y., Ohlendorf, D. H., and Matthews, B. W. (1982). *J. Mol. Biol.* **159**, 745–751.
Carroll, S. B., and Scott, M. P. (1985). *Cell (Cambridge, Mass.)* **43**, 113–126.
Chadwick, R., and McGinnis, W. (1987). *EMBO J.* **6**, 778–786.
Desplan, C., Theis, J., and O'Farrell, P. H. (1985). *Nature (London)* **318**, 630–635.
DiNardo, S., Kuner, J. M., Theis, J., and O'Farrell, P. H. (1985). *Cell (Cambridge, Mass.)* **43**, 59–69.
Doyle, H. J., Harding, K., Hoey, T., and Levine, M. (1986). *Nature (London)* **323**, 76–79.
Fjose, A., McGinnis, W. J., and Gehring, W. J. (1985). *Nature (London)* **313**, 284–289.
Frasch, M., Hoey, T., Rushlow, C., Doyle, H., and Levine, M. (1987). *EMBO J.* **6**, 749–759.
Fried, M. G., and Crothers, D. M. (1984). *J. Mol. Biol.* **172**, 241–262.
Galas, D., and Schmitz, A. (1978). *Nucleic Acids Res.* **5**, 3157–3170.
Garcia-Bellido, A. (1977). *Am. Zool.* **17**, 613–629.
Hafen, E., Kuroiwa, A., and Gehring, W. J. (1984). *Cell (Cambridge, Mass.)* **37**, 833–841.
Hall, M. N., and Johnson, A. D. (1987). *Science* **237**, 1007–1012.
Harding, K., Wedeen, C., McGinnis, W., and Levine, M. (1985). *Science* **229**, 1236–1242.
Harding, K., Rushlow, C., Doyle, H. J., Hoey, T., and Levine, M. (1986). *Science* **233**, 953–959.

Heberlein, U., England, B., and Tjian, R. (1985). *Cell (Cambridge, Mass.)* **41**, 965–977.
Hochschild, A., and Ptashne, M. (1986). *Cell (Cambridge, Mass.)* **44**, 681–687.
Ingham, P. W., and Martinez-Arias, A. (1986). *Nature (London)* **324**, 592–597.
Johnson, A. D., and Herskowitz, I. (1985). *Cell (Cambridge, Mass.)* **42**, 237–247.
Johnson, A. D., Poteete, A. R., Lauer, G., Sauer, R. T., Ackers, G. K., and Ptashne, M. (1981). *Nature (London)* **294**, 217–223.
Kassis, J. A., Poole, S. J., Wright, D. K., and O'Farrell, P. H. (1986). *EMBO J.* **5**, 3583–3589.
Kornberg, T., Siden, I., O'Farrell, P. H., and Simon, M. (1985). *Cell (Cambridge, Mass.)* **40**, 45–53.
Kuner, J. M., Nakanishi, M., Ali, Z., Drees, B., Gustavson, E., Theis, J., Kauvar, L., Kornberg, T., and O'Farrell, P. H. (1985). *Cell (Cambridge, Mass.)* **42**, 309–316.
Kunkel, T. A. (1985). *Proc. Natl. Acad. Sci. U.S.A.* **82**, 488–492.
Laughon, A., and Scott, M. P. (1984). *Nature (London)* **310**, 25–31.
Levine, M., and Harding, K. (1987). *Oxford Survey Eukaryotic Genes* **5** (in press).
Levine, M., Hafen, E., Garber, R. L., and Gehring, W. J. (1983). *EMBO J.* **2**, 2037–2046.
Macdonald, P. M., Ingham, P. W., and Struhl, G. (1986). *Cell (Cambridge, Mass.)* **47**, 721–734.
McGinnis, W., Levine, M. S., Hafen, E., Kuroiwa, A., and Gehring, W. J. (1984). *Nature (London)* **308**, 428–433.
McKay, R. (1981). *J. Mol. Biol.* **145**, 471–488.
Maxam, A., and Gilbert, W. (1980). *In* "Methods in Enzymology" (L. Grossman and K. Moldave, eds.), Vol. 65, pp. 499–560. Academic Press, New York.
Miller, A. M., MacKay, V. L., and Nasmyth, K. (1985). *Nature (London)* **314**, 598–603.
Nusslein-Volhard, C., Kluding, H., and Jurgens, G. (1985). *Cold Spring Harbor Symp. Quant. Biol.* **50**, 145–154.
Pabo, C. O., and Lewis, M. (1982). *Nature (London)* **298**, 443–447.
Poole, S. J., Kauvar, L. M., Drees, B., and Kornberg, T. (1985). *Cell (Cambridge, Mass.)* **40**, 37–43.
Ptashne, M., Jeffrey, A., Johnson, A. D., Maurer, R., Meyer, B. J., Pabo, C. O., Roberts, T. M., and Sauer, R. T. (1980). *Cell (Cambridge, Mass.)* **19**, 1–11.
Regulski, M., McGinnis, N., Chadwick, R., and McGinnis, W. (1987). *EMBO J.* **6**, 767–777.
Rushlow, C., Harding, K., and Levine, M. (1987a). "Hierarchical Interactions among Pattern Forming Genes in *Drosophila*," in press. Banbury Conference on Developmental Toxicology, Cold Spring Harbor Laboratory, Cold Spring Harbor, New York.
Rushlow, C., Doyle, H., Hoey, T., and Levine, M. (1987b). *Genes & Dev.* (submitted).
Scott, M. P., and O'Farrell, P. H. (1986). *Annu. Rev. Cell Biol.* **2**, 49–80.
Scott, M. P., and Weiner, A. J. (1984). *Proc. Natl. Acad. Sci. U.S.A.* **81**, 4115–4118.
Shepherd, J. C. W., McGinnis, W., Carrasco, A. E., DeRobertis, E. M., and Gehring, W. J. (1984). *Nature (London)* **310**, 70–71.
Studier, F. W., and Moffatt, B. A. (1986). *J. Mol. Biol.* **189**, 113–130.
Wakimoto, B. T., Turner, F. R., and Kaufman, T. C. (1984). *Dev. Biol.* **102**, 147–172.
Zoller, M. J., and Smith, M. (1984). *DNA* **3**, 479–488.

22
Identification and Characterization of Two Mouse Genes, *En-1* and *En-2*, with Sequence Homology to the *engrailed* and *invected* Genes of *Drosophila*

GAIL R. MARTIN AND ALEXANDRA L. JOYNER[1]
Department of Anatomy
University of California, San Francisco
San Francisco, California 94143

INTRODUCTION

The classic genetic approach to understanding how development is controlled is to first identify mutations that perturb the process and then to use information gained from study of the phenotypes of mutant embryos to determine the functions of the genes represented by the mutant alleles. In organisms that are particularly well suited to genetic investigation, such as the invertebrate species *Drosophila melanogaster*, this approach has been particularly successful. The coupling of classic genetic techniques with recombinant DNA technology has resulted in a rapidly accelerating understanding of how development proceeds in *Drosophila*. Basically, two sets of

[1]. Present address: Samuel Lunenfeld Research Institute, Division of Molecular and Developmental Biology, Mt. Sinai Hospital, 600 University Avenue, Toronto, Canada M5G 1X5.

genes function, one during oocyte maturation and the other during early embryogenesis, to establish a body plan composed of repeating segmental units. Subsequently, a third set of genes, the homeotic genes, function to provide each segment with a specific identity (1–5). A particularly intriguing finding is that a substantial proportion of these *Drosophila* genes have been found to contain a common 180-base pair (bp) sequence known as the homeo box (5–7).

Attempts to identify genes that regulate mammalian development by this classic approach have been hindered by the difficulty of identifying mutations that affect early embryonic development. Mammalian embryos, which are obtainable in relatively small numbers, have long gestation periods, and are inaccessible during development *in utero*, are poorly suited to the type of genetic analysis that has been successful in *Drosophila*. Moreover, there is no straightforward means of isolating the mammalian genes represented by mutant alleles that do perturb development, except in those cases in which the mutation has been caused by insertion of foreign DNA (8–10). However, recombinant DNA technology has made possible an alternative approach, known as "reverse genetics." Beginning with a cloned gene, it is possible to cause abnormal expression of that gene and potentially to use information gained from a study of the consequences of that perturbation to determine the function of the gene. In applying "reverse genetics" to the problem of development in mammals, the fundamental question is how to identify genes for further study that are likely to play significant roles in the control of developmental processes.

One means of identifying such genes is based on the assumption that there has been conservation during evolution of the basic mechanisms of embryonic development, and of the genes that control the process. Thus, it is thought that in vertebrates one set of genes, similar to those identified in *Drosophila*, functions to establish a basic body plan composed of repeating units, and another set of genes subsequently acts to specialize and diversify the cells in each of those units. If this premise is correct, it should theoretically be possible to isolate such genes from the mammalian genome by virtue of their sequence homology with the genes that control development in flies. Although the validity of the fundamental premise that in the early vertebrate embryo there exist repeating units that are functionally homologous to the segmental units of the early *Drosophila* embryo is still hotly debated (11), there is some experimental evidence in support of the idea that the genes which control development in flies may have been conserved during the evolution of vertebrates. This comes from the finding that the homeo box, now known to be present in more than 20 *Drosophila* genes that control pattern formation during embryogenesis (5), has been found to be highly conserved in numerous genes in vertebrates (12,13).

The specific approach we took to identifying genes that might control developmental processes in mammals was to ask whether the *engrailed (en)* gene of *Drosophila* has been conserved during mammalian evolution. This gene, first identified by traditional mutational analysis, is involved in the processes that subdivide the insect embryo into separate developmental units called compartments. Each segment of the fly is composed of an anterior and a posterior compartment (14); in mutants that lack *engrailed* function, posterior compartments develop abnormal patterns, and the compartment and segment borders are not maintained (15–17). The discovery that only posterior compartment cells are affected by mutations at the *en* locus has suggested that this gene functions as a binary developmental switch to "select" a posterior developmental pathway (18).

Studies of genomic and cDNA clones of the *en* gene indicate that the major embryonic transcript contains a 1700-bp open reading frame that includes a homeo box at its 3' end (19–21). *In situ* RNA hybridization (21–24) and antibody localization of the *en* protein product (25) show that *en* is expressed only in the cells of the posterior compartments during the early stages of embryonic development. These results are consistent with the view of the function of the *engrailed* locus as deduced from mutational analysis.

Molecular studies have also led to the identification of a gene adjacent to *en*, now designated *invected (inv)*, which has extensive sequence homology with it (20). Sequence comparisons of *en* and *inv* cDNA clones (26) have revealed two regions of homology between the genes. One region includes a stretch that encodes 17 contiguous amino acids that are identical in the two genes. This region is separated, by sequences encoding 5 (*en*) or 26 (*inv*) nonconserved amino acids, from a downstream homeo box that is very similar in *en* and *inv* but quite divergent from the homeo boxes found in other *Drosophila* genes. Additional homologous sequences encoding 31 amino acids are found immediately downstream of the homeo box. *In situ* RNA hybridization analysis of *inv* expression in developing *Drosophila* embryos has shown that, like *en*, early in development *inv* is expressed selectively in the cells of the posterior compartments (26). These results suggest that the *en* and *inv* genes have related functions during embryogenesis. As yet, however, no mutations of *inv* are known that would suggest its role in development.

We present here a description of the studies in which we identified and isolated two mouse genes with significant sequence homology to the *engrailed* and *invected* genes of *Drosophila*, determined their stage and tissue specificities of expression during mouse embryogenesis, and mapped them in the mouse genome. These results have been previously described by Joyner and co-workers in Refs. 27 and 28.

RESULTS

Isolation of Mouse Genomic and cDNA Clones with Sequence Homology to the *Drosophila* Engrailed and Invected Genes

To investigate the possibility that the mouse genome contains sequences homologous to those in the *Drosophila en* gene, Southern blot hybridization analysis of mouse genomic DNA digested separately with four different restriction endonucleases was carried out using a 1.4-kilobase (kb) probe derived from a *Drosophila en* cDNA clone. Under low stringency conditions this probe hybridized to at least two restriction fragment bands in each

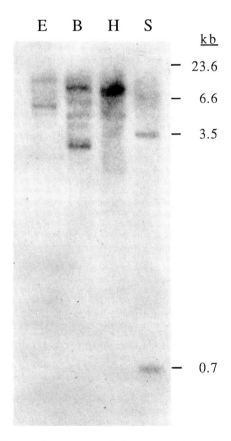

Fig. 1. Mouse DNA contains two genomic fragments with homology to a *Drosophila* engrailed cDNA probe. Southern blots of mouse genomic DNA (10 µg) digested with various restriction endonucleases (E, *Eco*RI; B, *Bam*HI; H, *Hin*dIII; and S, *Sst*I) were hybridized under conditions of low stringency to a 1.4-kb *Drosophila engrailed* cDNA probe (27).

digest of mouse DNA (Fig. 1). This suggested that the mouse genome contains two different genes with sequence homology to some portion of the *Drosophila en* gene.

These two genes, termed *En-1* and *En-2*, were isolated from a mouse genomic library (27,28) and their sequence homology with the *Drosophila en* gene confirmed (see below). cDNA clones representing both *En-1* and *En-2* mRNAs were obtained from a 12.5-day mouse embryo cDNA library (28). Figure 2 shows partial restriction maps of the *En-1* and *En-2* genomic DNAs that were isolated and illustrates the relationship, as far as it is known at

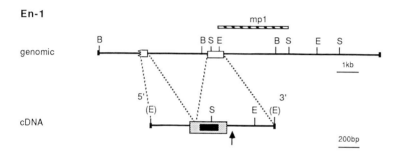

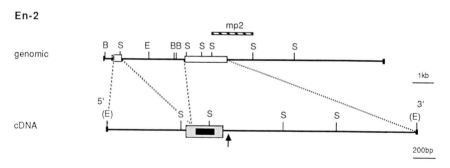

Fig. 2. Restriction maps of *En-1* and *En-2* genomic and cDNA clones. Partial restriction maps of the cloned *En-1* and *En-2* genomic regions are shown. The open boxes represent the known positions of exons. The hatched bars marked mp1 and mp2 represent the subcloned fragments used as probes for the mapping studies. In the illustration of each cDNA clone, the horizontal line indicates the DNA insert, the ends of which contain synthetic *Eco*RI sites (E). The stippled box represents the presently known limits of the "conserved" region, which is homologous to sequences in the *Drosophila engrailed* and *invected* genes; this region includes the homeo box, which is depicted as a filled rectangle. In each cDNA, the vertical arrow indicates the end of the open reading frame that includes the homeo box. The direction of transcription of both genes was determined as previously described (see Ref. 28). *Bam*HI; H, *Hind*III; S, *Sst*I; E, *Eco*RI.

present, of the longest *En-1* (1.2 kb) and *En-2* (3 kb) cDNA clones available to the genomic DNAs from which they were ultimately derived.

The region in each of the two mouse genes that shares significant sequence homology with the *Drosophila en* gene (see Fig. 2) was localized by hybridization analysis and subsequently sequenced (27,28). The data show that in the two mouse genes these regions of homology with *Drosophila en* are remarkably similar [295/375 nucleotides (79%) are identical]. Moreover, the sequences that have been conserved between the two mouse genes and the *Drosophila en* gene are the same as those that have been conserved between *engrailed* and its near neighbor in the *Drosophila* genome, *invected*. In each gene the sequence of this region constitutes an open reading frame that contains a homeo box and ends with a signal for the termination of protein synthesis.

Although the homology between the two mouse genes and the two *Drosophila* genes is readily apparent from nucleotide sequence comparisons, it is even more striking when the deduced amino acid sequences are compared. The 124 amino acids that can be encoded by the *En-1* and *En-2* conserved regions are shown in comparison with the corresponding sequences of the *Drosophila en* and *inv* genes in Fig. 3. Although the four sequences are not perfectly colinear, they can be aligned by making certain adjustments (see legend to Fig. 3). It is obvious from this comparison that there is a region which is remarkably conserved between the two mouse and two *Drosophila* genes extending over most of the sequences shown. The conserved region begins with a stretch of 17 contiguous amino acids that are identical in all four genes (positions numbered 4–20 in Fig. 3). Each of the four genes contains an intron that interrupts the nucleotide sequence coding for the last of these 17 amino acids (position 20). The first 9 of the next 13 amino acids in the sequence (positions 21–29) represent a conserved domain in which 7 are identical in *En-1* and *En-2*, and, of these 7, 4 are also identical in *en* and 6 are identical in *inv*. The remaining 4 amino acids in this stretch (positions 30–33) represent a region in which the most variation occurs in the four genes; the two *Drosophila* genes each contain additional amino acids in this region. The next 60 amino acids in the sequence, delineated by a box in Fig. 3, constitute the homeo box in which 56 of 60 amino acids (93%) are identical in the two mouse genes and 42 of 60 residues (70%) are identical in all four genes. In the two *Drosophila* genes, but not the mouse genes, the homeo box is interrupted by an intervening sequence. Immediately downstream of the homeo box there is a region (amino acids 94–114) in which 18 of 21 amino acids (86%) are identical in the two mouse genes and 15 of 21 (71%) are identical in all four genes. Although the predicted amino acid sequences of *en* and *inv* are homologous for an additional 10 amino acids, these residues are not conserved in either mouse gene.

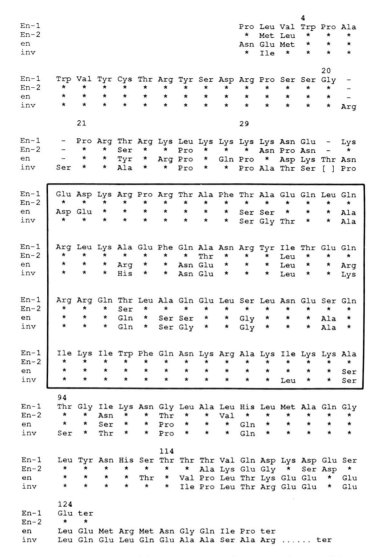

Fig. 3. Sequence comparison of the region conserved in *En-1* and *En-2* and the *Drosophila engrailed (en)* and *invected (inv)* genes. The predicted sequences were aligned to achieve the best match. To accomplish this a space, denoted by a dash (–), was inserted between amino acids that are actually contiguous. In addition, 22 amino acids in the *inv* sequence, denoted by brackets [ ], are not shown. The dots represent 6 amino acids in the *inv* sequence immediately upstream from the stop codon. The homeo box is delineated by a box. Asterisks denote amino acids in the *En-2*, *en* or *inv* sequences that are identical to the amino acid in the corresponding position in the *En-1* sequence. The *en* and *inv* sequences were determined by Poole *et al.* (20) and Coleman *et al.* (26).

Stage and Tissue Specificity of *En-1* and *En-2* Expression during Embryogenesis

In order to determine whether *En-1* and *En-2*, like their *Drosophila* counterparts, are expressed during embryonic development, we carried out a Northern blot hybridization analysis of poly(A)$^+$ RNA from PSA-1 teratocarcinoma cells, which serve as an *in vitro* model system for the mouse embryo at the peri-implantation stages of development (3.5 to 6.5 days of gestation; Ref. 29), and from whole mouse embryos at the mid- and late-gestation stages of development. In RNA from undifferentiated teratocarcinoma stem cells a relatively low abundance 1.8-kb *En-1* transcript was detected. Following differentiation, an additional, larger (2.8 kb) *En-1* transcript became detectable. Both transcripts were observed in RNA from embryos at 9.5 through 17.5 days of gestation. However, whereas the less abundant, smaller transcript appeared to be expressed at relatively constant levels in all samples tested, the levels of the more abundant, 2.8-kb transcript appeared to peak during the mid-gestation stages of development (Fig. 4, top).

In contrast to the pattern observed for *En-1*, a single 3.7-kb *En-2* RNA was readily detected at approximately equivalent levels in all teratocarcinoma samples tested (Fig. 4, bottom). In samples of poly(A)$^+$ RNA from embryos, the pattern of *En-2* expression was found to be more complex: the 3.7-kb *En-2* RNA was detected in all embryo samples tested, but two smaller transcripts, approximately 2.5 and 1.8 kb in length, became detectable by 15.5 days and were found at relatively high levels in the samples from embryos at 16.5 and 17.5 days of gestation. The increase in abundance of these two smaller transcripts coincided with a decrease in the abundance of the 3.7-kb transcript.

To examine the tissue specificity of *En-1* and *En-2* transcripts, conceptuses at 12.5 days of gestation were dissected into seven fetal and two extraembryonic parts, and total RNA was extracted from each sample. Northern blot hybridization analysis showed that the 2.8-kb *En-1* RNA was most abundant in the sample containing the posterior portion of the brain. This sample consisted primarily of midbrain, but also included cerebellum, pons, and medulla. As shown in Fig. 5, the 2.8-kb *En-1* RNA was approximately 3- to 5-fold more abundant in the posterior brain sample than in those from the anterior portion of the brain, the spinal cord, face, limbs, or carcass; samples derived from the viscera appeared to be completely negative. Although there were several faint bands of hybridization in the lanes containing RNA from yolk sac and from placenta, neither of these extraembryonic tissue samples contained detectable amounts of the 2.8-kb *En-1* RNA. The 1.8-kb

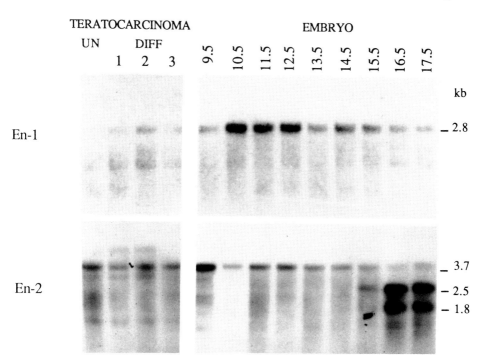

Fig. 4. Expression of *En-1* and *En-2* RNA during mouse embryogenesis. Northern blots of poly(A)+ RNA (5 μg) isolated from PSA-1 teratocarcinoma stem cells maintained in the undifferentiated state (un) and at three stages of differentiation (diff 1, 2, 3; see Ref. 27) and from whole embryos at the days of gestation indicated were analyzed by hybridization under conditions of high stringency to (top) a single-stranded 140-bp *En-1* homeo box-containing probe (see Ref. 27) or (bottom) a double-stranded 3-kb *En-2* cDNA probe (see Ref. 28). The results of rehybridizing the blot shown in at bottom with a probe for β-actin indicated that the apparent peak in abundance of the 3.7-kb *En-2* RNA at 9.5 days of gestation was probably an artifact due to differences in the amount of RNA present in each lane.

En-1 RNA seen in all poly(A)+ RNA samples from teratocarcinoma cells and embryos was not observed in any of the total RNA samples from fetal tissues, possibly because it is expressed in most tissues at levels too low to detect under the conditions we employed.

The data in Fig. 5 show that the 3.7-kb *En-2* transcript is also abundant in the RNA sample from the posterior portion of the brain and that no *En-2* RNA is detectable in the other samples. Abundant expression of both *En-1* and *En-2* in the posterior portion of the brain continues during development, since the 2.8-kb *En-1* and 3.7-kb *En-2* transcripts were also readily detectable in total RNA samples from this portion of the brain of 17.5 day

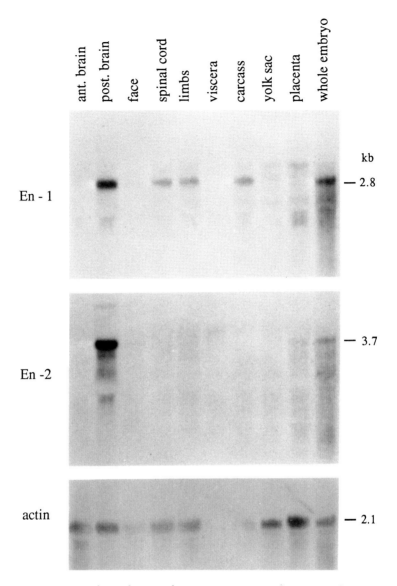

Fig. 5. Tissue specificity of *En-1* and *En-2* expression at mid-gestation. The tissues indicated were dissected from mouse fetuses at 12.5 days of gestation. The brain was removed from the head and divided into the anterior portion, containing the two frontal lobes, and the posterior portion consisting of midbrain, cerebellum, pons, and medulla; the remainder of the head constituted the "face" sample. The spinal cord sample was free of the spinal column but included the spinal ganglia. The "viscera" sample contained a pool of internal organs, and the "carcass" sample contained what remained of the fetus after the other tissues were removed. In addition, yolk sac and placenta samples were dissected from the conceptuses. A Northern blot of total RNA (15 μg) from each tissue was analyzed by reiterative hybridization and stripping using *En-1* and *En-2* cDNA insert probes and a probe for β-actin.

fetuses and of newborn mice; these transcripts were also detected in poly(A)$^+$ RNA from whole adult brains (data not shown).

Genetic Mapping of the *En-1* and *En-2* Loci

The approximate map positions of *En-1* and *En-2* were determined by means of recombinant inbred (RI) strain analyses using probes derived from *En-1* and *En-2* genomic DNA (mp1 and mp2, respectively, see Fig. 2). The data from such an analysis (28) indicated that *En-1* is located on mouse chromosome 1, in the vicinity of *Dominant hemimelia (Dh)*, a mutation that causes abnormalities in skeletal development and spleenlessness (30), and that *En-2* is located on mouse chromosome 5, in the vicinity of *Hemimelic extratoes (Hx)* and *Hammertoe (Hm)*, a pair of very closely linked mutations that cause similar abnormalities in skeletal development (30).

DISCUSSION

We have shown that the mouse has two genes, *En-1* and *En-2*, which have conserved the majority of the sequences that are homologous in the two *Drosophila* genes, *engrailed* and *invected*. The region of homology, containing a centrally located homeo box, can code for 107 amino acids (positions 4–29 and 34–114, Fig. 3), of which 78 (73%) are identical in all four genes. No other regions of significant homology have been found between the two *Drosophila* genes (26), and sufficient relevant data for the two mouse genes are not yet available. The conserved sequences represent a carboxy-terminal protein domain with a molecular mass of approximately 12,000 daltons. The existence of such extensive homology suggests that the conserved domain in the products of all four genes performs a similar biochemical function. Although this function may be related to that of other homeo box-containing genes, it is likely to have some unique features specified by the conserved sequences outside of the homeo box.

Although the two mouse genes might be the homologs of the *Drosophila engrailed* and *invected* genes with respect to their specific biochemical functions, it remains to be determined whether *En-1* and *En-2* play some role in the control of embryogenesis in mammals that is in any way analogous to the developmental functions of their counterparts in *Drosophila*. Our studies establish that both genes are expressed in PSA-1 teratocarcinoma cells maintained in the undifferentiated state and at three different stages of differentiation. Such undifferentiated cells are thought to be equivalent in many respects to the inner cell mass (ICM) cells of the preimplantation mouse blastocyst (3.5–4.5 days of gestation), and the differentiating cultures at stages 1, 2, and 3 appear to represent or have features in common with mouse embryos at stages prior to (4.5–6.5 days of gestation), during (6.5–8.5 days), and after (8.5–10.5 days) gastrulation, respectively (31–33). The data thus imply that both genes are expressed early in embryogenesis, during periods when fundamental cell lineage decisions are being made and when the basic embryo body plan is being established. The finding that, later in development, both *En-1* and *En-2* are expressed most abundantly in the posterior portion of the brain suggests that both genes have some function in the development of the central nervous system. Expression of both *engrailed* and *invected* has also been detected in the central nervous system of developing *Drosophila* embryos (23,24,26). Although it is an attractive idea that there are similarities in the developmental functions of the mouse and *Drosophila* genes, at present this is highly speculative. Some insight into this question may be gained from *in situ* RNA hybridization and protein localization studies of *En-1* and *En-2* products on sections of mouse embryos, which will serve to define the spatial distribution and cell type specificity of the expression of *En-1* and *En-2* during mouse embrygenesis.

The data reported here show that *En-1* and *En-2* are unlinked, whereas the *engrailed* and *invected* genes are adjacent to each other in the *Drosophila* genome (26). Thus, any functional significance in the close linkage of the *Drosophila* genes has not been conserved in the evolution of the mammalian *engrailed*-like genes. Both *En-1* and *En-2* map in the vicinity of mutations that are known to cause developmental abnormalities, but both *En-1* sequences on chromosome 1 and *En-2* sequences on chromosome 5 have now been separated by recombination from their respective nearby mutant loci, *Dominant hemimelia (Dh)* and *Hemimelic extratoes (Hx)* (data not shown). These results suggest, but do not conclusively demonstrate, that *Dh* and *Hx* are not mutant alleles of *En-1* or *En-2*, respectively. Further experiments aimed at determining the actual physical distance between the mammalian *engrailed*-like genes and their neighboring mutant loci are in progress, and it may yet be found that there is a relationship between *En-1* and *Dh*, and/or between *En-2* and *Hx/Hm*.

Now that we have succeeded in identifying *En-1* and *En-2* as candidates for genes that play a role in the control of mammalian development, we plan to carry out more detailed analyses of the cell and tissue specificities of their expression and to study the biochemical properties of their protein products. Moreover, we hope to realize the full potential of the reverse genetics approach by studying the consequences of perturbing the expression of these two genes during embryogenesis. The information gained from such studies should ultimately help to elucidate their functions in the developing mammalian embryo.

ACKNOWLEDGMENTS

This work was supported by funds from National Institutes of Health Grant HD-20959. A.L.J. was supported by a postdoctoral fellowship from the Medical Research Council of Canada. Figures 3 and 5 were previously published by Joyner and Martin (Ref. 28).

REFERENCES

1. Nusslein-Volhard, C. (1979). "Determinants of Spatial Organization" (S. Subtelny and I. R. Konigsberg, eds.), pp. 185–211. Academic Press, New York.
2. Nusslein-Volhard, C., and Weischaus, E. (1980). *Nature (London)* **287**, 795–801.
3. Lewis, E. B. (1978). *Nature (London)* **276**, 565–570.
4. Kaufman, T. C., Lewis, R., and Wakimoto, B. (1980). *Genetics* **94**, 115–133.
5. Gehring, W. J., and Hiromi, Y. (1986). *Annu. Rev. Genet.* **20**, 147–173.
6. McGinnis, W., Levine, M. S., Hafen, E., Kuriowa, A., and Gehring, W. J. (1984). *Nature (London)* **308**, 428–433.
7. Scott, M. P., and Weiner, A. J. (1984). *Proc. Natl. Acad. Sci. U.S.A.* **81**, 4115–4119.
8. Jaenisch, R., Harbers, K., Schnieke, A., Lohler, J., Chumakov, I., Jahner, D., Grotkopp, D., and Hoffman, E. (1983). *Cell (Cambridge, Mass.)* **32**, 209–216.
9. Wagner, E. F., Covarrubias, L., Stewart, T. A., and Mintz, B. (1983). *Cell (Cambridge, Mass.)* **35**, 647–655.
10. Woychik, R. P., Stewart, T. A., Davis., L. G., D'Eustachio, P., and Leder, P. (1985). *Nature (London)* **318**, 36–40.
11. Raff, E. C., and Raff, R. A., Brown, S. D. M., and Greenfield, A. J. (1985). *Nature (London)* **313**, 185–186.
12. McGinnis, W., Garber, R. L., Wirz, J., Kuroiwa, A., and Gehring, W. J. (1984). *Cell (Cambridge, Mass.)* **37**, 403–408.
13. Martin, G. R., Boncinelli, E., Duboule, D., Gruss, P., Jackson, I., Krumlauf, R., Lonai, P., McGinnis, W., Ruddle, F., and Wolgemuth, D. (1987). *Nature (London)* **325**, 221–222.
14. Garcia-Bellido, A., Ripoll, P., and Morata, G. (1973). *Nature (London) New Biol.* **245**, 251–253.
15. Lawrence, P. A., and Morata, G. (1976). *Dev. Biol.* **50**, 321–337.
16. Kornberg, T. (1981). *Proc. Natl. Acad. Sci. U.S.A.* **78**, 1095–1099.
17. Lawrence, P. A., and Struhl, G. (1982). *EMBO J.* **1**, 827–833.
18. Garcia-Bellido, A. (1975). *Ciba Found. Symp.* **29**, (new series), 161–182.
19. Kuner, J. M., Nakanishi, M., Ali, Z., Drees, B., Gustavson, E., Theis, J., Kauvar, L., Kornberg, T., and O'Farrell, P. H. (1985). *Cell (Cambridge, Mass.)* **42**, 309–316.
20. Poole, S. J., Kauvar, L. M., Drees, B., and Kornberg, T. (1985). *Cell (Cambridge, Mass.)* **40**, 37–43.
21. Fjose, A., McGinnis, W. J., and Gehring, W. J. (1985). *Nature (London)* **313**, 284–289.
22. Kornberg, T., Siden, I., O'Farrell, P., and Simon, M. (1985). *Cell (Cambridge, Mass.)* **40**, 45–53.
23. Ingham, P., Martinez-Arias, A., Lawrence, P., and Howard, K. (1985). *Nature (London)* **317**, 634–636.
24. Weir, M., and Kornberg, T. (1985). *Nature (London)* **318**, 433–439.

25. DiNardo, S., Kuner, J. M., Theis, J., and O'Farrell, P. H. (1985). *Cell (Cambridge, Mass.)* **43,** 56–69.
26. Coleman, K. G., Poole, S. J., Weir, M., Soeller, W. C., and Kornberg, T. (1987). *Genes & Dev.* **1,** 19–28.
27. Joyner, A. L., Kornberg, T., Coleman, K., Cox, D., and Martin, G. R. (1985). *Cell (Cambridge, Mass.)* **43,** 29–37.
28. Joyner, A. L., and Martin, G. R. (1987). *Genes & Dev.* **1,** 29–38.
29. Martin, G. R. (1980). *Science* **209,** 768–776.
30. Green, M. C., ed. (1981). "Genetic Variants and Strains of the Laboratory Mouse." Gustav Fischer, Stuttgart.
31. Martin, G. R., and Evans, M. J. (1975). *Proc. Natl. Acad. Sci. U.S.A.* **72,** 1441–1445.
32. Martin, G. R., and Evans, M. J. (1975). *Cell (Cambridge, Mass.)* **6,** 467–474.
33. Martin, G. R., Wiley, L. M., and Damjanov, I. (1977). *Dev. Biol.* **61,** 230–244.

23
Structure and Developmental Expression of Murine Homeo Box Genes

PETER GRUSS, CAROLA DONY, BERND FÖHRING,
AND MICHAEL KESSEL
Max Planck Institute of Biophysical Chemistry
Department of Molecular Cell Biology
D-3400 Göttingen, Germany

INTRODUCTION

The discovery of sequence conservation between homeotic genes of *Drosophila* (1) and other organisms (2,3), including mouse and man, has stimulated considerable discussion of the intriguing possibility that similar molecular mechanisms may underlie the generation of cellular diversity and pattern formation in many different organisms. In order to test this assumption, a detailed structural and functional analysis of murine homeo box-containing genes is required.

Presently, the family of murine genes consists of approximately 15 members [see Martin *et al.* (4) for references], some contained in clusters (5–7). As initially deduced by comparison with yeast mating-type genes (8,9), the helix–turn–helix motif present in the homeo domain most likely represents a DNA-binding domain. Experimental evidence at least points to specific DNA binding of *engrailed* and *fushi tarazu* (10). In the case of murine genes, some indications of a nuclear location (11) and DNA binding (12) are available.

The idea that homeo box genes act in specifying positional information during murine development can only be put to the test by carefully monitoring their activity and then carrying out a functional analysis. Colberg-Poley *et al.* (13) noted differential activity of murine homeo box genes in their study of the expression profile of Hox 1.1 (m6) during differentiation of F9 cells into parietal endoderm (14,15). Mouse embryos have also been studied at different stages of development. Northern analysis carried out in several laboratories revealed transient expression of *Hox 1.1, Hox 1.2* (5), *Hox 1.3* (16), *Hox 1.4* (7,17,18), *Hox 1.5* (19), *Hox 2.1, Hox 2.2* (6,20,21), and *Hox 3.1* (22,23). *Hox 1.1, Hox 1.3,* and *Hox 2.1* transcripts were reported to show temporal and spatial restriction. They peak at 11–12 days postcoitum (pc), *Hox 3.1* being most abundant in spinal cord (5,6,20,23). Recently, *Hox 1.4* expression has also been demonstrated in day-7.5 embryos, later becoming detectable in spinal cord and mesonephric kidney (24), and in adult male animals in spermatocytes and spermatids (17,18). The spatial and temporal expression profile of homeo box genes found in studies carried out in humans concurs with that found in other mammals (25).

While these results suggest a strong similarity between corresponding mammalian and *Drosophila* homeo box genes, firm conclusions cannot be made without more extensive information on the temporal and spatial expression of murine homeo box genes. Besides being a means of overcoming the limitation on the amount of material available in early murine embryos, *in situ* hybridization techniques can provide information about the activity of a gene even to the single cell level. Here we discuss results of an *in situ* analysis of the spatial and temporal expression profile of *Hox 1.3*.

RESULTS AND DISCUSSION

Structure of Murine Homeo Box Genes

The family and nomenclature of murine homeo box genes was introduced by Martin *et al.* (4). Up to now all six members of the *Hox 1* cluster, as well as *Hox 2.1* and *Hox 3.1*, have been sequenced. Since a high degree of structural homology might be indicative of a functional conservation, we have compared the murine homeo box sequences with each other and also with *Antennapedia (Antp)* and *engrailed (en)* of *Drosophila* (Fig. 1). In Fig. 1, the amino acid sequence of the homeo box domain (61 amino acids), as obtained from a conceptual translation of the respective nucleic acid sequence, was taken for computer-assisted comparative analysis. In order to facilitate recognition of the various degrees of homology, we have introduced both numbers and shading. The figure in each square represents the number of amino acids which are identical with the *Antp* homeo domain. Additionally, background shading was used which reflects the degree of homology. Thus, 100% homol-

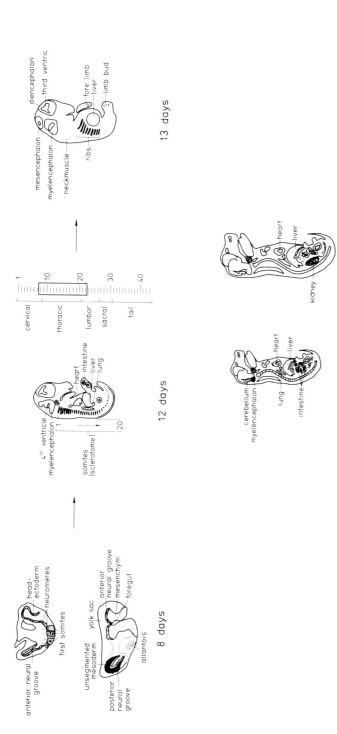

Fig. 1. The sequences of 10 murine homeo domains, each 61 amino acids long, were compared with each other and arranged according to their homology to the *Antennapedia* homeo domains of the *Drosophila* homeo domains and the *engrailed* gene products. The elements of the resulting matrix are the numbers of homologous amino acids. They are underlayed with a gray tone which reflects the number of homologous amino acids in order to facilitate the recognition of groups. The primary structures for the *Antp* (2), *Hox 1.1* (13), *Hox 2.1* (20), *Hox 1.2* (5), *Hox 1.4* (7), *Hox 3.1* (22), *Hox 1.5* (26), *En1* (27), *En2* (27), and *En* (28) homeo domains have been published elsewhere. The *Hox 1.3* sequence is from our unpublished results, and the *Hox 1.6* sequence was communicated to us prior to publication by D. Duboule.

ogy is represented by the number 61 on a completely black square. Using this means of representation it becomes evident that with the sequences employed for comparison we can detect at least two different families. One family of murine genes is highly homologous to Antp (Hox 1.1, 2.1, 1.2, 1.4, 1.3, 3.1, 1.5). It is possible that both these members and Antp originate from a common ancestral precursor. Similarly, the murine engrailed genes (En1, En2) are highly homologous to Drosophila engrailed but less related to Antp. Again, one can postulate a common precursor, distinctly different from Antp. The analysis presented in Fig. 1 also allows the interesting comparison of all members of the Hox 1 cluster. It can be deduced that the highest degree of homology with Antp can be found with Hox 1.1 and the lowest with Hox 1.6. Thus, there seems to be a gradient of decreasing homology to Drosophila Antp, beginning with Hox 1.1 and ending with Hox 1.6. This result could be indicative of evolutionary duplication events where Hox 1.1 represents the most ancestral homeo box, whose sequence was then modified according to what processes needed to be controlled.

Recently, two papers (11,29) reported the structure of two murine homeo box transcription units. Based on these data, it appears that at least Hox 1.1, Hox 2.1, and also Hox 1.3 (M. Fibi et al., unpublished data) are simply structured genes containing two exons and one intron. In all cases (and in agreement with corresponding genes from Drosophila) the homeo box domain is located on the last exon and preceded immediately by the intron. Interestingly, patches of homology can also be found on comparing sequences outside of the homeo box. One stretch of six amino acids is often found preceding the homeo box by 5 to 16 amino acids. Additionally, and depending on the genes compared, homologous amino acids are present at the amino terminus and at the carboxy-terminal end of the homeo domain. Thus, it appears as if the respective protein is constructed by combining different protein domains, which is possibly indicative of an at least mechanistically related function. In order to gain further insight into the role of these genes present in the murine genome, a detailed functional analysis is mandatory.

Functional Analysis of Hox 1.3

Our interest was mostly focused on the functional analysis of the first three genes of the Hox 1 cluster. Initially, we used the Northern technique to analyze the presence of stable RNA in teratocarcinoma cells and also in mouse embryos at various developmental stages. Our results demonstrated the differential expression of the Hox 1.1 and 1.2 genes. Specifically, using F9 cells initially, we showed that differentiation of F9 cells in parietal endoderm was accompanied by induction of stable Hox 1.1 and 1.3 RNA (30,31;

B. Zink et al., personal communication). Interestingly, this induction occurred in F9 cells at the posttranscriptional level, most likely by stabilizing the RNA. Similarly, differential expression was observed after examining the RNA of mouse embryos. For *Hox 1.1* and also *Hox 1.3* the highest level of stable RNA was observed at days 12–14 of gestation. Additionally, stable RNA was detected in some organs of adult animals, such as kidney and brain (31). However, for precise determination of the temporal and spatial expression profile of these genes, the Northern technique is inadequate. Thus, *in situ* analysis was employed which can yield information on the expression down to the single cell level.

A schematic representation of our results (32) is shown in Fig. 2. In Fig. 2 parasagittal sections from various stages of mouse embryogenesis are depicted. All tissues found to have detectable levels of *Hox 1.3* RNA are marked in black. Interestingly, a distinct spatial and temporal expression can

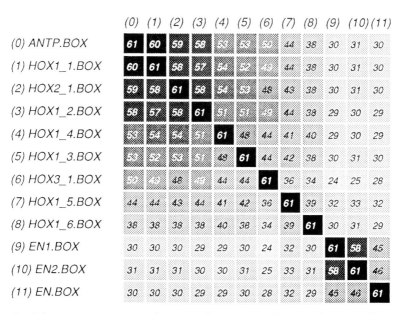

Fig. 2. Schematic representation of mouse embryo sections. Parasagittal sections of 8-, 12-, 13-, and 18-day-old embryos are drawn schematically. The presence of *Hox 1.3* RNA as detected by *in situ* analysis (32) is indicated by drawing it black. The boxed region to the right of the 12-day section accentuates the somites/sclerotomes found to be positive for *Hox 1.3* expression.

be seen. This expression seems largely to be limited to the thoracic region of the murine body. Beginning at day 8, a signal is present in the unsegmented mesoderm but not in the first seven cervical somites. This pattern is retained in 12-day embryos in which somites have given rise to sclerotomes (as well as dermatomes and myotomes). Again, at this stage the cervical sclerotomes are negative, but sclerotomes of the thoracic area express *Hox 1.3* RNA. Sclerotomes subsequently yield embryonal ribs and vertebrae. In 13-day-old embryos we again discover *Hox 1.3* transcripts in thoracic vertebrae and embryonal ribs. Thus, the expression of *Hox 1.3* beginning at day 8 in noncondensed somites follows a developmental lineage. This lineage is represented by presomitic mesoderm, thoracic somites, thoracic sclerotomes, ribs. It is interesting to speculate that the presence of the *Hox 1.3* in early mesoderm might be involved in this process of pattern formation.

The expression of *Hox 1.3* is not limited, however, to somite-derived structures alone. It can also be found in regionally specific mesodermal components of lung, gut, intestine, and kidney. In all these organs mesoderm plays a major role in determining the character of the tissue. All structures mentioned thus far with the exception of the intestine and kidney cease to express *Hox 1.3* RNA at later stages of mouse development. Thus, in 18-day embryos mesodermal components of intestine and kidney express this gene, while ribs, vertebrae, lung, and gut do not.

Besides mesodermal components of the embryo only one other organ of ectodermal derivation expresses *Hox 1.3* RNA, namely, distinct parts of the brain. Specifically, as first seen around day 12 of gestation, certain hindbrain regions representing myelencephalon carry *Hox 1.3* RNA that persists and can also be seen at day 18. In this context it should be noted that, in *Drosophila*, the presence of homeo box-containing transcripts has been noted in spinal ganglia. In mammalian embryogenesis also a presumed mesodermal component, the notochord, plays a major role in the induction of the neurulation process. Thus, it is conceivable that early mesodermal induction mechanisms could also lead to induction of *Hox* genes in ectodermal derivatives.

Taking all things together in summary, the highly distinct temporal and spatial expression profile of *Hox 1.3* is compatible with the notion that this gene is involved in certain developmental control processes of the mouse. Our present efforts are directed toward functional analysis using appropriate vectors for introduction into embryonal stem cells and mouse zygotes.

ACKNOWLEDGMENTS

We thank Rosemary Drescher for expert and speedy preparation of the manuscript. This research was supported by the Deutsche Forschungsgemeinschaft (SFB 236) and the Max Planck Society. This manuscript was prepared in 1987.

REFERENCES

1. Gehring, W. J. (1985). *Cell (Cambridge, Mass.)* **40,** 3–5.
2. McGinnis, W., Levine, M. S., Hafen, E., Kuroiwa, A., and Gehring, W. J. (1984a). *Nature (London)* **308,** 428–433.
3. Scott, M. P., and Weiner, A. J. (1984). *Proc. Natl. Acad. Sci. U.S.A.* **81,** 4115–4119.
4. Martin, G., Boncinelli, E., Duboule, D., Gruss, P., Jackson, I., Krumlauf, R., Lonai, P., McGinnis, W., Ruddle, F., and Wolgemuth, D. (1987). *Nature (London)* **325,** 21–22.
5. Colberg-Poley, A. M., Voss, S., Chowdhury, K., Stewart, C. L., Wagner, E. F., and Gruss, P. (1985a). *Cell (Cambridge, Mass.)* **43,** 39–45.
6. Hart, C. P., Awgulewitsch, A., Fainsod, A., McGinnis, W., and Ruddle, F. (1985). *Cell (Cambridge, Mass.)* **43,** 9–18.
7. Duboule, D., Baron, A., Mahl, P., and Galliot, B. (1986). *EMBO J.* **5,** 1973–1980.
8. McGinnis, W., Garber, R. L., Wirz, J., Kuroiwa, A., and Gehring, W. J. (1984b). *Cell (Cambridge, Mass.)* **37,** 403–408.
9. Porter, S. D., and Smith, M. (1986). *Nature (London)* **320,** 766–768.
10. Desplan, C., Theis, J., and O'Farrell, P. (1985). *Nature (London)* **318,** 630–635.
11. Kessel, M., Schulze, F., Fibi, M., and Gruss, P. (1987). *Proc. Natl. Acad. Sci. U.S.A.* (in press).
12. Fainsod, A., Bogarad, L. D., Ruusala, T., Lubin, M., Crothers, D., and Ruddle, F. H. (1986). *Proc. Natl. Acad. Sci. U.S.A.* **83,** 9532–9536.
13. Colberg-Poley, A. M., Voss, S., Chowdhury, K., and Gruss, P. (1985b). *Nature (London)* **314,** 713–718.
14. Strickland, S., and Mahdavi, V. (1978). *Cell (Cambridge, Mass.)* **15,** 393.
15. Hogan, B. L. M., Barlow, D. P., and Tilly, R. (1983). *Cancer Surv.* **2,** 115.
16. Zink, B., Kessel, M., Colberg-Poley, A. M., and Gruss, P. (1987). Submitted.
17. Wolgemuth, D. J., Englmyer, E., Duggal, R. N., Gizang-Ginsberg, E., Mutter, G. L., Ponzetto, C., Vivano, C., and Zakeri, Z. F. (1986). *EMBO J.* **5,** 1229–1235.
18. Rubin, M. R., Toth, L. E., Patel, M. D., D'Eustachio, P., and Nguyen-Huu, M. C. (1986). *Science* **233,** 663–667.
19. Ruddle, F. H., Hart, C. P., and McGinnis, W. (1985). *Trends Genet.* **1,** 48–51.
20. Jackson, I. J., Schofield, P., and Hogan, B. (1985). *Nature (London)* **317,** 745–748.
21. Hauser, C. A., Joyner, A. L., Klein, R. D., Learned, T. K., Martin, G. R., and Tjian, R. (1985). *Cell (Cambridge, Mass.)* **43,** 19–28.
22. Breier, G., Bucan, M., Francke, U., Colberg-Poley, A. M., and Gruss, P. (1986). *EMBO J.* **5,** 2209–2215.
23. Awgulewitsch, A., Utset, M. F., Hart, C. P., McGinnis, W., and Ruddle, F. H. (1986). *Nature (London)* **320,** 328–335.
24. Gaunt, S. F., Miller, R. J., Powell, D. J., and Duboule, D. (1986). *Nature (London)* **324,** 662–663.
25. Mavilio, F., Simeone, A., Giampaolo, A., Faiella, A., Zappavigna, V., Acampora, D., Poiana, G., Russo, G., Peschle, C., and Bopncinelli, E. (1986). *Nature (London)* **324,** 664–666.
26. McGinnis, W., Hart, C. P., Gehring, W. J., and Ruddle, F. H. (1984). *Cell (Cambridge, Mass.)* **38,** 675–680.
27. Joyner, A. L., Kornberg, T., Coleman, K. G., Cox, D. R., and Martin, G. R. (1985). *Cell (Cambridge, Mass.)* **43,** 29–37.
28. Poole, S. J., Kauvar, L. M., Drees, B., and Kornberg, T. (1985). *Cell (Cambridge, Mass.)* **40,** 37–43.
29. Krumlauf, R., Holland, P. W. H., McVey, J. H., and Hogan, B. L. M. (1987). *Development* **99,** 603–617.

30. Colberg-Poley, A. M., Voss, S. D., and Gruss, P. (1986). *Chem. Scr. R. Acad., Nobel Conf.* **26**, 299–303.
31. Colberg-Poley, A. M., Voss, S., and Gruss, P. (1987). *Oxford Surv. Eukaryotic Genes* (N. Maclean, ed.), in press.
32. Dony, C., and Gruss, P. (1987). *EMBO J.* (in press).

PART V

SMALL MOLECULES IN DIFFERENTIATION AND DEVELOPMENT

24
Molecular Dissection of Pattern Formation in Vertebrate Limbs: Concepts and Experimental Approaches

GREGOR EICHELE AND CHRISTINA THALLER
Department of Cellular and Molecular Physiology
Harvard Medical School
Boston, Massachusetts 02115

BACKGROUND

The developing chick limb is an excellent system to investigate, at a biochemical and molecular biological level, some of the fundamental processes that underly embryonic development. Here we review some of our recent work regarding the role of vitamin A derivatives in limb pattern formation.

All vertebrate limbs develop from small buds protruding from the embryonic flank. Figure 1 illustrates how a chick wing is formed. In early stages (Hamburger–Hamilton stages 19 to 23 correspond to day 3 to 4 of embryonic development; 1) the bud-shaped wing anlage consists of an ectodermal hull filled with apparently undifferentiated mesenchyme. At this time many of the critical events that eventually lead to the complexity of a limb are taking place. However, at present the language in which instructions of early development are written is poorly understood. At stage 24 limb mesenchyme cartilage condensation begins (Fig. 1), and over a period of about 3 days the different elements of the limb evolve. At day 10 of development, a small but complete wing can be seen. The chick wing consists of humerus, radius, and

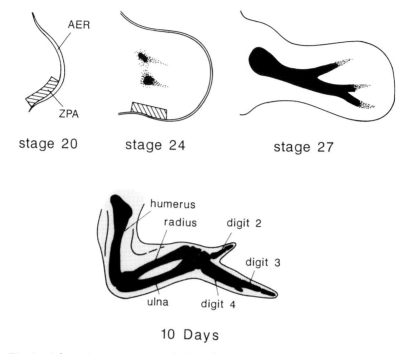

Fig. 1. Schematic representation of selected stages of chick wing development. At Hamburger–Hamilton stage 20 (3.5 days of incubation) the wing anlage is a 1 mm wide bud that consists of apparently undifferentiated mesenchyme. The ZPA (zone of polarizing activity) is marked by a hatched area. The AER (apical ectodermal ridge) forms a rim around the early limb bud. At stage 24, cartilage begins to condense. Rudiments for humerus and the forearm elements are well demarcated at stage 27 (5.5 days of incubation). At Day 10 of development all skeletal elements are formed. The handplate consists of three digits, numbered 2, 3, and 4.

ulna and a rather intricate hand plate with three digits designated 2, 3, and 4 (Fig. 1). All vertebrate limbs display such a stereotypic pattern, indicating that they have evolved from a common ancestral form. Consistent with this evolutionary relatedness, embryonic limb development in vertebrates is highly conserved.

Among vertebrate animals the chick is most suitable for studying limb morphogenesis, because the embryo is easily accessible and thus amenable to manipulations while still developing. Therefore, a substantial body of experimental work has been done on the chick limb. Of particular relevance to our own studies is the mesenchymal tissue that extends along the posterior bud margin, known as the zone of polarizing activity (ZPA, see Fig. 1). In their seminal experiment, Saunders and Gasseling grafted a block of ZPA tissue to the anterior margin of a host bud (2). They found that such buds with two ZPAs develop into wings with a mirror-symmetrical digit pattern duplication (*432234* pattern). The ZPA seems to be a true inducer, since additional digits derive from host and not from transplant tissue (3,4). Moreover, the ZPA is not species specific, since ZPA tissue obtained from any amniote animal examined to date induces duplications when grafted into a chick wing or leg bud (5,6).

It has been suggested that limb pattern formation along the anteroposterior axis depends on a gradient of a diffusible morphogen (7,8). This morphogen is thought to be first released from the ZPA and then diffuses into the adjacent tissue, thereby forming a concentration gradient. Thus, cells closest to the ZPA would see a high concentration of morphogen and form a digit *4*, cells in between would see less morphogen and form a digit *3*. The cells farthest away would be exposed to an even lower concentration and interpret this to form a digit *2* (7,8). By grafting a ZPA to the anterior margin, the morphogenetic cue is produced in a mirror-symmetrical fashion, by the endogenous ZPA and by the grafted one. The structure that arises consequently will have bilateral symmetry. This rather straightforward model poses an intriguing question, namely, that of the chemical nature of the postulated ZPA morphogen.

A promising advance toward resolving this question was an unexpected discovery by Tickle and colleagues (9) and Summerbell (10). They found that local application of all-*trans*-retinoic acid from a piece of DEAE-cellulose paper (9) or newsprint (10) to the chick limb bud produces pattern duplications which are morphologically indistinguishable from those obtained by ZPA grafts. As discussed below, there is now increasing evidence that ZPA and all-*trans*-retinoic acid share a common mechanism of action.

RETINOIC ACID MIMICS ZPA TISSUE AND INDUCES PATTERN DUPLICATIONS IN A HIGHLY SPECIFIC WAY

Retinoic acid is applied locally in order to mimic the localized ZPA. Typically, a positively charged Dowex bead of 250 μm diameter is presoaked in a dimethyl sulfoxide solution of all-*trans*-retinoic acid. The bead is washed briefly and implanted *in ovo* at the anterior margin of a wing bud of a stage 20 embryo (11). The implant slowly releases the absorbed retinoic acid and thereby acts as a local source for a period of about 24 hr (12). Studies in model systems show that the concentration of retinoic acid in the soaking solution is proportional to the amount of retinoic acid in the bead which, in turn, determines the amount of retinoic acid released into the bud tissue (11,12). After the bead is implanted, the embryos are left to develop for 7 more days. At that time the cartilage pattern of the limb can readily be visualized by staining the embryo with Alcian green dye.

Dose–response studies showed that the pattern changes are dose dependent (9–11). Beads soaked in solvent only give normal 234 patterns. The addition of small amounts of retinoic acid (soaking concentration ~0.2 μm/ml) causes the formation of an additional digit 2, resulting in a 2234 pattern. Higher doses of retinoic acid first lead to an additional digit 3, that is, to a 32234 pattern and finally to a duplication with a 432234 pattern or, more frequently, to a 43234 pattern (Fig. 2). This happens at soaking concentrations of 5–15 μg/ml. The average concentration of retinoic acid in the tissue necessary to induce such a duplication with additional digits 3 and 4 is about 25 nM (11). Hence, retinoic acid is effective at a concentration range typical for hormonelike substances. The well-known teratogenic effects of retinoids can also be induced by local retinoic acid release, but very high doses (soaking concentration >100 μg/ml) are needed (10,11). As for retinoic acid, ZPA tissue grafts show a graded response: implantations of 30 to 40 ZPA cells yield an additional digit 2, while 80 cells induce extra digits 2 and 3 and 150 cells yield a 432234 duplication (3).

One of the convenient features of the beads releasing retinoic acid is that they can easily be removed after implantation. Kinetic studies have shown that the remaining retinoic acid is cleared with a half-time of 15 to 20 min (13). This makes it possible to explore the time course of digit induction. To generate the time–response curves shown in Fig. 3, beads were first soaked

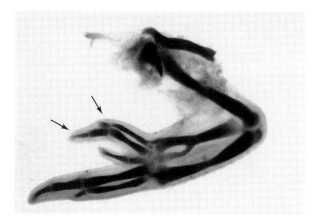

Fig. 2. Cartilage pattern of a 10-day-old chick wing that developed after local application of all-*trans*-retinoic acid to the anterior wing bud margin at Day 3.5 of development. The treated wing (ventral aspect) displays a 43234 digit pattern. The additional digits are those located at the anterior side of the handplate and are marked with arrows. For the normal digit pattern, see Fig. 1.

in either retinoic acid or a synthetic retinoic acid analog (TTNPB; (E)-4-[2-(5,6,7,8,-tetrahydro-5,5,8,8-tetramethyl-2-naphthalenyl-1-propenyl]benzoic acid) and implanted into wing buds of stage 20 embryos. Subsequently, the implants were removed after the period of treatment indicated on the time axis. Then incubation was continued until day 10. As can be seen in Fig. 3, treatments for up to 10 hr result in the normal 234 pattern. Thereafter, additional digits are formed, a digit 2 always first, followed by a 3, and last, at about 18 hr, a digit 4. These data indicate that digit induction is biphasic. There exists a "priming phase," during which retinoic acid induces a process required but not sufficient to specify additional digits. The commitment is reversible, since removal of the bead results in a normal pattern. During the second time span, referred to as a "duplication phase," digits are irreversibly specified at an average rate of one new digit every 1 to 2 hr. It is important to realize that neither during the priming nor the duplication phase is there any indication of terminal differentiation of mesenchyme into digit cartilage. Thus, at a time when digits are not yet formed as visible structures retinoic acid will provide a determinative signal to limb bud cells to later become a particular digit. With respect to timing there is again good

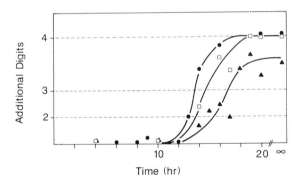

←stage 20→ ←stage 21→ ← stage 22→ ←stage 23

Fig. 3. Time–response curves showing the induction of additional digits as a function of the period of exposure to retinoic acid (□ and ▲) or the synthetic retinoid TTNPB (●) (modified from Ref. 13). Beads were presoaked in 33 (▲) or 333 μM (□) all-*trans*-retinoic acid or 14 μM TTNPB and then implanted into wing buds of stage 20 embryos (zero time point). Subsequently, the beads were removed at the time points indicated on the abscissa. The digit patterns that formed were scored numerically (for details, see Ref. 11). Normal digit patterns coincide with the abscissa. Patterns with an additional digit 2 (on the average) are represented by the first horizontal line; the second line represents patterns that have additional digits 2 and 3. The third line represents patterns of the type 432234, 43234, or 4334, all with an extra digit 4 at the most anterior position. The time–response curves are biphasic, consisting of a priming phase (0 to 10 hr) and a duplication phase (11 to 19 hr). TTNPB, (E)-4-[2-(5,6,7,8-Tetrahydro-5,5,8,8-tetramethyl-2-haphthalenyl)-1-propenyl]benzoic acid.

correspondence between retinoic acid and the ZPA. Smith (14) found that removing a ZPA 10 hr after it was grafted resulted in normal limbs. Waiting for 12 hr often gave an additional digit 2, while an incubation period of 15 hr sometimes lead to a pattern with an additional digit 3.

Although digit pattern duplications are the most conspicuous sign of retinoid treatment, there is also a remarkable effect on the pattern of feather anlagen, which is particularly obvious in wing specimens with 4334 digit patterns. Figure 4 displays the normal left wing and the retinoic acid-treated contralateral right wing. Note the gradual anteroposterior increase of the length of the feather anlagen in the untreated wing (Fig. 4, left). Anteriorly feather buds are tiny, but posteriorly they are long, filamentous structures. In the treated wing (Fig. 4, right) such long filaments are present posteriorly as well as anteriorly. These changes in the skin pattern represent, of course, an example of a mesodermal–ectodermal interaction. These highly specific effects at low retinoic acid concentrations, and the similar results seen with retinoic acid treatment and ZPA grafting, indicate that retinoic acid could be the hypothetical signaling substance of the ZPA.

RETINOIC ACID CAN INDUCE DIGIT FORMATION IN THE ABSENCE OF ZPA TISSUE

Because the ZPA is still present in the buds that received a retinoic acid-releasing implant, one can argue that the signal to form additional digits in

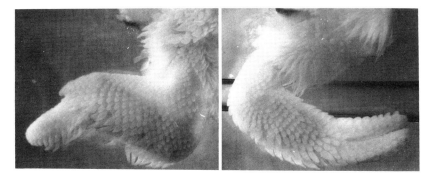

Fig. 4. The photograph at left shows a normal untreated left wing of a 10-day-old embryo. The picture at right displays the corresponding right wing that developed following retinoid treatment. The treated wing has a 4334 digit pattern. The forearm of the control specimen displays a highly asymmetrical pattern of feather anlagen; anteriorly the anlagen are tiny, but at the posterior margin feather rudiments are long and filamentous. In contrast, the feather anlagen of the retinoid-treated wing clearly displays mirror-symmetry.

the above experiments primarily stems from the ZPA and not from retinoic acid. In this case, retinoic acid would act as a permissive agent and not be a true morphogen. To examine this question, we have excised the entire posterior half of the wing bud, thereby removing the ZPA and the prospective tissue that would form ulna and digits (see fate map in Fig. 5). Usually this mutilation results in a limb that has only a humerus, a radius, and occasionally a digit 2 (Fig. 5, center). What happens if retinoic acid is applied to such an anterior half-bud (Fig. 5, right)? The fate map (15) clearly shows that in normal development, the anterior region contributes little tissue to the hand. Yet, the striking result is that retinoic acid-treated anterior half-buds develop into a complete extremity (Fig. 5, right). Note, that the rescued digit pattern has a reversed polarity (the term "rescue" should indicate that

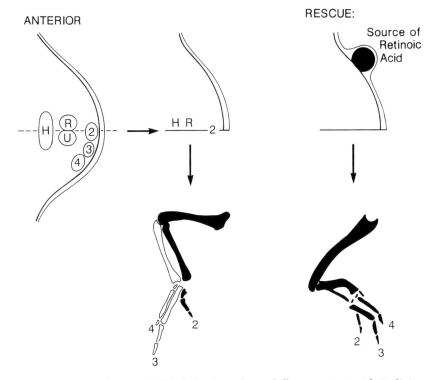

Fig. 5. Rescue of anterior half-buds by slow release of all-*trans*-retinoic acid. (Left) Fate map of stage 20 limb bud tissue (after Ref. 15). (Center) After removal of the posterior limb bud half including the ZPA, the limb formed has a severely truncated handplate, often all digits are missing, and sometimes a digit 2 is formed. (Right) Treatment of an anterior half-bud with all-*trans*-retinoic acid results in a complete hand that exhibits a reversed polarity, that is, the most anterior digit is now a digit 4 and not a digit 2 as in the normal wing. Abbreviations: H, humerus; R, radius; U, ulna; 2, digit 2; 3, digit 3; 4, digit 4.

tissue that normally does give rise to little terminally differentiated structures does so under the influence of retinoic acid). In the normal hand (Fig. 1), the anteroposterior digit sequence is *234*. The experimental limb displays a *432* sequence. We conclude that retinoic acid, by itself, will (1) generate a hand, and (2) determine the polarity of the pattern. There is again a remarkable similarity between ZPA grafts and retinoic acid implants: when ZPA tissue is grafted into anterior half-buds, a pattern with reversed polarity will also develop (2).

LIMB BUDS CONTAIN ENDOGENOUS RETINOIC ACID

One of the crucial questions to ask is whether the above experiments reflect a physiological function of retinoic acid in limb development, or whether we are seeing a pharmacological effect. To examine this question we have carried out two types of experiments. First, we have searched for retinoic acid in the early limb rudiment and determined its concentration in the tissue. Second, we have investigated whether retinoic acid is enriched in the ZPA (16).

To search for endogenous all-*trans*-retinoic acid, limb buds of stage 21 embryos were collected. The buds were homogenized, spiked with a known amount of tritiated all-*trans*-retinoic acid (to serve as an internal standard), and extracted with a mixture of ethyl acetate and methyl acetate. The extract was first fractionated by reversed-phase high-performance liquid chromatography (HPLC). Figure 6a displays a sample chromatogram showing the absorbance at 350 nm, where retinoids characteristically absorb. Note that peak A_2 coelutes with the added radioactive all-*trans*-retinoic acid standard. The fractions containing this peak were pooled and rechromatographed on a normal-phase HPLC column. Only one major peak is found that again coelutes with the added tracer retinoic acid (data not shown). Alternatively, peak fractions A_2 from the initial reversed-phase column were pooled and rechromatographed on a second reversed-phase column using a more polar solvent. Again, tracer retinoic acid and the "unknown" substance comigrate (not shown). In a next step, the putative retinoic acid was methylated. The product, reapplied to the same column, coelutes with the methylated internal standard retinoic acid as a single peak (Fig. 6b). Molecular weight determination by gas chromatography–mass spectroscopy (GC/MS) confirm that this peak is all-*trans*-methyl retinoate. We conclude that early chick limb buds contain all-*trans*-retinoic acid.

The chromatograms, in combination with the recovery calculated from the radiolabeled internal standard retinoic acid, afford quantitation of retinoic acid. We found that a limb bud of a stage 21 embryo, the stage when retinoic

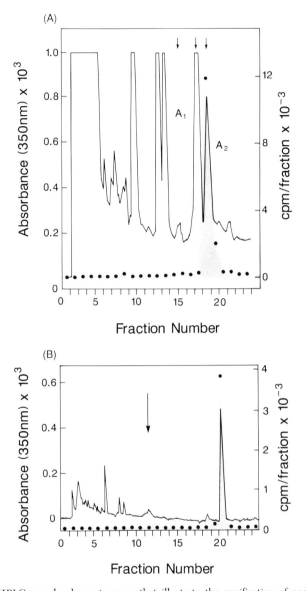

Fig. 6. HPLC sample chromatograms that illustrate the purification of endogenous all-trans-retinoic acid from chick limb buds. (a) Fractionation of limb bud extract on a reversed-phase column. Limb buds from stage 21 chick embryos were homogenized after addition of all-trans-[^{3}H]-retinoic acid as an internal standard and were extracted with organic solvent. The concentrated extract was applied to a C_{18} column and eluted with acetonitrile/methanol/2% acetic acid (6:2:2). Peak A_2 comigrates with the internal standard all-trans-retinoic acid. Peak A_1 is all-trans-retinol. (b) all-trans-[^{3}H]Methylretinoate and methylated endogenous all-trans-retinoic acid copurify. The methyl esters were generated by diazomethane treatment of a limb bud extract that was prepurified on two different C_{18} colums. Solid lines show the absorbance measured at 350 nm; shaded areas with black dots, the elution profile of internal standards as determined by scintillation counting. The arrows mark the elution position of 13-cis-retinoic acid, all-trans-retinol, and all-trans-retinoic acid (left to right in Fig. 6a) and all-trans-retinoic acid (Fig. 6b).

acid can induce extra digits, contains 6.3 pg retinoic acid. This corresponds to a mean tissue concentration of 25 nM. The dose–response experiments reviewed above have revealed that the amount of retinoic acid needed in the tissue to induce a complete set of additional digits is also about 25 nM. This match of concentration is good evidence that endogenous retinoic acid plays an important role in the specification of the limb pattern. Fractionated extracts of stage 21 buds exhibit a second major peak (A_1 in Fig. 6a) that represents all-*trans*-retinol. The amount of retinol present in a stage 21 bud is 140 pg, which corresponds to a tissue concentration of approximately 600 nM, or 25 times that of retinoic acid (16).

To test whether extracted retinoic acid can induce additional digits, duplication bioassays with purified extracts of whole stage 21 chick embryos were performed. Extract obtained from several hundred embryos was purified by HPLC, and fractions containing retinoic acid were concentrated. Slow release beads were soaked in the concentrate and implanted in the usual position. In most cases *43234* and *32234* digit pattern duplication developed (16). Thus, chick embryo extracts contain an endogenous morphogenetic substance which is retinoic acid, as we now know.

Is retinoic acid enriched in the ZPA? To determine the distribution of retinoic acid, stage 21 limb buds were dissected into a smaller posterior part, which contains the ZPA, and a larger anterior piece (see Fig. 1 for location of the ZPA). The two tissue samples were analyzed separately, and retinoic acid was quantitated by the same methods as used for intact buds. Six independent analyses were carried out with 800 to 1000 bud parts each. The amount of all-*trans*-retinoic acid in the anterior and posterior tissue segments is rather similar (3.52 ± 0.25 pg anteriorly and 2.95 ± 0.34 pg posteriorly). However, anteriorly the mean DNA content (a measure of the amount of tissue) is 1.9 ± 0.11 μg per bud, whereas the smaller posterior part contains only 0.62 ± 0.02 μg DNA. Hence, there is an enrichment of retinoic acid in the ZPA-containing, posterior segment. This enrichment can be numerically expressed as the ratio of picograms retinoic acid per microgram DNA posterior, divided by picograms retinoic acid per microgram DNA anterior. The average ratio is 2.6 ± 0.4. This means that in the tissue part which contains the ZPA, all-*trans*-retinoic acid is enriched 2.6-fold. There is a high point of retinoic acid in the ZPA, but the rest of the bud also contains appreciable amounts of retinoic acid. Therefore, retinoic acid forms a concentration gradient.

Can a gradient of the kind we have measured in the limb bud provide a cue for pattern formation? To examine this question, the concentration gradient of endogenous retinoic acid was compared with that established after a duplication-inducing dose of retinoic acid had been applied from an anteriorly positioned slow release bead (16). In this case the enrichment factor is

3.3 (±0.3). Although the biologically effective gradient of exogenously applied retinoic acid is somewhat steeper than the endogenous gradient, they are in the same range.

Additional evidence for the importance of a graded rather than a uniform concentration profile comes from earlier work (13). In these studies retinoid was distributed either uniformly throughout the bud or in the form of a concentration gradient. It was found that the second distribution, the gradient, is more effective in inducing additional digits than the uniform concentration profile.

FUTURE PROBLEMS

How is the retinoic acid gradient generated? There are two obvious ways. One possibility is that the ZPA enzymatically generates retinoic acid from its biosynthetic precursor retinol. Several observations are consistent with this mechanism. First, there are substantial amounts of retinol present; the average concentration of retinol in limb buds is 25 times higher than that of retinoic acid. Second, we found that limb bud tissue can specifically convert retinol to retinoic acid (21). Third, retinol itself is essentially evenly distributed (16), so the retinoic acid gradient must arise during one of the subsequent oxidation steps. Fourth, if retinoic acid is locally released from a bead it will diffuse deeply into the tissue (17). Likewise, retinoic acid produced in the ZPA could readily spread across the limb bud. An alternative mechanism of gradient formation assumes an even rate of synthesis of retinoic acid throughout the bud but a preferential degradation of retinoic acid anteriorly.

Another yet unresolved question is how do cells interpret and respond to the retinoic acid gradient. Almost certainly, this interpretation ultimately involves changes in the expression pattern of specific genes. A substantial body of work demonstrates that retinoids have the capacity to specifically affect gene expression (18,19). Of particular interest is the recent identification of retinoic acid-inducible genes that are activated in murine teratocarcinoma cells within a few hours after retinoic acid treatment (20). Whether this *in vitro* induction is a model of the mode of action of retinoic acid in the limb bud remains to be seen.

ACKNOWLEDGMENTS

This work was supported in part by grant HD 20209 from the National Institutes of Health to G.E.

REFERENCES

1. Hamburger, V., and Hamilton, H. (1951). *J. Morphol.* **88,** 49.
2. Saunders, J. W., Jr., and Gasseling, M. T. (1968). *In* "Epithelial–Mesenchymal Interactions" (R. Fleischmajer and R. E. Billingham, eds.), pp. 78–97. Williams & Wilkins, Baltimore, Maryland.
3. Tickle, C. (1981). *Nature (London)* **289,** 295.
4. Honig, L. S. (1983). *Dev. Biol.* **97,** 424.
5. Tickle, C., Shellswell, G., Crawley, A., and Wolpert, L. (1976). *Nature (London)* **259,** 396.
6. Fallon, J. F., and Crosby, G. M. (1977). *In* "Vertebrate Limb and Somite Morphogenesis" (D. A. Ede, J. R. Hinchliffe, and M. Balls, eds.), pp. 55–69. Cambridge Univ. Press, Cambridge.
7. Wolpert, L. (1971). *Curr. Top. Dev. Biol.* **6,** 183.
8. Tickle, C., Summerbell, D., and Wolpert, L. (1975). *Nature (London)* **254,** 199.
9. Tickle, C., Alberts, B. M., Lee, J., and Wolpert, L. (1982). *Nature (London)* **296,** 564.
10. Summerbell, D. (1983). *J. Embryol. Exp. Morphol.* **78,** 269.
11. Tickle, C., Lee, J., and Eichele, G. (1985). *Dev. Biol.* **109,** 82.
12. Eichele, G., Tickle, C., and Alberts, B. M. (1984). *Anal. Biochem.* **142,** 542.
13. Eichele, G., Tickle, C., and Alberts, B. M. (1985). *J. Cell Biol.* **101,** 1913.
14. Smith, J. C. (1980). *J. Embryol. Exp. Morphol.* **60,** 321.
15. Hinchliffe, J. R., Garcia-Porrero, J. A., and Gumpel-Pinot, M. (1981). *Histochem. J.* **13,** 643.
16. Thaller, C., and Eichele, G. (1987). *Nature (London)* **327,** 625.
17. Eichele, G., and Thaller, C. (1987). *J. Cell Biol.* **105,** 1917.
18. Hauser, C. A., Joyner, A. L., Klein, R. D., Learned, T. K., Martin, G. R., and Tjian, R. (1985). *Cell (Cambridge, Mass.)* **43,** 19.
19. Colberg-Poley, A. M., Voss, S. D., Chowdhury, K., Stewart, C. L., Wagner, E. F., and Gruss, P. (1985). *Cell (Cambridge, Mass.)* **43,** 39.
20. LaRosa G. J., and Gudas L. J. (1987). *Proc. Natl. Acad. Sci. U.S.A.* **85,** 329.
21. Thaller, C., and Eichele, G. (1988). *Development* **103,** 473.

25
Differentiation of a Multipotent Human Intestinal Cell Line: Expression of Villin, a Structural Component of Brush Borders[1]

D. LOUVARD, M. ARPIN, E. COUDRIER, B. DUDOUET,
J. FINIDORI, A. GARCIA, O. GODEFROY, C. HUET,
E. PRINGAULT, S. ROBINE, AND C. SAHUQUILLO MERINO

Institut Pasteur
75724 Paris Cedex 15, France

THE INTESTINAL EPITHELIUM: A MODEL SYSTEM FOR STUDY OF CELL DIFFERENTIATION AND THE DEVELOPMENT OF A POLARIZED CYTOSKELETON

The intestinal epithelium is a valuable tissue for the study of cell differentiation in that it is spatially organized around its proliferative units, the crypts. Each crypt consists of a clone of cells (Ponder *et al.*, 1) produced by a group of undifferentiated proliferative cells whose progeny express at least four dramatically different phenotypes (Leblond and Cheng, 2). The determinants of phenotype commitment and expression in this tissue are unknown. The temporal sequence of differentiative changes occurring in each cell type can be readily defined, since the orderly upward migration of cells

---

[1]. Based on the Opening Address delivered by D. Louvard at the symposium, "Molecular Mechanisms in Cellular Growth and Differentiation," held at Arden House on the Harriman Campus of Columbia University.

from crypt to villus arranges the cells along this axis in order of age (Madara and Trier, 3). A major unfulfilled goal of intestinal culture systems is the controlled recreation of the crypt–villus axis of epithelial differentiation in the absence of the complex connective tissue lamina propria.

Like all simple epithelia, the intestine provides a valuable model for the study of cell polarity. Fully differentiated intestinal cells are more strictly polarized than many other epithelial cell types, perhaps because they are designed to face the threats of the intestinal lumen, a situation in which missorting of membrane components could compromise the epithelial barrier. Also unique to this epithelium is the dramatic conversion that occurs during fetal development from a nonpolarized, stratified cell layer to a simple columnar epithelium (Trier and Moxey, 4; Colony and Neutra, 5). This process provides the opportunity to follow normal, *in vivo* assembly of tight junctions and *de novo* formation of apical membrane domains (Madara *et al.*, 6).

The apical brush border of intestinal absorptive cells, designed to expand the membrane surface area available for hydrolysis and transport, provides a unique model for the study of cytoskeleton assembly and cytoskeleton–membrane interactions. The molecular architecture of this structure has been defined in considerable detail using normal intestinal cells, but synchronously differentiating *in vitro* systems should provide an opportunity to define the factors that control assembly and maintenance of this unique, polarized structure.

In attempts to reproduce the complex stages of epithelial intestinal differentiation, investigators have used three general strategies: (1) separation and culture of normal epithelial cells; (2) *in vitro* maintenance of intestinal explants; and (3) culture of neoplastic epithelial cells derived from intestinal adenocarcinomas. The goal in establishing intestinal cell culture systems is to obtain cells and tissues that mimic as closely as possible their counterparts *in vivo*. Ideally, epithelial cell cultures should allow the process of normal cell differentiation and other functions to be recapitulated, manipulated, and observed under controlled conditions. This has proved particularly difficult to achieve for the intestinal epithelium, where the normal *in vivo* state is a complex dynamic process with continuous cell proliferation and cell loss in which individual cells survive only for a few days. Attempts at establishing primary, differentiated cell cultures or short-term differentiated monolayers from adult intestine have been uniformly unsuccessful. Successful long-term primary culture of undifferentiated intestinal epithelial cells was achieved, however, using benign human tumor cells (Friedman *et al.*, 7), rat fetal epithelial cells (Negrel *et al.*, 8), and collagenase-dissociated cells from weaning rat small intestine (Quaroni *et al.*, 9; Quaroni and May, 10). Selected epithelial colonies from the latter cultures were serially passaged, providing nonneoplastic, proliferative lines of intestinal epithelial cells. These lines, now maintained for nearly a decade, can form monolayers of cuboidal, polarized cells, but they consistently fail to differentiate in monolayer culture or to express brush border enzymes. Dozens of human adenocarcinoma cell lines are available since adenocarcinoma of the colon is a common human malignancy (Fogh and Trempe, 11). The growth properties of many lines in culture and in transplants have been described and reviewed (Fogh et al., 12). We have focused our interest on the HT29 cell line because it is currently the only example of an epithelial intestinal cell for which the differentiation features can be manipulated in culture.

A PLURIPOTENT HUMAN INTESTINAL CELL LINE THAT CAN BE INDUCED TO DIFFERENTIATE IN CULTURE

The HT29 cell line does not show polarity or other undifferentiated characteristics of intestinal cells under standard conditions (in media containing glucose and normal serum), and until recently these cells were useful only for studies of general cell features such as ubiquitous receptors and carbohydrate metabolism (Rousset et al., 13; Rousset, 14). When grown in the absence of glucose, however, HT29 cells exhibit a high degree of differentiation (Pinto et al., 15), surpassing in some respects all other neoplastic lines developed to date (Rousset, 14). Phenotypes resembling terminally differentiated absorptive cells appeared in highly polarized, confluent HT29 cell monolayers when galactose, inosine, or uridine was substituted for glucose as carbon sources (Pinto et al., 15; Wice et al., 16), or even in the total absence of these additives (Zweibaum et al., 17). Conditions permissive for differentiation also include the presence of supplemental human transferrin and nonessential amino acids and seeding a high density so that confluence is reached within a few days after plating. Differentiated HT29 cells express four brush border enzymes: alkaline phosphatase, sucrase-isomaltase, aminopeptidase N, and dipeptidyl peptidase IV (Pinto et al., 15; Zweibaum et al., 18), all of which are typical of normal fetal colon. HT29 cells grown in the absence of glucose form highly polarized, differentiated monolayers on either permeable filters or uncoated impermeable glass. Since differentiation can be manipulated in culture, HT29 monolayers offer unique opportunities to reproduce and dissect the process of intestinal cell differentiation and to identify factors that govern the physiological functions of terminally differentiated intestinal cells.

The value of the HT29 cell line has been further extended by the isolation of clones and subclones. Some of these, such as the multipotent clone designated HT29-18, retain partially differentiated characteristics such as the

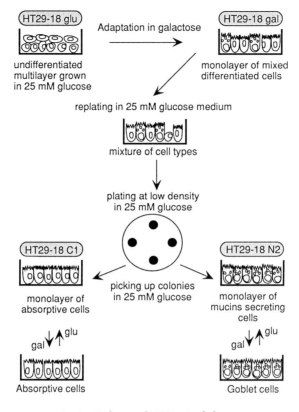

Fig. 1. Isolation of HT29-18 subclones.

ability to form monolayers when returned to glucose-containing medium (Huet et al., 19). Subcloning of differentiated HT29-18 cells in media containing glucose by limiting dilution produced colonies that, when returned to glucose-free medium, expressed a single phenotype, either globlet cells or absorptive cells (Fig. 1) (Huet et al., 19).

ORGANIZATION OF INTESTINAL BRUSH BORDER CYTOSKELETON

The molecular architecture of the enterocyte brush border has been extensively studied as a model system for membrane–cytoskeleton interactions and assembly of actin filaments and filament bundles. The isolated cytoskeleton of intestinal brush borders can be divided into two distinct areas, the microvilli and the terminal web, which is a cytoskeleton-rich region in the apical cytoplasm beneath the microvillus. Although these two areas overlap structurally, each one performs different functions within the bush border. Each microvillus contains a bundle of 20 to 30 actin microfilaments anchored in the plasma membrane at the tip of the microvilli. Transmission electron micrographs of specimens fixed in presence of 15–20 mM $MgCl_2$ showed that the axial bundles of microfilaments are bridged laterally by short, thin microfilaments disposed regularly along the longitudinal axis of the microvilli. These bridges form an α-helix with a pitch of 33 nm around the axial bundle. The major polypeptides found in the microvilli microfilaments have been recently studied and characterized (Mooseker, 20). Known proteins and their contribution to the organization of intestinal brush border microvilli are schematically represented in Fig. 2.

The core of the microvilli is composed of microfilaments made of β- and γ-actin isoforms. All these filaments show the same polarity when they are incubated with proteolytic fragment of myosin S1 or the heavy meromyosin (HMM). Their nucleation point is thought to be located near the tip of the microvilli. The rootlets of these filaments are plunging in the terminal web. Two polypeptides of 68 kDa (Fimbrin) and 95 kDa (villin) bind the actin filaments along its entire length. In contrast, polypeptides of 80 kDa and 110 kDa are distributed only along the actin filaments in the microvillus part, whereas tropomyosin is found essentially associated with the rootlets of actin microfilaments.

The terminal web cytoskeleton consists of three distinct filamentous domains oriented perpendicularly to the microvilli core bundles. Two of the filamentous networks are associated with the junctional complex of the cell. The base or rootlet ends of the microvilli core are bundled together with the filamentous meshwork in which they are embedded; this meshwork constitutes the third filamentous domain. Finally, 10-nm filaments made of prekeratin are attached to the desmosomes of the maculae adherens. They form a thick, dense disk of filaments oriented perpendicularly to the microvilli core bundle. Above the maculae adherens, the zonula adherens junction is associated with a circumferential ring of actin filaments of mixed polarity, as indicated by their decoration with the myosin fragments. This organization is similar to the stress fiber organization in cultured cells. In addition to actin these filaments contain myosin and tropomyosin. They appear to be closely associated with the membrane zonula adherens junctions, possibly through a complex that includes α-actinin and vinculin. The latter two proteins are thought to mediate a similar interaction between actin and plasma membrane in adhesion plaques of fibroblasts. The third filamentous domain of the terminal web is made of the rootlet end of the microvilli, core bundle, and the filamentous meshwork in which they are embedded. It is referred to

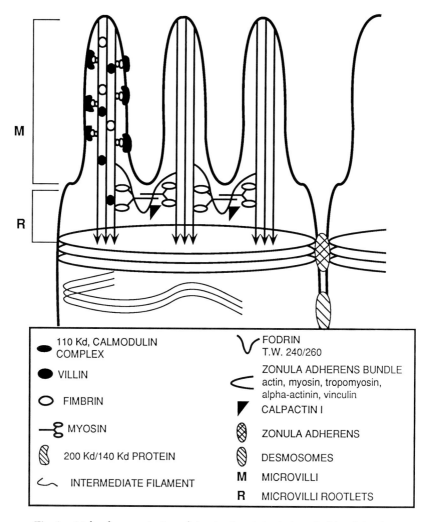

Fig. 2. Molecular organization of structural proteins in intestinal brush borders.

as the interrootlet domain. Two actin-binding proteins have been characterized in the rootlet bundles, villin and tropomyosin. The rootlets are embedded in a dense meshwork of fine filamentous material which presumably cross-links adjacent rootlets. Some of these filaments are members of a ubiquitous class of long flexible cross-linking proteins related to erythrocyte spectrin. Another filamentous protein is myosin, which is of the typical nonmuscle variety. Two other proteins have been localized by light microscopic immunocytochemistry in the terminal web; they are caldesmon and a

36-kDa substrate for the pp60 *Sarc* tyrosine kinase, which has been named calpactin I.

VILLIN, A CALCIUM-REGULATED ACTIN-BINDING PROTEIN

Villin is a globular actin-binding protein. Unlike fimbrin it has been found in large amounts, mainly in cells having a well-organized brush border (intestine and kidney proximal tubule cells) or in cells which have the same embryological origin as intestinal mucosa: (duct cells of pancreas and biliary duct cells). Villin is a monomeric calcium-binding protein in solution (three Ca^{2+} binding sites per molecule). Calcium binding induces pronounced conformational changes of villin. When the calcium concentration is below 10^{-6}–10^{-7} M, villin acts as a bundling factor. At micromolar concentrations of calcium, villin leads to the formation of short filaments by preventing monomer addition and forming at the barbed end what is known as a functional cap. At concentrations above 10 μM, villin severs actin microfilaments into short fragments. Different actin binding sites have been mapped on the protein using a combination of protease treatments. Using V8 protease, Glenney *et al.* (21) cleaved villin into two fragments with apparent molecular weights of 90,000 and 8.5,000. The largest fragment, called the villin core, has Ca^{2+}- dependent nucleation and severing actin-binding activities but does not bundle actin filaments. The smallest fragment, called the *head piece*, consists of 76 amino acids located at the carboxy terminus of villin. This small polypeptide (head piece) binds to actin irrespective of the presence of Ca^{2+} in the medium. Its primary sequence has been determined by Glenney *et al.* (21). In addition, Matsudaira *et al.* (22) have shown that a 45-kDa tryptic fragment derived from the amino terminus of villin retains severing activity. Further cleavages of this peptide with V8 protease yielded a 13-kDa subfragment that contained a Ca^{2+}- binding site and bound to F actin in a Ca^{2+}-dependent fashion. Polyclonal and monoclonal antibodies raised against pig villin display good immunological cross-reactivity between species such as chicken and *Xenopus* (Dudouet *et al.*, 23; Fiegel *et al.*, 24). All these antibodies recognized the head piece domain, suggesting sequence homologies among species.

Accordingly, the primary sequence of this domain in humans displays 65% homology with the primary sequence of the carboxy-terminal domain of chicken villin (Pringault *et al.*, 25). Moreover, 57% homology and 45% identity have been found between the amino acid sequences of human villin and human gelsolin (Kwiatkowski *et al.*, 26); (Arpin *et al.*,, 27). Gelsolin (90 kDa) is a Ca^{2+}-regulated actin-binding protein whose function is to sever, nucle-

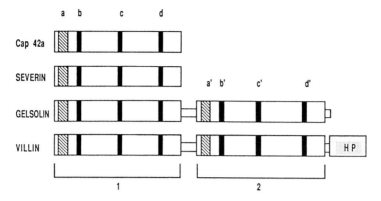

Fig. 3. Schematic representation of the structural organization of villin and gelsolin. In each of the large domains 1 and 2, hatched areas a and a' represent two homologous motifs while dark areas correspond to three motifs identical with each other and repeated. The carboxy-terminal end of villin, unique to this protein, is indicated by HP (Arpin et al., 27).

ate, and block actin filaments. However, unlike villin, gelsolin has no bundling activity. The amino acid sequences of both proteins display two duplicated domains (labeled 1 and 2 in Fig. 3). The same internal motifs aa', bb', cc', dd', with identical characteristic sequences, are found in each of the domains of both proteins. In addition, the motifs bb', cc', dd' present similarities to one another. However, one important difference between villin and gelsolin resides in the carboxy-terminal part of the molecule. The head piece region of villin represents an extension which is present only in this protein (Fig. 3).

VILLIN IS AN ORGAN-SPECIFIC MARKER IN ADULTS AND EMBRYOS

In contrast to the other cytoskeleton proteins found in intestinal brush borders, villin displays remarkable tissue-specific expression. Using antibodies for this protein, several groups have shown by immunochemical and immunocytochemical procedures the occurrence of this protein in some cells of the urogenital and gastrointestinal tracts (Table I). It has been shown that cells which have a brush border contain a large amount of this protein accumulated at their apex. By ultrastructural immunolabeling it has also been shown that villin is localized along the entire length of the axial microfilament bundles (Dudouet et al., 23). Villin has been also found in undifferentiated crypt cells of the intestinal mucosa, which do not yet have an

TABLE I
Villin Detection in Organs and Cell Lines[a]

Gastrointestinal tract	+	Other tissues	
Small intestine	+	Lung	
Large intestine	+	Bronchial epithelium	−
Gallbladder		Alveolar epithelium	−
Pancreas	+	Tissue homogenate	−
Duct	−	Skin	
Acini	+	Epidermas	ND
Tissue homogenate		Dermis	ND
Liver	+	Sweat glands	−
Large bile duct	−	Tissue homogenate	−
Hepatocytes	−	Choroid plexus	−
Tissue homogenate	−		
Stomach			
Esophagus		Cell lines derived from	
		human colonic adenocarcinoma	
Urogenital tract			
Kidney		HT29-18	+
Proximal tubule	+	$CaCo_2$	+
Distal tubule	−	Cell lines derived from	
Cortex homogenate	+	kidney	
Epididymis		Porcine kidney proximal	
Ductus epididymidis	−	tubule (LLCPKI)	+
Ductuli efferentes	+	Canine kidney distal tubule	
Tissue homogenate	−	(MDCK)	−
Testis	−		
Oviduct	−	Other cell lines	
Uterus endometrium	−	HeLa (human uterus)	−
Bladder urothelium	−	HEP2 (human cervix)	−
Prostate glands	−		
Human embryonic kidney	+		

[a] These results were usually obtained by a combination of three different methodologies: immunolocalization of villin by immunofluorescence; immunodetection of the villin by immunoreplica procedure; or enzyme-linked immunosorbent assay (ELISA). In some cases, these results were derived from only one technique (23, 27–31). ND, no data available.

organized brush border but express lower but significant amounts of villin (Robine et al., 28). In the embryo, the primitive human gut (8 weeks of gestation) or the rat gut (14 days of gestation), which contains a multilayer of rapidly dividing precursor enterocyte cells, also expresses villin (Table I). Cell lines established from normal or tumor cells derived from intestine or from kidney proximal tubular cells continue to express this protein (Table I). Finally, epithelial cells lining the pancreatic and bile ducts also express villin. Taken together, these observations emphasize two facts: (1) villin can be expressed early by precursors (at the embryonic or adult stages) of cells displaying, once fully differentiated, a brush border; (2) adult differentiated cells lacking a brush border but derived from the same embryonic precursors express this protein (i.e., pancreatic and bile duct cells).

VILLIN EXPRESSION DURING INTESTINAL DIFFERENTIATION

Differentiation of intestinal cells is a continuous and gradual process throughout the life span. Morphogenesis of the brush border also occurs in intestinal epithelia cells of the adult. When the cells migrate up from the crypt to the villus tip, microvilli are gradually and synchronously elongated (Leblond and Cheng, 2). Studies on the differentiation of intestinal cells indicate that high levels of membrane-bound hydrolases are expressed in the differentiated cells of the upper part of the villi. In contrast, villin is expressed in the immature enterocytes from the crypt as well as in the mature enterocytes from the tip of the villi. However, there is a gradient of villin expression along the crypt–villus axis. The amount of villin in the mature enterocyte is 10-fold higher than the amount in the immature crypt cells (Robine et al., 28). All the cells along the villi express mRNAs coding for villin. The level of villin mRNA is low in the crypt, reaches a maximum at the bottom of the villus, but decreases significantly toward the tip of the villus (Boller et al., 32).

The clone HT29-18 isolated in our laboratory from parent cells grown in glucose (Godefroy et al., 33) and able to differentiate in presence of galactose into several cell types similar to those found in intestinal mucosa (Huet et al., 19) (Fig. 1) has been used as a model to follow villin expression. On differentiation of HT29 cells grown in the presence of galactose we observed a significant increase in the amount of villin, whereas the amount of total actin is similar in both conditions. Villin is colocalized with actin microfilaments in the apex of the differentiated cells, whereas it is dispersed throughout the cytoplasm in undifferentiated cells. The rate of synthesis, measured by incorporation of [^{35}S]methionine followed by a chase with cold methionine, is also increased in differentiating cells, whereas the stability of the protein is the same in both conditions of culture (Dudouet et al., 23). Moreover, we have shown that the 10-fold increase of the amount of protein is correlated with a large increase in the amount of mRNAs encoding villin. These results suggest that the regulation of expression of villin is due to an increase in the rate of transcription of villin gene(s) or to the stabilization of villin mRNAs (Pringault et al., 25). Both tissue-specific expression of the villin gene and its induction during terminal differentiation emphasize the strict regulation of villin gene activity during embryonic development and in adult life.

SUMMARY AND CONCLUSIONS

Using a pluripotent cell line (HT29) that can be manipulated in culture, the differentiation pattern of HT29 cells has been investigated and compared with intestinal differentiation *in vivo*. We focused our study on the establishment of a polarized cytoskeleton based on actin microfilaments, assembled during the terminal differentiation of this cell line. The apical domain of intestinal cells is generated *de novo*, leading to the formation of the so-called brush border. We showed that the expression of villin, a calcium-regulated actin-binding protein found in the brush border, is organ specific. Furthermore, its level of expression in intestinal cells can be correlated with the differentiation stage of the cells under investigation. The unique structural properties of villin may account for its expression in a few epithelial cell types as well as its localization at the apex of differentiated cells. Since cDNA clones for this protein are now available, this new tool can be used to investigate villin function *in vivo* under various conditions after introduction of the villin gene in selected cell types. Our ultimate goal is to understand the role played by structural proteins in the assembly and regulation of the architecture of the intestinal brush border. The availability of suitable intestinal cell lines and molecular tools allowing analysis of this process should be useful in the future for the understanding of the rules governing the biogenesis of this important cell surface specialization.

ACKNOWLEDGMENTS

This work was supported by grants from the "Institut National de la Santé et de la recherche médicale" (No. 86-7008), the "Fondation pour la Recherche Médicale," the "Association poiur la Recherche sur le Cancer" (No. 6379), the CNRS UA-1149, and the Ligue Nationale Française contre le Cancer."

REFERENCES

1. Ponder, B. A. J., Schmidt, C. H., Wilkinson, M. M., Wood, M. J., Monk, M., and Reid, A. (1985). *Nature (London)* **313,** 689–691.
2. Leblond, O. P., and Cheng, H. (1976). *In* "Stem Cells of Renewing Cell Populations"

(A. B. Cairnie, R. K. Lala, and D. G. Osmond, eds.), pp. 7–31. Academic Press, New York.
3. Madara, J., and Trier, J. S. (1987). In "Physiology of the Gastrointestinal Tract" (L. Johnson, ed.), Vol. 2, pp. 1209–1250. Raven, New York.
4. Trier, J. S., and Moxey, P. C. (1979). Ciba Found. Symp. **70**, 3–29.
5. Colony, P. C., and Neutra, M. R. (1983). Dev. Biol. **97**, 349–363.
6. Madara, J. L., Neutra, M. A., and Trier, J. S. (1981). Dev. Biol. **86**, 170–178.
7. Friedman, E. A., Higgins, P. J., Lipkin, M., Shinya, H., and Gelb, A. M. (1981). In Vitro **17**, 632–644.
8. Negrel, R., Rampal, R., Nano, J. L., Cavenel, C., and Ailhaud, G. (1983). Exp. Cell Res. **143**, 427–437.
9. Quaroni, A., Wands, J., Trielstad, R. S., and Isselbacher, K. J. (1979). J. Cell Biol. **80**, 248–265.
10. Quaroni, A., and May, R. J. (1980). In "Methods in Cell Biology" (G. C. Harris, B. F. Trump, and G. D. Stoner, eds.), Vol. 21B, pp. 403–427. Academic Press, New York.
11. Fogh, J., and Trempe, G. (1975). In "Human Tumor Cells in Vitro" (J. Fogh, ed.), pp. 115–141. Plenum, New York.
12. Fogh, J., Fogh, J. M., and Orfeo, T. (1977). J. Natl. Cancer Inst. **59**, 221–226.
13. Rousset, M., Laburthe, M., Chevalier, G., Boissard, C., Rosselin, G., and Zweibaum, A. (1981). FEBS Lett. **126**, 38–40.
14. Rousset, M. (1986). Biochimie **68**, 1035–1040.
15. Pinto, M., Appay, M. D., Simon-Assmann, P., Chevalier, G., Dracopoli, N., Fogh, J., and Zweibaum, A. (1982). Biol. Cell **44**, 193–196.
16. Wice, B. M. L., Trugnan, G., Pinto, M., Rousset, M., Chevalier, G., Sussaulx, E., Lacroix, B., and Zweibaum, A. (1985). J. Biol. Chem. **260**, 139–146.
17. Zweibaum, A., Pinto, M., Chevalier, G., Dussaulx, E., Triadou, N., Lacroix, B., Haffen, K., Brun, J. L., and Rousset, M. (1985). J. Cell. Physiol. **122**, 21–29.
18. Zweibaum, A., Hauri, H. P., Sterchi, E., Chantret, I., Haffen, K., Bamat, J., and Sordat, B. (1984). Int. J. Cancer **34**, 591–598.
19. Huet, C., Sahuquillo-Merino, C., Coudrier, E., and Louvard, D. (1987). J. Cell Biol. **105**, 345–358.
20. Mooseker, M. S. (1985). Annu. Rev. Cell Biol. **1**, 209–241.
21. Glenney, J. R., Geisler, N., Kaulfus, P., and Weber, K. (1981). J. Biol. Chem. **256**, 8156–8161.
22. Matsudaira, P., Jakes, R., and Walker, J. E. (1985). Nature (London) **315**, 248–250.
23. Dudouet, B., Robine, S., Huet, C., Sahuquillo-Merino, C., Blair, L., Coudrier, E., and Louvard, D. (1987). J. Cell Biol. **105**, 359–369.
24. Figiel, A., Schilt, J., Dudouet, B., Robine, S., and Dauca, M. (1987). Differentiation **36**, 116–124.
25. Pringault, E., Arpin, M., Garcia, A., Finidori, J., and Louvard, D. (1986). EMBO J. **5**, 3119–3124.
26. Kwiatkowski, D. D. J., Stossel, T. P., Orkin, S. H., Mole, J. E., Colten, H. R., and Yin, H. L. (1986). Nature (London) **323**, 455–458.
27. Arpin, M., Pringault, E., Finidori, J., Garcia, A., Jeltsch, J. M., Vanderkerckhove, J., and Louvard, D. (1988). J. Cell Biol. (in press).
28. Robine, S., Huet, C., Moll, R., Sahuquillo-Merino, C., Coudrier, E., Zweibaum, A., and Louvard, D. (1985). Proc. Natl. Acad. Sci. U.S.A. **82**, 8488–8492.
29. Bretscher, A., Osborn, M., Wehland, J., and Weber, K. (1981). Exp. Cell Res. **135**, 213–219.

30. Reggio, H., Coudrier, E., and Louvard, D. (1982). "Membranes in Growth and Development," pp. 89–105. Alan R. Liss, New York.
31. Drenckhahn, D., and Mannherz, H. G. (1983). *Eur. J. Cell Biol.* **30**, 167–176.
32. Boller, K., Arpin, M., Pringault, E., and Reggio, H. (1988). *Differentiation* (in press).
33. Godefroy, O., Huet, C., Blair, L., Sahuquillo-Merino, C., and Louvard, D. (1988). *Biol. Cell* **63**, 41–55.

Index

A

Actin
  abl and *mos* transcription in spermatogenesis, 178
  villin as binding protein, 346–348
  villin expression during intestinal differentiation, 350
Adenocarcinoma, 343
Adhesion protein
  collagens, 76
  gene expression in primary liver cell cultures, 86–87
ADP-ribosylation, 51–52, 56
Adults
  morphogenesis of brush border in intestinal epithelium, 350
  normal skin cells and TGFα, 122
  villin as organ-specific marker in, 348–350
α-IR3
  studies of type I somatomedin receptor, 5, 7
  two site hypothesis for type I somatomedin receptor, 7
Albumin
  endothelial cells, 246
  as tracer, 228–229, 231, 235, 237, 239–241, 243–245
Amino acids
  dpp polypeptide, 254–255
  human villin and gelsolin, 348
  mouse and *Drosophila en* and *inv* genes, 310
  nucleotide and deduced sequence of the 2HS epitope of human laminin receptor, 113–115
  phospho-*engrailed* protein from HS-EN cells, 274, 277
  sequence of homeo box domain, 320

Angiogenesis
  embryonic stem cells, 211–213
  growth and development, 207–208
Anion transporter protein
  plasmalemma vesicles, 228
  v-*erb*A oncogene and expression in maturing cells, 143
Ankyrin, 56
Antennapedia (Antp)
  homology to murine homeo box genes, 320, 322
  studies of homeotic mutants, 259–260
ANT-C gene, 262
AtT20 mouse pituitary corticotrope cell line, 37
Autophosphorylation
  tyrosine 416, 31–32
  tyrosine 527, 33

B

Baly, D. L., 10
Basement membrane
  components, 108–109
  interaction with metastatic tumor cell, 118
  three-step hypothesis of invasion, 108
Bithorax gene complex (BX-C)
  function as regulators of "downstream" genes, 261
  as gene cluster, 259
  Hox sequences, 262
  wild-type functions, 260
Blastoderms, 268
Blood-brain barrier (BBB) endothelium, 219, 221
Blood vessels
  blood island induction, 208
  cell types and structure, 207

embryonic stem cells as an experimental model system for early vascular development, 211–217
vascular regression in embryo, 216–217
vascular wall development, 214–215
vasculogenesis, 208, 210
Brain
  en genes and development of central nervous system, 316
  expression of *Hox* 1.3 RNA, 324
  membrane and effects of detergents on G proteins, 54
  tissue specificity of *en* transcripts, 312–313, 315
Bundgaard, M., 244–245

C

cAMP-dependent protein kinase, 29
Campos-Ortega, J. A., 255
Capillaries, permeability, 226–227
Carcinoma cells
  laminin and metastasis, 109
  pp60$^{C\text{-}SRC}$ activity, 39
CCAAT box, 81
Cell junctions, continuous microvascular endothelium, 226
Cell line
  defined, 72
  pluripotent human intestinal and differentiation in culture, 343–344
Cell strain, defined, 72
C-*erb*A, 138–140
CHAPS, 53
Chicken
  limb as research model, 329, 330–331
  limb buds and endogenous retinoic acid, 336, 338–339
  retinoic acid and ZPA tissue, 331–336
  wing anatomy, 329–330
Chou, P. Y., 116
Cis-overexpression (COE) effect, 262–263
Clark, E. L., 214
Clark, E. R., 214
Clayton, D. F., 82
Colberg-Poley, A. M., 320
Collagen
  adhesion protein, 76
  gene expression in primary liver cell cultures, 86–87
  permissive and nonpermissive for responsiveness to specific hormones, 98
  type and extracellular matrices, 76
  type IV and basement membranes, 108
Colloidal gold-albumin complexes, 228–229, 231, 235, 237, 241
Conrad, H. E., 78–79, 94
Crypt-villus axis, 342
c-src gene, 25–26
Culture, cell
  classic method, 72–73
  conditions and matrix synthesis in liver cells, 79
  differentiated and intestinal epithelium, 342–343
  extracellular matrix, 75–76, 78–80
  gene expression in primary liver cultures, 81–83
  gene expression studies in hepatoma cells, 90
  hormonally defined media (HDM), 73–75
  pluripotent human intestinal cell line and differentiated, 343–344
  types of, 71–72
  utility and validity for gene expression studies, 71
Cushman, S. W., 10
Cyanogen bromide, 115
Cyclic AMP, 46
Cysteine
  residues and EGF receptor, 132
  structural aspects of G proteins, 49
Cytoskeletal elements, signal transduction, 55
Cytoskeleton
  intestinal epithelium as model system for development of polarized, 341–343
  organization of intestinal brush border, 344–346

D

Darnell, J. E., 82
Decapentaplegic (dpp) gene
  Hin region and function, 250, 252–253
  organization, 249–250
  polypeptide, 254–255
Deformed (Dfd) gene
  binding of proteins, 299–300
  DNA-binding activities, 289, 291, 296
  eve+ gene activity, 280
DeLarco, J. E., 121

Depolymerization, theories of hormone action, 47
Desplan, C., 285, 302
Detergents, effects of on G proteins, 52–54
Dexamethasone, 84
Diabetes mellitus
  glycated albumin, 241
  insulin and hexose transport, 9
Diacylglycerol, 65
DNA
  binding activities of homeo box proteins, 289, 291, 296
  human placental and transformants, 171
  isolation of human laminin receptor, 110–111
  overlapping domain of laminin receptor inserts, 111, 113
  v-*erb*A and binding of, 145–146
Dominant hemimelia (*Dh*) gene, 315, 316
Drohan, W., 111
Drosophila
  affinis and homeotic mutants, 259
  cellular blastoderm stage of embryo, 268
  conservation of genes during evolution, 315–316
  en promoter joined to *lacZ* coding sequences, 269
  en protein expression, 274
  eve gene and development of embryo, 279–280
  eve protein function, 298
  homeo box-containing transcripts in spinal ganglia, 324
  identification and function of gene products, 249
  inv gene and development, 307
  melanogaster
    classic approach to genetics, 305–306
    en intron and sequence divergence from *D. virilis*, 297
    homeotic mutants, 259, 267
    invected protein, 277
    murine *en* genes, 322
    serine 12 and pp60$^{C\text{-}SRC}$, 28
    structure of EGF receptor, 133
    study of *en* gene in, 261–262
  virilis
    en intron and sequence divergence from *D. melanogaster*, 297
    en protein, 274, 277

E

Electrophoresis, S6 phosphorylation, 20
Embryos and embryogenesis
  angiogenesis, 208
  cellular blastoderm stage, 268
  dpp phenotypes, 252–253
  eve gene and development of *Drosophila*, 279–280
  evolutionary conservation of basic mechanisms of development, 306, 315–316
  functional analysis of *Hox* 1 cluster, 322–324
  mammalian and classic approach to genetics, 306
  metazoan cells and cellular interactions, 69
  stage and tissue specificity of *en* expression, 312–313, 315
  vascular regression, 216–217
  vasculogenesis, 210
  villin as organ-specific marker in, 348–350
Embryonic stem cells (ESCs), 211–217
Enat, R., 74
Endothelial cells
  albumin, 246
  differentiation in embryo, 210, 217–219, 221
  embryonic angiogenesis, 211–213
  growth and development, 207–208
  molecular basis for vascular wall development, 214–215
  regulatory mechanisms governing hepatocellular functions, 70
  structure of blood vessels, 207
  vascular and FGF, 169–170
  vascular regression in embryo, 216–217
Endothelium
  attempts at structural–functional integration, 227–229, 231, 235, 237, 239–241, 243–245
  general organization, 224
  multiplicity of functions, 223–224
  structural basis of permeability in continuous, 245–246
Engrailed (*en*) gene
  binding of proteins, 299–300
  characterization, 307
  conservation during mammalian evolution, 307, 315–316
  DNA-binding activities, 289, 291, 296

eve+ gene activity, 280
genetic mapping of loci, 315
importance to embryonic and subsequent development, 268
isolation of mouse genomic and cDNA clones, 308–310
patterns of expression, 281
promoter, 268–269
protein as phosphoprotein, 269, 274, 277–278
regulation of by the eve protein, 296–298
stage and tissue specificity of expression during embryogenesis, 312–313, 315
study of in *Drosophila*, 261–262
transcription unit and high-affinity *eve* binding sites, 285, 287
Epidermal growth factor (EGF)
biological effects of TGFα, 122
characterization, 131
properties of receptor mutants, 133–134
S6 kinase activation, 20–21
S6 phosphorylation, 19, 20
structure of receptor, 132–133
Epithelial cells, autocrine regulation by TGFα and TGFβ, 125–126
Epithelium, intestinal as model system for study of cell differentiation, 341–343
Erythroblasts
contribution of v-*erb*A to transformation of, 141–144
distinct and measurable effects of v-*erb*A on, 138
effects of mutant v-*erb*A and v-*erb*B genes on, 140–141
mechanism for cooperativity between v-*erb*A and v-*erb*B, 147
Escherichia coli
en protein, 274
eve protein, 280–281, 296
expression of full-length homeo box proteins, 284–285
Even-skipped (*eve*) gene
comparison of protein binding, 299–300
development of *Drosophila* embryo, 279–280
DNA binding studies, 280–281, 289, 291, 296
en transcription unit and high-affinity binding sites, 285, 287
high-affinity *eve* binding sites near gene, 287, 289

mechanism of protein binding, 298–299
patterns of expression, 281
regulation of by *eve* protein, 296–298
Evolution
conservation of basic mechanisms of embryonic development, 306
dpp polypeptide, 255
EGF receptor domains, 133
engrailed gene of *Drosophila*, 307
structure of murine homeo box genes, 320, 322
vertebrate limbs, 330
Extracellular matrix
effects of culture conditions on synthesis in liver cells, 79
embryonic vasculogenesis and angiogenesis in vascular, 215
hormones and synergy of effects, 90–91
insoluble regulatory signals needed for normal liver gene expression, 75–76, 78–79
studies with extracts, 88–89
substrata and hepatoma cells, 91–93

F

Fasman, G. D., 116
Feather anlagen, 334
Fibroblast growth factor (FGF)
broad spectrum of mitogenicity, 169–170
determination of roles in normal and neoplastic processes, 171
distinct from SGF as biological entity, 201–202
embryonic angiogenesis, 213
FGF-5 expression in tumor cell lines, 170–171
FGF-5 gene and secreted growth factor, 168
Leydig cells and protein secretion, 198
oncogene-transformed cell lines, 150
proliferation of Leydig and Sertoli cells, 192, 194
Sertoli cells and protein secretion, 198
SGF induction of extracellular accumulation of sulfated glycoprotein, 194–195
tumors and *ras*-transformed cells, 151–152
Fibroblasts
effects of v-*erb*A, 138

Fibroblasts (cont.)
  growth of in culture, 158
  neuronal pp60^{C-SRC}, 36, 37
  v-erbB and mitogenic signaling, 147
Focus assays, 157–158
Footprint assays, 302–303
Frokjaer-Jensen, J., 244–245

G

Galactose, 350
Garcia-Bellido, A., 261
Gasseling, M. T., 330–331
Gatmaitan, Z., 73, 74
Gelbart, W. M., 252
Gelsolin, 348
Gene clusters, bithorax gene complex, 259
Gene expression
  DNA and v-erbA protein, 146
  hypotheses on mechanisms of matrix and hormonal regulation of, 94–95
  in normal liver, 80–90
  retinoids, 339
  studies and defined hormone and matrix conditions, 95
  studies in hepatoma cells, 90–94
  utility and validity of cell culture for studies on, 71
  v-erbA and erythroblasts, 141–142
Genetic mutation
  en gene and embryonic development, 268
  eve and homeo box genes, 280
  genetic organization governing insect development, 267
  Hin region of dpp, 250, 252
Genetics
  classic approach, 305–306
  reverse as alternative approach, 306
Germ cells
  development of mammalian testes, 187–188
  expression of mitogenic activity, 188
  expression of mos in male, 177–178
Ghitescu, L., 229
Glenney, J. R., 347, 348
Globin system, 93
Glucocorticoids, 83
Glucose, insulin and redistribution of transporter proteins, 10
Glycated albumin, 241

Glycosaminoglycans
  biological role of heparan sulfate PG, 94–95
  heparin and FGFs, 169
  liver gene expression, 87
  proteoglycans in liver, 78–79
  synergies with hormones, 89–91
Glycosylation, proteins in mature eggs and zygotes, 181
Glycosylphosphatidylinositol, 66–67
Gospodarowicz, D., 202–203
G proteins (GTP-binding proteins)
  depolymerization theory, 47
  effects of detergents on, 52–54
  hormone and neurotransmitter activation, 51–52
  as polymeric structures, 55–56
  programmable messenger theory, 46–47
  signal transduction and cytoskeletal elements, 55
  structure and functions, 47–51
  theories of hormone action, 46
Growth factors
  oncogene-transformed cell lines, 150
  role in tumor growth, 154
  see also Epidermal growth factor; Fibroblast growth factor; Herparin-binding growth factor; Platelet-derived growth factor; Seminiferous growth factor; Transforming growth factor
Guguen-Guillouzo, C., 82
Guillouzo, A., 82

H

Hammertoe (Hm) gene, 315
Haptotaxis, 116
Harrison, L. C., 5
Heberlein, U., 302
Hemimelic extratoes (Hx) gene, 315, 316
Heparan sulfate proteoglycan, 66
  basement membranes, 108
  biological role, 94–95
  proteoglycans and glycosaminoglycans in liver, 78–79
Heparin-binding growth factor (HBGF)
  immunological properties, 200
  induction of specific peptides by SGF, 196, 198–200
Hepatocyte
  classic cell culture conditions, 81
  hormonally defined medium, 73, 75

regulatory mechanisms governing hepatocellular functions, 70
Hepatomas
  gene expression studies in, 90–94
  hormonally defined medium, 73–75
HER2 protein, 132
Hexose transport, 9
High density lipoprotein (HDL), 73
Hintz, R. L.
  atypical insulin receptors, 7
  IGF-II receptors, 4, 7
Hippocampus, 35–36
Hogness, D., 262
Holley, R. W., 123, 124
Homeo box genes
  conservation in vertebrate genes, 306
  DNA-binding activities of four different, 289, 291, 296
  expression of full-length in *Escherichia coli*, 284–285
  footprint assays, 302–303
  functional analysis, 322–324
  immunoprecipitation assays, 302
  murine development studies, 319–320
  plasmid constructions, 301
  preparation of protein extracts, 301–302
  structure of murine, 320, 322
Homeobox (*Hox*) sequences
  conservation of ANT-C and BX-C, 262
  studies of gene clusters, 260
Homeotic genes. *See* Gene clusters
Hormonally defined media (HDM)
  gene expression in hepatoma cells, 91
  hepatocytes and hepatomas, 73–75
  influence of serum-free on gene expression in cultured liver cells, 82–86
Hormones
  depolymerization theory, 47
  extracellular matrix and synergy of effects, 89–91
  G proteins and cytoskeletal elements, 55
  programmable messenger theory, 46–47
  second messenger hypothesis, 45–46
Horuk, R., 10
HS-EN cell line, 274

I

IGF-I
  membrane redistribution of transferrin receptors, 12
  subtypes of receptor, 5
IGF-II
  Man-6-9 receptor and phosphorylation in intact cells, 15–16
  Man-6-P receptor and general receptor structure, 13–14
  Man-6-P receptor and phosphorylation, 17
  as model for study of phosphorylation/dephosphorylation reactions, 12–13
  primary structure deduced from cDNA, 17
Immunoglobin, 93
Immunoprecipitation assays, 302
Inositol phosphate glycans
  as enzyme modulators, 62–64
  produced by insulin-sensitive phosphatidylinositol glycan-specific phospholipase C activity, 65–66
Insect, genetic organization of development, 267
Insulin
  cysteine-rich domain and receptor, 132
  hexose transport and cell metabolism, 9
  inositol phosphate glycans as enzyme modulators, 62–64
  intracellular second messengers and phosphorylation, 59
  and Man-6-P receptors in 32P-labeled control and insulin-treated adipocytes, 14–15
  molecular mechanisms and receptor recycling, 12
  nutrient uptake and membrane redistribution of transporters and receptors, 17
  PDGF and FGF activity, 160
  recycling process between plasma membrane and intracellular endosomal membranes, 10–11
  time frames and mechanisms of signal transduction, 59
  turnover of glycosylphophatidylinositol, 66–67
Insulin-sensitive enzyme modulators
  inositol phosphate glycans, 62–64
  purification and biological characterization, 61–62
Insulin-sensitive phosphatidylinositol glycan, 65–66
Intestine
  organization of brush border cytoskeleton, 344–346

INDEX 361

pluripotent human cell line and differentiation in culture, 343–344
stem cells, 70
villin expression during differentiation, 350–351
Invected (inv) gene
  characterization, 307
  conservation of during evolution, 315
  isolation of mouse genomic and cDNA clones, 308–310
Irish, V. F., 252

J

Jaye, M., 111
Jonas, H. A., 5, 7
Joyner, A. L., 307

K

Keratinocytes
  inhibition of proliferation by TGFβ, 126
  negative regulation by TGFβ, 125
Kinases
  activation of S6, 20–21
  characteristics of S6, 21–22
  phosphorylation of IGF-II/Man–6–9 receptors in intact cells, 15–16
Kleinman, H. K., 89
Kono, T., 10
Kornberg, T., 262
Kull, F. C., 5

L

*Lac*Z coding sequences, 269
Laminin
  components of basement membranes, 108–109
  expression as early marker for vascular maturation, 215
  interaction between metastatic tumor cell and basement membrane, 118
  role of in tumor cell metastasis, 109–110
Laminin receptor
  expression of mRNA, 116–118
  interaction with tumor cell, 109
  isolation of human cDNA, 110–111
  monoclonal antibodies to human, 110

nucleotide and deduced amino acid sequence of the 2HS epitope of human, 113–115
overlapping domain of cDNA inserts, 111, 113
synthetic peptides of the putative ligand binding domain, 116
LAN-5 human neuroblastoma line, 37
Larner, J., 60
Leydig cells
  development of mammalian testes, 187–188
  proliferation of, 192, 194
  properties of testicular cell lines, 191–192
  protein secretion in response to growth factors, 198
Limbs
  buds and endogenous retinoic acid, 336, 338–339
  development of vertebrate, 329–330
  zone of polarizing activity (ZPA), 330–331
Lipshitz, H., 262
Liver
  culture conditions and matrix synthesis in liver cells, 79
  embryonic angiogenesis, 212
  epithelial-mesenchymal dyad, 70
  gene expression in normal, 80–90
  matrix types produced by cells, 76
  proteoglycans and glycosaminoglycans in liver, 78–79
  stem cells, 70–71

M

Mammose 6-phosphate receptors, 4
Man-6-9 receptors, 15–16
Man-6-P-linked lysosomal enzymes, 13
Man-6-P receptor
  IGF-II and general receptor structure, 13–14
  IGF-II and phosphorylation, 14–15, 17
  primary structure deduced from cDNA, 17
Marceau, N., 83
Marquardt, H., 122
Martin, G., 320
Mato, J. M., 64
Matrigel, 89
Matsudaira, P., 348

Metastasis
    approaches to study of molecular genetics, 107–108
    complex multistep process, 107
    laminin and laminin receptor in metastasis, 109–110
    tumor cell interaction with basement membrane, 118
Metazoan cells, 69–71
Michalopoulos, G., 82
Microvascular endothelium
    continuous, 225–227
    general organization, 224
Microvilli
    organization of intestinal brush border cytoskeleton, 344–345
    villin expression during intestinal differentiation, 350
Mitogen, activity and mammalian testes, 188
Moffatt, B. A., 284
Monoclonal antibodies, human laminin receptor, 110
Monomeric albumin, 237, 239–241, 244
Morgan, D. O., 5, 7
Morphogen, ZPA and limb development, 331
Mouse
    conservation of *en* and *inv* genes, 315
    family of genes, 319–320
    functional analysis of *Hox* 1 cluster, 322–324
    genetic mapping of *en* loci, 315
    isolation of genomic and cDNA clones with sequence homology to *Drosophia en* and *inv* genes, 308–310
    stage and tissue specificity of *en* expression during embryogenesis, 312–313, 315
    structure of homeo box genes, 320, 322
    teratocarcinoma cells and retinoic acid, 339
Müllerian inhibiting substance, 124

N

Neoplasia, study of EGF receptor, 131
Neoplastic cells, transfection studies, 93
Neufeld, G., 202–203
Neuroblastomas, 39
Neuronal cells
    pp60^{C-SRC}, 35–37

src and *ras* protooncogene and differentiation, 176
Nissley, S. P., 4
Nucleotides, 113–115

O

Octylglucoside, 53–54
Oncogenes
    cell proliferation, 149–150
    detection of activated by focus assay, 158
    expression of growth factors, 154
    identification of cellular, 175
    mechanisms of activation, 175–176
    see also Protooncogenes
Oocytes, 178–179
Organogenesis, 211–213

P

Parenchymal cells, 75
Peptides
    induction of specific by SGF, 196, 198–200
    see also Polypeptides
Pericytes
    regulation of cell differentiation, 215
    structure of blood vessels, 207
Pertussis toxin, 51–52, 56
pH, v-*erb*B-transformed erythroblasts, 144
Phosphatases, 21
Phosphatidylinositol (PI)
    glycan-specific phospholipase C activity, 65–66
    glycosylated derivative of inositol phosphate, 63
Phosphoamino acid, 15
Phospholipase C, 65–66
Phosphopeptides, 20
Phosphoprotein, *en* protein, 269, 274, 277–278
Phosphorylation
    cascade hypothesis, 60
    engrailed protein, 277–278
    IGF-II/Man-6-P receptors in ^{32}P-labeled control and insulin-treated adipocytes, 14–15
    insulin and intracellular second messengers, 59

Phosphorylation (*cont.*)
  neuronal and fibroblast forms of pp60^{C-SRC}, 36–37
  pathways involved in S6, 22–23
  PDGF and pp60^{C-SRC}, 34–35
  pp60^{C-SRC} and polymavirus middle T antigen, 38
  proteins in mature eggs and zygotes, 181
  regulation of pp60^{C-SRC} and pp60^{V-SRC}, 27–29, 31–32, 40–41
  stimulation of S6, 19–20
  TPA-induced kinase C and EGF receptor, 134
  tyrosine 527, 32–33
Pitot, H. C., 82
Plasmalemma
  continuous endothelium, 225
  differentiated microdomains on the luminal, 225–226
  monomeric albumin and vesicles, 239
  preferential binding of A-Au, 229, 231, 237
  vesicles and transport of anionic plasma proteins, 228
Plasmids, homeo box proteins, 301
Platelet-derived growth factor (PDGF), 34–35
Polyergin, 123–124
Polymavirus, 37–38
Polymers, structure of G proteins, 55–56
Polypeptides
  dpp, 254
  see also Peptides
pp60^{C-SRC}
  biological activity, 27
  characterization, 25
  complexity of phosphorylation process, 40–41
  elevation of protein kinase activity in transformed cells, 39–40
  functions, 26
  neuronal, 35–37
  PDGF and phosphorylation, 34–35
  polymavirus middle T antigen, 37–38
  serine and phosphorylation, 27–29
  tyrosine 416 and autophosphorylation, 31–32
  tyrosine 527 and phosphorylation, 32–33
pp60^{V-SRC}
  biological activity, 27
  Ser-17 phosphorylation, 29
  tyrosine 416 and autophosphorylation, 31
  tyrosine 527 and phosphorylation, 32–33
  vanadate and protein kinase activity, 39
Protein kinase C
  engrailed protein, 277–278
  PDGF and phosphorylation of pp60^{C-SRC}, 34
  relation between insulin action and, 65
Proteins
  engrailed as phosphoprotein, 269, 274, 277–278
  preparation of homeo box extracts, 301–302
  synthesis and S6 phosphorylation, 19–20
Protein-tyrosine kinases, 26
Proteoglycans
  biological role of heparan sulfate PG, 94–95
  glycosaminoglycans in liver, 78–79
  liver gene expression, 87
  synergies with hormones, 89–90
Protooncogenes
  abl and reproductive processes, 176–177
  abl compared to *mos* and actin, 178
  mos expression in male germ cells, 177–178
  mos expression of oocytes, 178–179
  posttranscriptional processing of *mos* in zygotes, 179–181
  normal cellular role, 176
  ras, *src* and neuronal differentiation, 176
  unique patterns of transcription during spermatogenesis in mouse, 181

R

Radioreceptor assays, 4
RAS proteins, 48
Ras-transformed cells, 151–152
Reagan, F. P., 210
Rechler, M. M., 4
Reid, L. M., 88
Retinoic acid
  digit formation in absence of ZPA tissue, 334–336
  generation of concentration gradient, 339
  limb buds and endogenous, 336, 338–339
  ZPA tissue and pattern duplications, 331–334
Retinol, 339
Retroviruses, *c-src* sequences, 26–27

Rhodopsin-rich plates, 48–49
Ricca, G., 111
Ringold, G. M., 84
RNA, expression of laminin receptor, 116–118
Robinson, J., 78
Roth, R. A., 5

S

Saunders, J. W., Jr., 330–331
Schizosaccharomyces pombe, 277
Sclerotomes, 324
Seminiferous growth factor (SGF)
  biochemical characterization, 189–191
  distinct from FGF as biological entity, 201–202
  immunological properties of heparin-binding growth factors, 200
  induction of extracellular accumulation of sulfated glycoprotein, 194–196
  induction of specific peptides by, 196, 198–200
  proliferation of Leydic and Sertoli cells, 192, 194
  testicular cell lines and expression of distinct pleiotypic effects, 191–192
Serine
  IGF-II/Man-6-P receptor and casein kinase II substrates, 16
  neuronal and fibroblast forms of pp60^{C-SRC}, 36
  phosphorylated residues of IGF-II/Man-6-P receptors, 15
  phosphorylation of pp60^{C-SRC}, 27–29, 34, 35, 38
Sertoli cells
  development of mammalian testes, 187–188
  expression of mitogenic activity, 188
  proliferation of, 192, 194
  properties of testicular cell lines, 191–192
  protein secretion and growth factors, 198
  SGF induction of extracellular accumulation of sulfated glycoprotein, 194
Signal transduction
  α-subunits of G proteins, 56
  cytoskeletal elements and G proteins, 55
Skin, normal adult cells and TGFα, 122
Smith, J. C., 334
Smith, M., 301

Somatomedin receptors
  characterization of types I and II, 3–4
  studies of type I with α-IR3, 5, 7
  two site hypothesis for type I, 7
Spectrin, G polymers, 56
Spermatogenesis
  cell types and synchronous renewal, 187–188
  mos transcription compared to *abl* and actin, 178
Staphylococcus aureus, 302
Stem cells
  embryonic as an experimental model system for early vascular development, 211–217
  gastrointestinal system, 70
  liver, 70–71
Stewart, P. A., 221
Striatum, neuronal pp60^{C-src}, 35–36
Studier, F. W., 284
Sulfated glycoprotein, 194–196
Summerbell, D., 331
Suzuki, Y., 10
SV40, 81

T

TATA box, 81
Technau, G. M., 255
Temperature, effects of detergent on G proteins, 53
Terminal web, 345–346
Testes
  development of mammalian, 187–188
  expression of mitogenic activity, 188
  identification of cell type expressing mos, 177
3T3 cells
  phosphatases and S6 kinase activity, 21
  purification of S6 kinase from vanadate-stimulated, 22
Threonine
  IGF-II/Man-6-P receptor and casein kinase II substrates, 16
  phosphorylated residues of IGF-II/Man-6-P receptors, 15
Thyroid hormone receptor
  gene expression, 141
  v-*erb*A as ligand-independent form of, 145
Tickle, G., 331
Tissue culture plastic, 82–86

Todaro, G. J., 121
Transducin, structure of G proteins, 48–49
Transfection, studies into hepatoma cells and hepatocytes, 92–94
Transferrin receptors
  effects of growth factors on number expressed on cell surface, 10
  insulin action on system, 17
  recycling process, 11–12
Transformation assays, 171
Transforming growth factor (TGF)
  autocrine regulation of epithelial cells, 125–126
  role of TGFα in oncogenesis by induction of autocrine growth, 131
  TGFα characterization, 121–123
  TGFβ characterization, 123–125
Tumor cells
  angiogenesis factors, 212
  basement membrane and three-step hypothesis of invasion, 108
  elevation of pp60^{C-src} protein kinase activity in, 39–40
  expression of FGF-5, 170–171
  interaction of metastatic with basement membrane, 118
  laminin, laminin receptor, and metastasis, 109–110
  ras-transformed cells and FGF, 151–152
  response in culture compared to normal cells, 79
Tumor promoter phorbol ester (TPA), 134
Tyrosine
  416 autophosphorylation of pp60^{C-src}, 31–32, 33, 38
  527 phosphorylation of pp60^{C-src} and pp60^{V-src}, 32–33, 38
  neuronal and fibroblast forms of pp60^{C-src}, 36, 37
  PDGF-induced phosphorylation, 34–35

V

Vanadate (sodium orthovanadate)
  pp60^{V-src} protein kinase activity, 39
  S6 kinase activity, 21
Vascular wall, development, 214–215
Vasculogenesis
  defined, 208
  development of blood vessels, 208, 210
  embryonic stem cells, 211

V-erbA
  binding of DNA, 145–146
  characterization, 137–138
  compared to c-erbA, 138–140
  contribution to transformation of erythroblasts, 141–144
  effects of mutant genes on erythroblasts, 140–141
  ligand-independent form of thyroid hormone receptor, 145
  mechanism for cooperativity with V-erbB, 147
V-erbB
  characterization, 137–138
  effects of mutant genes on erythroblasts, 140–141
  erythroblasts and synthesis of anion transporter, 143–144
  mechanism for cooperativity with V-erbB, 147
Vertebrates, limb development, 329–330
Villin
  as calcium-regulated actin-binding protein, 346–348
  crypt axis, 342
  expression during intestinal differentiation, 350–351
  as organ-specific marker in adults and embryos, 348–350

W

Wardzala, L. S., 10
Weiner, F. R., 83
Wiley, M. J., 221

Y

Yeast, phosphorylation at Tyrosine 527, 33

Z

Zerknullt (zen) gene, 289, 291, 296
Zoller, M. J., 301
Zone of polarizing activity (ZPA)
  limb development, 330–331
  retinoic acid, 331–336, 338
Zygotes
  binding of proteins, 299–300
  posttranscriptional processing of mos in, 179–181